ACCLAIM FOR KAY REDFIELD JAMISON'S

Exuberance

"Our foremost chronicler of the mind's darkest weather . . . demonstrates that exuberance is necessary for creativity and discovery."
—*Salon*

"[Jamison's] prose leaps from the page." —*Discover*

"Extraordinary grace. . . . A literary stylist of the first order, with prose of luminescent eloquence." —*Seattle Post-Intelligencer*

"A book to dive into for inspiration. . . . Jamison makes a compelling, memorable case for finding joy in the world."
—*The Star-Ledger* (Newark)

"Important. . . . Captivating. . . . [Jamison's] sentences stop one with their beauty and power. . . . [She] plunges into the light, celebrating exuberance as a powerful life force that fuels discoveries, inspires us, and even helps us to survive." —*The Baltimore Sun*

"A hymn to exuberance in all the places it may be found. . . . A well-written and engaging study." —*The Wall Street Journal*

"A delight to read. . . . Offers a compelling exploration of the beguiling emotion essential to the hope for joy and discovery that keeps us going."
—*The Oregonian*

"[Jamison's] lyrical stream flows abundantly, her generosity of spirit is everywhere, and her wisdom has never been greater. A delight of a book." —Antonio Damasio, author of *Descartes' Error*

"[Jamison's] energy and intelligence, and her wide-ranging knowledge of literature and history keep the book buoyant and entertaining." —*Psychology Today*

"Jamison's wise and spirited book celebrates the leaders, the scientists, the artists who take risks, transporting the rest of us to real and imaginary places we might never experience on our own. . . . [She also] eloquently describes the border that enthusiasm shares with fanaticism." —*The Boston Herald*

"Kay Jamison's unquenchable prose matches the luxuriant lives of her doers and shakers."
—James Watson, Nobel laureate and author of *The Double Helix*

"Nobody understands better than Kay Redfield Jamison the line between high spirits and madness. . . . [She] grants us new license to be ambitious, to be wildly in love with the world's mysteries and our pursuit of them." —*Elle*

"Jamison's exquisite prose carries us through this thought-provoking book as if we were surfing on splendid waves—the moments of inspiration just keep coming. This is one of the best books of [the year]." —*Spirituality and Health*

"In a witty and seamless meditation, [Jamison] explores the long-neglected positive trait of exuberance, a normal but highflying button-busting appetite for life." —*Time*

"Jamison has by now produced an impressive and thorough investigation of moods and mood disorders studied from all angles, including the most personal." —*The Washington Post Book World*

Kay Redfield Jamison

Exuberance

Kay Redfield Jamison is Professor of Psychiatry at the Johns Hopkins University School of Medicine and codirector of the Johns Hopkins Mood Disorders Center. She is also Honorary Professor of English at the University of St. Andrews in Scotland. She is the author of the national bestsellers *An Unquiet Mind: A Memoir of Moods and Madness*, *Night Falls Fast: Understanding Suicide*, and *Touched with Fire: Manic-Depressive Illness and the Artistic Temperament*. She is coauthor of the standard medical text on manic-depressive illness and author or coauthor of more than one hundred scientific papers about mood disorders, creativity, and psychopharmacology. Dr. Jamison, the recipient of numerous national and international scientific awards, was Distinguished Lecturer at Harvard University in 2002 and the Litchfield lecturer at the University of Oxford in 2003. She is a John D. and Catherine T. MacArthur Fellow.

ALSO BY KAY REDFIELD JAMISON

Night Falls Fast:
Understanding Suicide

An Unquiet Mind:
A Memoir of Moods and Madness

Touched with Fire:
Manic-Depressive Illness and the Artistic Temperament

TEXTBOOKS

Manic-Depressive Illness
(with Frederick Goodwin)

Abnormal Psychology
(with Michael Goldstein and Bruce Baker)

EXUBERANCE

The Passion for Life

BY

Kay Redfield Jamison

VINTAGE BOOKS
A DIVISION OF RANDOM HOUSE, INC.
NEW YORK

FIRST VINTAGE BOOKS EDITION, SEPTEMBER 2005

 Published in the United States by Vintage Books, a division of Random House, Inc., New York, and in Canada by Random House of Canada Limited, Toronto. Originally published in hardcover in the United States by Alfred A. Knopf, a division of Random House, Inc., New York, in 2004.

Owing to limitations of space, all acknowledgments for permission to reprint previously published material may be found preceding the index.

The Library of Congress has cataloged the Knopf edition as follows:
Jamison, Kay R.
Exuberance / by Kay Redfield Jamison.
p. cm.
Includes index.
1. Joy. 2. Enthusiasm. I. Title.
BF575.H27J36 2004
152.4'2—dc22
2004046561

Vintage ISBN: 0-375-70148-6

Book design by Virginia Tan

www.vintagebooks.com

Printed in the United States of America
10 9 8 7 6 5 4 3 2 1

For my father,
Marshall Jamison, Ph.D.,
whose exuberance
has been deeply and wonderfully contagious

and

Daniel Auerbach, M.D.,
to whom I owe so much of my life and work

In Memoriam

Richard Wyatt, M.D.,

my husband, colleague, and closest confidant

and

Dr. Harriet Braiker,
David Mahoney, Dr. Anthony Storr,
Judy Hoyer, and Katharine Graham,

irreplaceable friends

Contents

EXUBERANCE

CHAPTER ONE

"Incapable of Being Indifferent"

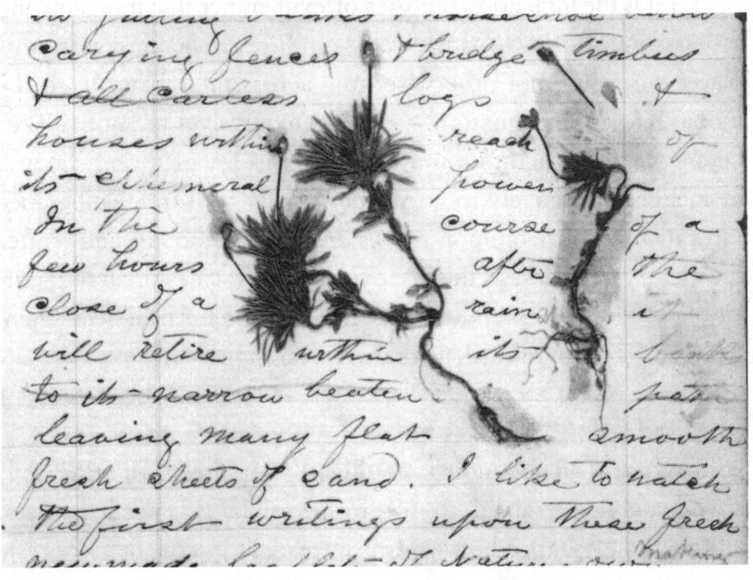

Carying fences & bridge timbers
& all careless logs &
houses within reach of
its ephemeral power
In the course of a
few hours after the
close of a rain it
will retire within its banks
to its narrow beaten path
leaving many flat smooth
fresh sheets of sand. I like to watch
the first writings upon these fresh

It is a curious request to make of God. *Shield your joyous ones,* asks the Anglican prayer: *Shield your joyous ones.* God more usually is asked to watch over those who are ill or in despair, as indeed the rest of the prayer makes clear. "Watch now those who weep this day," it goes. "Rest your weary ones; soothe your suffering ones." The joyous tend to be left to their own devices, the exuberant even more so.

Perhaps this is just as well. Those inclined toward exuberance have enjoyed the benign neglect of my field. Psychologists, for reasons of clinical necessity or vagaries of temperament, have chosen to dissect and catalogue the morbid emotions—depression, anger, anxiety—and to leave largely unexamined the more vital, positive ones. Not unlike God, if only in this one regard, my colleagues and I have tended more to those in the darkness than to those in the light. We have given sorrow many words, but a passion for life few.

Yet it is the infectious energies of exuberance that proclaim and disperse much of what is marvelous in life. Exuberance carries us places we would not otherwise go—across the savannah, to the moon, into the imagination—and if we ourselves are not so exuberant we will, caught up in the contagious joy of those who are, be inclined collectively to go yonder. By its pleasures, exuberance lures us from our common places and quieter moods; and—after the victory, the harvest, the discovery of a new idea or an unfamiliar place—it gives ascendant reason to venture forth all over again. Delight is its own reward, adventure its own pleasure.

Exuberance is an abounding, ebullient, effervescent emotion. It is kinetic and unrestrained, joyful, irrepressible. It is not happiness, although they share a border. It is instead, at its core, a more restless, billowing state. Certainly it is no lulling sense of contentment: exuberance leaps, bubbles, and overflows, propels its energy through troop and tribe. It spreads upward and outward, like pollen toted by dancing bees, and in this carrying ideas are moved and actions taken. Yet exuberance and joy are fragile matter. Bubbles burst; a wince of disapproval can cut dead a whistle or abort a cartwheel. The exuberant move above the horizon, exposed and vulnerable.

Exuberance keeps occasional company with grief, though grief may command the greater mention. Blake's belief that "Under every grief & pine / Runs a joy with silken twine" is a received theme in folklore. Our greatest joys and sorrows ripen on the same vine, says the American proverb. Danger and delight grow on one stalk, maintains the English one. Intense emotions inhabit a correspondent territory: joy may be our wings and sorrow our spurs, but the boundaries between the moods are open. Wings and high moods are shivery things; the joyous do indeed need shielding.

Exuberance is a vital emotion; it demands not only defense but exposure, for despair far more than joy has found sympathy with poets and scholars. Joy lacks the *gravitas* that suffering so effortlessly commands. Joy without reflection is evanescent; without counterweight, it has no weight at all. Or so one would think.

Yet joy is essential to our existence. Exuberance, joy's more energetic relation, occupies an ancient region of our mammalian selves, and one to which we owe in no small measure our survival and triumphs. It is a material part of our pursuits—love, games, hunting and war, exploration—and it is a vibrant force to signal victory, proclaim a time to quicken, to draw together, to exult, to celebrate. Exuberance is ancient, material, and profound. "The Greeks understood the mysterious power of the hidden side of things," wrote Louis Pasteur. "They bequeathed to us one of the most beautiful words in our language—the word 'enthusiasm'—*en theos*—a god within. The grandeur of human actions is measured by the inspiration from which they spring. Happy is he who bears a god within, and who obeys it."

Like many essential human traits, exuberance is teeming in some and not to be caught sight of in others. For a few, exuberance is in the blood, an irrepressible life force. It may ebb and flow, but the underlying capacity for joy is as much a part of the person as having green eyes or a long waist. For them, as the psalm promises, a full joy cometh in the morning. Not so for most others. Exuberance is a more occasional thing, something to be experienced only at splendid moments of love or attainment, or known in youth but lost with time. The nonexuberant lack fizz and risibility: they need to be lifted up on the enthusiasm of others; roused by dance or drug; impelled by music. They do not kindle of their own accord.

Variation in temperament is necessary. Exuberance, indiscriminately apportioned, is anarchical. If all were effervescent, the world would be an exhausting and chaotic place, driven to incoherence by competing enthusiasms or becalmed by indifference to the day-to-day requirements of life. Our species, like most, is well served by a diversity of temperaments, a variety of energies and moods. Exuberance is a fermenting, pushing-upward-and-forward force, but sometimes fixity is critical to survival. The joyous, and the not so, need one another in order to survive.

I believe that exuberance is incomparably more important than we acknowledge. If, as it has been claimed, enthusiasm finds the opportunities and energy makes the most of them, a mood of mind that yokes the two is formidable indeed. Exuberant people take in the world and act upon it differently than those who are less lively and less energetically engaged. They hold their ideas with passion and delight, and they act upon them with dispatch. Their love of life and of adventure is palpable. Exuberance is a peculiarly pleasurable state, and in that pleasure is power.

"Why should man want to fly at all?" asked Charles Lindbergh. "People often ask these questions. But what civilization was not founded on adventure, and how long could one exist without it? Some answer the attainment of knowledge. Some say wealth, or power, is sufficient cause. I believe the risks I take are justified by the sheer love of the life I lead." Man's exuberant spirit of adventure, Lindbergh argued elsewhere, is beyond his power to control. "Our earliest records," he said, "tell of biting the apple and baiting the dragon, regardless of hardship or of danger, and from this inner drive, perhaps, progress and civilization developed. We moved from land to sea, to air, to space, era on era, our aspirations rising."

Psychologists, who in recent years have taken up the study of positive emotions, find that joy widens one's view of the world and expands imaginative thought. It activates. It makes both physical and intellectual exploration more likely, and it provides reward for problems solved or risks taken. Through its positive energies, it heals as well. One joy, the Chinese believe, scatters a hundred griefs, and certainly it can be an antidote to fatigue and discouragement. Into those set back by failure, joy transfuses hope.

Exuberance is also, at its quick, contagious. As it spreads pell-mell throughout a group, exuberance excites, it delights, and it dispels tension. It alerts the group to change and possibility. Ted Turner, who would know, believes a leader is someone with the ability to "create infectious enthusiasm." This is a defining quality of great teachers, statesmen, and adventurers. Put to good use, infectious enthusiasm is a wonderful thing; used badly, it is calamitous.

Mostly, exuberance is a bounty and a blessing. It has its dangers, and we shall examine them in depth, but it is, all told, an amazing thing. Amazing and, on occasion, transfiguring. This was indisputably the case in May 1903, when two bounding enthusiasts hiked together in Yosemite. One was the President of the United States, the other a Scottish celebrant of the American wilderness. They were both, by temperament, utterly incapable of being indifferent.

Life for Theodore Roosevelt, said one friend, was the "unpacking of endless Christmas stockings." This would have gotten no argument from Roosevelt, a man who well into his fifties delighted in Christmas as an occasion of "literally delirious joy," and who believed that the entirety of life was a Great Adventure. The man "who knows the great enthusiasms," he held, lays claim to the high triumphs of life.

Born in 1858 into one of New York City's wealthiest families,

Theodore Roosevelt seems to have burst into the world a full-throated exuberant. For this, he owed a considerable debt to his father. "I never knew any one who got greater joy out of living than did my father," he wrote, and the seasons of his childhood, so beholden to his father's love and enthusiasms, "went by in a round of uninterrupted and enthralling pleasures." From his earliest days he exulted in life. At the age of ten, he wrote to his mother with breathless enthusiasm: "What an excitement to have received your letter. My mouth opened wide with astonish [*sic*] when I heard how many flowers were sent in to you. I could revel in the buggie ones. I jumped with delight when I found you heard the mocking bird."

Roosevelt, years later, was still jumping. One debutante said he did not so much dance as "hop." Another recounted his "unquenchable gaiety" and his unerring ability at formal dinner parties to send her into such uncontrollable fits of laughter that she had no choice but to leave the table. His Harvard classmates depicted him as a fast-moving, rapid-talking enthusiast who often wore them out with his boisterous exuberance. He held his far-flung interests with delight and stocked his college rooms with piles of books, a large tortoise, sundry snakes, and a collection of lobsters. He zoomed, he bolted, he boomed and gesticulated wildly as he went.

Roosevelt's vivacity receded when his father died. He felt, he said, as though "I should almost perish." It was a devastating loss. For the rest of his life he would miss, though himself incorporate, his father's rare mixture of infectious joy and keen sense of public duty. "Sometimes, when I fully realize my loss," he wrote in his diary a few months after his father's death, "I feel as if I should go mad."

Stoked by a restless energy not uncommon in those with exuberant temperaments, Roosevelt drove his desolation into action. He rowed, hiked, hunted, boxed, and swam furiously during the fevered weeks following his father's death. With slight cause other than annoyance he impetuously shot and killed a neighbor's dog. He nearly drove his horse into the ground through reckless gallops in the Oyster Bay countryside and was no easier on himself: "He'll kill himself before he'll even say he's tired," remarked one doctor of Roosevelt's frenetic behavior. Yet through it all there remained an irrepressible sense of life: "I am of a very buoyant temper," he wrote his sister not long after their father died. It was a temper that would serve him well and ill, but mostly well.

In the years immediately following his father's death, Roosevelt fell passionately in love, married, graduated from law school, and published the first of the nearly forty books he would go on to write. In 1881 he was elected to the New York State Assembly, where, as he put it, he "rose like a rocket." An ardent reformer, and lustily so throughout his political life, he became a mercurial, unstoppable irritant to his fellow Republicans.

Roosevelt's life in politics was abruptly broken when, on St. Valentine's Day of 1884, both his wife and his mother died. "You could not talk to him about it," said a close friend. He drew a cross in his diary for the date of the fourteenth of February and wrote, "The light has gone out of my life." In a pitch of energy reminiscent of the period following his father's death, Roosevelt abruptly took off for the Dakota Badlands, where he lived out his conviction that "black care rarely sits behind a rider whose pace is fast enough." He hunted, wrote an improbable number of books, and ran a cattle ranch. The hard work ultimately made wide inroads into his grief. "We felt the beat of hardy life in our veins," he wrote later in his autobiography, "and ours was the glory of work and the joy of living." Despite his distress, he said, "I enjoyed life to the full."

He returned to the East, remarried, and threw himself back into politics with gusto. He became a gale force in Washington. President Benjamin Harrison, who had appointed him civil service commissioner, said that the crusading Roosevelt "wanted to put an end to all the evil in the world between sunrise and sunset." Rudyard Kipling found himself caught up in a gentler form of Roosevelt's persuasive energies and, like most, he was completely captivated. After dining with him one evening at the Cosmos Club in Washington, Kipling knew himself bewitched: "I curled up on the seat opposite and listened and wondered until the universe seemed to be spinning around and Theodore was the spinner."

Roosevelt loped onward from post to post. He served energetically as assistant secretary of the Navy, and then led the 1st U.S. Volunteer Cavalry, the "Rough Riders," during the Spanish-American War. His zest for war, as for life, knew few limits. He had, one journalist put it, enough "energy and enthusiasm to inspire a whole regiment." Roosevelt exulted that the war was "bully," "the great day" of his life. It was, he said, his "crowded hour." He seemed to relish his brushes with death as passionately as he loved the rest of life. He was recommended for the Medal of Honor and returned to politics a war hero, a legend. He was elected governor of New York and then, within a few years' time, vice president of the United States. When William McKinley was assassinated in September 1901, Roosevelt became, at the age of forty-two, the youngest president in American history. He was also the liveliest.

The new president's exuberance was captured by a reporter from the *New York Times:* "The President goes from one to another . . . always speaking with great animation, gesturing freely, and in fact talking with his whole being, mouth, eyes, forehead, cheeks and neck all taking their mobile parts. . . . A hundred times a day the President will laugh, and when he laughs he does it with the same energy with which he talks. It is usually a roar of laughter, and it comes nearly every five minutes . . . sometimes he doubles up in paroxysm. You don't smile with Mr. Roosevelt; you shout with laughter with him, and then you shout again while he tries to cork up more laugh[ter] and sputters: 'Come gentlemen, let us be serious.' " Another journalist wrote, "You go into Roosevelt's presence . . . and you go home and wring the personality out of your clothes."

The White House rang out not only with laughter but with the squeals of children and the clattering of their ponies going up and down the marble stairs of the presidential mansion. Roosevelt was frequently to be found chasing or being chased by his children and their animals around the White House grounds. "You must always remember," said a British diplomat, "the President is about six." Certainly Roosevelt did nothing to dispel that impression. His zeal was infectious. His magnetic force, said one observer, "surrounded him as a kind of nimbus, imperceptible but irresistibly drawing to him everyone who came into his presence." Roosevelt used his infectious enthusiasm, which was tethered to a highly disciplined intelligence, to render unprecedented reform through the actions of the federal government. Nowhere was this more obvious and lasting than in his drive to conserve the American wilderness.

Roosevelt's passion for the American land and for natural history went back to his childhood. His father, one of the founders of the American Museum of Natural History, enthusiastically encouraged his young son's collecting and stuffing of animal specimens for the "Roosevelt Museum of Natural History," which the boy set up in the family's New York City mansion. He wrote in his autobiography that he had "fully intended to make science my life-work," an ambition reflected by his being first in his class in zoology at Harvard. When Roosevelt returned to New York from the Dakota Badlands, he formed the Boone & Crockett Club to promote preservation of big game animals and to encourage forest and land conservation. It was an effective group, later influential in establishing Yellowstone Park and in saving great stretches of timberlands.

By the time Roosevelt became president, America's natural resources had been stripped. Native bison herds were decimated; only eight hundred or so of the original sixty million animals survived. Many other species of birds and mammals were on the edge of extinction and nearly half of all forest lands had been logged. "Ever since man in recognizably human shape made his appearance on this planet," wrote Roosevelt, "he has been an appreciable factor in the destruction of other forms of animal life."

Roosevelt acted quickly to stop the destruction. "There can be no greater issue than that of conservation in this country," he declared. "We do not intend that our natural resources shall be exploited by the few against the interests of the many, nor do we intend to turn over to any man who will wastefully use them by destruction, and leave to those who come after us a heritage damaged by just so much." He ended his remarks in revival fervor: "We stand at Armageddon, and we battle for the Lord."

Roosevelt engaged not only the Lord, but the Congress and the American public as well. With characteristic vigor he set out on a campaign of persuasion. Action must be taken, it must be bold, and it must be now. His outrage was visceral, and his persuasiveness nearly absolute. "He is doubtless the most vital man on the continent, if not the planet, to-day," observed the naturalist John Burroughs. "He is many-sided, and every side throbs with his tremendous life and energy. . . . His interest in the whole of life, and in the life of the nation, never flags for a moment. His activity is tireless."

Roosevelt's exuberant campaign was manifestly successful. He doubled the number of national parks, created 150 national forests, added nearly 150 million acres of timber to the government reserves, set up more than fifty federal wildlife preserves, initiated thirty major irrigation programs, and established sixteen national monuments. One journalist commented that if Roosevelt continued to create reserves "there would be little ground left to bury folks on." The president's own summing up of his conservation efforts was different but equally succinct: "During the seven and a half years closing on March 4, 1909, more was accomplished for the protection of wildlife in the United States than during all the previous years, excepting only the creation of the Yellowstone National Park." He had been audacious in his use of the presidency and deft in his employment of conviction and enthusiasm.

Roosevelt's passion for saving the wilderness stayed with him. "Wild beasts and birds are by right not the property of the people alive to-day, but the property of the unborn generations, whose belongings we have no right to squander," he wrote a few years before he died. "It is barbarism to ravage the woods and fields, rooting out the mayflower and breaking branches of dogwood as ornaments." These were sustaining passions—"A grove of giant redwoods or sequoias," he believed, "should be kept just as we keep a great and beautiful cathedral"—and these passions and great enthusiasms made Roosevelt the activist he was. Without them it is unimaginable that the nation's wilderness would be as vast and wonderful as it now is.

Shortly before Theodore Roosevelt went to Norway to accept his Nobel Peace Prize in 1910 he gave a lecture at the Sorbonne, in Paris. It was his most eloquent tribute to the centrality of exuberance in action: "It is not the critic who counts; not the man who points out how the strong man stumbles, or where the doer of deeds could have done them better. The credit belongs to the man in the arena, whose face is marred by dust and sweat and blood; who strives valiantly . . . who knows the great enthusiasms, the great devotions; who spends himself in a worthy cause; who at best knows in the end the triumph of high achievement, and who at the worst, if he fails, at least fails while daring greatly, so that his place shall never be with those cold and timid souls who have . . . known neither victory nor defeat."

John Muir was different from Theodore Roosevelt in a thousand particulars. An immigrant Scot who knew neither a loving father nor a privileged upbringing, he preferred the company of mountains to that of men, had no desire to govern, and did not willingly put down in cities. But, like Roosevelt, he had a passion for wild places and a persuasive exuberance that vaulted passion into action.

Muir was born in 1838 on the east coast of Scotland. His lasting childhood influences, he said, were the sea and hills and his father's restless, often cruel, evangelical Presbyterianism. Nature was Muir's deliverance: "When I was a boy in Scotland," he wrote, "the natural inherited wildness in our blood ran true on its glorious course as invincible and unstoppable as stars. . . . Kings may be blessed; we were glorious, we were free—school cares and scoldings, heart thrashings and flesh thrashings alike, were forgotten in the fullness of Nature's glad wildness."

"Glorious" was a term Muir would invoke time and again in his accounts of nature, despite his conscious attempts to eradicate it from his writing. "Glorious" and "joy" and "exhilaration": no matter how often he scratched out these words once he had written them, they sprang up time and again to dominate his descriptions of the world as he experienced it. His exultant roots were deep, and never, as a writer or as a speaker, was he fully able to bridle the abounding delight of his language.

The Muir family emigrated to America in 1849. Young Muir worked at a brutal pace on their Wisconsin farm until he was able to escape, first to Madison and then, as a student, to the University of Wisconsin. A chance lesson in botany from a fellow classmate who queried him on the unlikely similarities between a straggling pea vine and the hardwood locust tree sent Muir "flying to the woods and meadows in wild enthusiasm." Before long, his college room, like Roosevelt's at Harvard, was brimming with life but instead of lobsters and tortoises he kept gooseberry bushes, wild plum, posies, and peppermint plants. He collected the specimens and studied them late into the night: "My eyes never closed on the plant glory I had seen," he later recalled.

Muir left Madison without a degree and wandered off, as he put it, into the "University of the Wilderness." This "glorious botanical and geological excursion" would last for the next fifty years. On the flyleaf of his journal, with the expansiveness that would characterize his life and writings, he inscribed: "John Muir, Earth-Planet, Universe."

When Muir was thirty years old he saw Yosemite for the first time. He fell in love. Everything, he wrote, was "glowing with Heaven's unquenchable enthusiasm. . . . I tremble with excitement in the dawn of these glorious mountain sublimities, but I can only gaze and wonder." Watching the sun come up between the mountain peaks and over the rock domes of Yosemite, he proclaimed what he experienced: "Our camp grove fills and thrills with the glorious light. Everything awakening alert and joyful . . . every pulse beats high, every cell life rejoices, the very rocks seem to thrill with life. The whole landscape glows like a human face in a glory of enthusiasm." With good cause, Muir's *My First Summer in the Sierra* has been described as the journal of a "soul on fire."

The mountains, the trees, and the air were, he effused, "joyful, wonderful, enchanting, banishing weariness and a sense of time." The streams "sing bank-full"; they leap, shout in "wild, exulting energy . . . joyful, beautiful." Muir joined in the exuberance of nature: "I shouted and gesticulated in a wild burst of energy," he wrote. "Exhilarated with the mountain air, I feel like shouting this morning with excess of wild animal joy." He climbed to the top of a hundred-foot Douglas spruce in the midst of a winter gale and, clinging to the tree, joined in its "rocking and swirling . . . [its] wild ecstasy." Never, he said, had he enjoyed "so noble an exhilaration of motion. The slender [tree]tops fairly flapped and swished in the passionate torrent, bending and swirling backward and forward, round and round, tracing indescribable combinations of vertical and horizontal curves, while I clung with muscles firm braced, like a bobolink on a reed."

Whether he was climbing in the Sierra or exploring the crevices of an Alaska glacier, Muir's fiery, joyous relationship with nature burned on. A friend described Muir's reaction to seeing a thick spread of mountain flowers in Alaska: "Muir at once went wild. . . . From cluster to cluster of flowers he ran, falling on his knees, babbling in unknown tongues, prattling a curious mixture of scientific lingo and baby talk." He felt himself to be "doomed to be 'carried of the spirit into the wilderness,' " and at times expressed concern that he was out of control: "I feel as if driven with whips, and ridden upon," he wrote his sister. "I am swept onward in a general current that bears on irresistibly."

Muir's exuberance found a match only in Nature's: "Every summer my gains from God's wilds grow greater," he wrote his fiancée from Alaska. "This last seems the greatest of all. For the first few weeks I was so feverishly excited with the boundless exuberance of the woods and the wilderness, of great ice floods, and the manifest scriptures of the ice-sheet that modelled the lovely archipelagoes along the coast, that I could hardly settle down to the steady labour required in making any sort of Truth one's own."

"Manifest scriptures" is a telling phrase. Muir's capacity to exult in nature, and his feeling of unity with the mountains and trees, led inexorably to a deep mystical appreciation of the world as he experienced it. The great sequoias were not just magnificent, they were sacred. The light that streaked through the sequoia woods was oracular, the tree sap sacramental. In a rhapsodic letter to a friend, written with the sap of a sequoia tree, Muir let his devotions tumble out: "Do behold the King in his glory, King Sequoia! Behold! Behold! seems all I can say. Some time ago I left all for Sequoia and have been and am at his feet; fasting and praying for light, for is he not the greatest light in the woods, in the world? . . . I'm in the woods, woods, woods, and they are in *me-ee-ee*. The King tree and I have sworn eternal love. . . . I've taken the sacrament with Douglas squirrel, drunk Sequoia wine, Sequoia blood. . . . I wish I were so drunk and Sequoical that I could preach the green brown woods to all the juiceless world, descending from this divine wilderness like John the Baptist . . . crying, Repent, for the Kingdom of Sequoia is at hand!"

Having taken communion with the woods, Muir offered up their liniment: "There is a balm in these leafy Gileads—pungent burrs and living King-juice for all defrauded civilization; for sick grangers and politicians; no need of Salt rivers. Sick or successful, come suck Sequoia and be saved." The salvation of the wilderness was not an abstraction to John Muir. He understood nature, felt nature, and then illuminated her to those who did not. The slaying of the wilderness was to him personal and intolerable.

With an almighty energy, Muir threw himself into saving the great groves of the sequoias and the mountain ranges of Yosemite. He brought into words the beauty he had seen and viscerally knew. He became an interpreter of nature, a prophet. "He sung the glory of nature like another Psalmist," said Muir's editor, Robert Underwood Johnson, and "as a true artist, was unashamed of his emotions." Muir took Johnson camping in Yosemite in 1889 and together they drew up plans for a campaign to establish what is now Yosemite National Park. Muir's writings and compelling enthusiasm were essential forces leading to the preservation of that part of the Sierra. When the Sierra Club was founded in 1892 to preserve the American wilderness, Muir became its first president; he held that office until his death twenty-two years later.

Muir branded others with his own almost painful awareness of the wilderness, attempting to make them feel some measure of what he himself felt so acutely. One friend, in discussing their exploration of Alaska, said, "Muir was always discovering to me things which I would never have seen myself and opening up to me new avenues of knowledge, delight, and adoration. . . . How often have I longed for the presence of Muir to heighten my enjoyment by his higher ecstasy, or reveal to me what I was too dull to see or understand . . . for I was blind and he made me see!" Person after person acknowledged a debt of profound comprehension: "To have explored with Muir the great glacier which bears his name," proclaimed one, "to have wandered with him in Yosemite and the Kings River Canyon, is to have come, through his enthusiasm and vision, a little nearer the hidden mysteries of nature."

Emerson, who visited Muir in Yosemite, said that Muir's was the most original mind in America. Muir used this originality to advantage, pouring it into persuasive, exultant language. His speech was described by those who heard him as nonstop and magnetizing—it had a "spell of fire and enthusiasm and glowing vitality," said one acquaintance—and Muir talked to all who would listen about the need to conserve the wild. He barraged lawmakers with letters and petitions and poured his heart into writing the books that would eventually bring the world's attention to the spectacular Sierra Nevada and its great stands of sequoias.

One of those who paid attention was the president of the United States, Theodore Roosevelt. "I write to you personally to express the hope that you will be able to take me through the Yosemite," he requested of Muir. "I do not want anyone with me but you." Muir agreed, and in May 1903 he met the president and together they set off with packers and mules. For several days they hiked and camped in the Sierra, an experience they both recollected with great pleasure. Muir wrote to his wife, "I had a perfectly glorious time with the President and the mountains"; to a friend he said, "I fairly fell in love with him." Roosevelt, who declared he had never felt better in his life, was no less enthusiastic in his letter to Muir: "I trust I need not tell you, my dear sir, how happy were the days in the Yosemite I owed to you, and how greatly I appreciated them. I shall never forget our three camps; the first in the solemn temple of the great sequoias; the next in the snow storm among the silver firs near the brink of the cliff; and the third on the floor of the Yosemite, in the open valley fronting the stupendous rocky mass of El Capitan with the falls thundering in the distance on either hand."

It is not possible to know the extent to which Muir's contagious enthusiasm for Yosemite and the sequoia groves influenced Roosevelt's subsequent actions. Roosevelt was already committed to the idea of conservation and had been for many years, but there is no doubt that he felt an additional sense of urgency after hiking with Muir. Immediately following his trip to Yosemite, Roosevelt delivered what was to become one of his most famous speeches. "I have just come from a four days' rest in Yosemite," he said to a gathering of people in Sacramento. "I want [the trees] preserved because they are the only things of their kind in the world. Lying out at night under those giant sequoias was lying in a temple built by no hand of man, a temple grander than any human architect could by any possibility build, and I hope for the preservation of the groves of giant trees simply because it would be a shame to our civilization to let them disappear. They are monuments in themselves." He talked about the need to conserve American forest lands and then closed with a plea for the rights of future generations: "I ask that your marvelous natural resources be handed on unimpaired to your posterity. We are not building this country of ours for a day. It is to last through the ages."

Muir expressed the same desire differently. "Any fool can

destroy trees," he wrote. "They cannot run away; and if they could, they would still be destroyed. . . . It took more than three thousand years to make some of the trees in these Western woods—trees that are still standing in perfect strength and beauty, waving and singing in the mighty forests of the Sierra. Through all the wonderful, eventful centuries since Christ's time—and long before that—God has cared for these trees, saved them from drought, disease, avalanches, and a thousand straining, levelling tempests and floods; but He cannot save them from fools—only [the government] can do that."

Muir knew in his marrow that wilderness was a necessity, that "going to the mountains is going home." It was scripture: in wilderness "lies the hope of the world." To save nature was to save oneself. "The galling harness of civilization drops off," he said, "and the wounds heal ere we are aware."

John Muir and Theodore Roosevelt were exuberant men. Infectiously enthusiastic, stupendously energetic, they left the country a wilder and more beautiful place because of their vision and action. Both registered the mountains and lands so keenly that they could not but act with urgency when the wilderness was threatened. Their receptive natures allowed them to feel and see that which was essential in the lands, that which could not be gotten elsewhere. Neither was capable of doing nothing when there was much to be done. Their joy in the wild was contagious to those around them. Both were persuasive by temperament and able to convince others of what they felt to be a moral imperative. Conservation was in their blood, not just in their intellect. It was elemental.

For Muir, it was a single, sustained, and consistent life's passion to preserve the wilderness. For Roosevelt, there would be many other crusades to engage his energies over a long and diverse political life. But because Roosevelt was a politician, because he had so many other passions and commitments, he was in a wider arena, with a more powerful scope, and therefore better able to act on behalf of the lands they cherished in common. "All of us who give service, and stand ready for sacrifice, are the torch-bearers," wrote Roosevelt. "We run with the torches until we fall, content if we can pass them to the hands of the other runners. . . . These are the torch-bearers; these are they who have dared the Great Adventure."

John Muir spoke of a more inward journey: "I only went out for a walk," he wrote, "and finally concluded to stay out till sundown, for going out, I found, was really going in."

CHAPTER TWO

"This Wonderful Loveliness"

In the exuberance of nature begins our own. And nature is self-evidently exuberant. One pair of poppies, given seven years and the right conditions, will produce 820 thousand million million million descendants. A single pair of spiders over the same time period and under ideal circumstances will give rise to 427 thousand million million more spiders. The fertility and diversity of nature are staggering. In a sliver of Brazilian forest only a few kilometers square, scientists have counted more than 1,500 species of butterfly. Lichens, among nature's oldest and slowest of living things, grow nearly everywhere—together with blueberries under the snow, on stained-glass windows of churches and cathedrals, in deserts and in birch woodlands, on the backs of tortoises—and an individual community of lichen may survive longer than most human civilizations. (One in Swedish Lapland is thought by scientists to have begun its collective life nearly nine thousand years ago.) The 13,500 known species of lichen are as exuberant in color as they are diverse in form and habitat, showing themselves in scarlets, snow whites, emerald greens, blacks, and sulfur yellows.

The abundance of nature is most obviously seen in the diversity of species, those "endless forms most beautiful" before which Darwin stood in awe. There are a million and a half species of fungi, eight million of insects (including 350,000 varieties of beetles alone), and more than a million species of bacteria. Nowhere is the exuberance of nature more apparent than in the tropical rain forest. Tens of thousands of species of animals and plants live on or under the forest floor, twine around the trees, or perch in the branches. Vines and orchids and fruits and flowers proliferate wildly under the canopy of the high trees. The rain forest is, as E. O. Wilson puts it, a "green cathedral," a teeming marker to nature's fullness and variation. But it is not the only luxuriant place. John Muir wrote of the layered richness of the Yosemite wilderness: "How deeply with beauty is beauty overlaid! The ground covered with crystals, the crystals with mosses and lichens and low-spreading grasses and flowers, these with larger plants leaf over leaf with every changing colour and form, the broad palms of the firs outspread over these, the azure dome over all like a bell-flower, and star above star." Each place he looked, everywhere in the mountains and valleys, Muir saw examples of nature's exuberance.

Nature's lavishness is not limited to the rain forests or to the Sierra Nevada, of course, although in those places it is spectacular. It is also evident in the dispersal of plants and animals across the earth and throughout the seas, in the snowflakes that fall by the skyful, distinct yet innumerable, and in the thickness of the starfields. Astronomers live among numbers that are unimaginable, fabulous, and terrifying. The universe, they believe, contains at least 10^{21} stars—1,000,000,000,000,000,000,000—and is home to more than 125 billion galaxies. The Milky Way alone, scientists reckon, contains 10^{41}—100,000,000,000,000,000,000,000,000,000, 000,000,000,000—grams of diamond dust; or, differently put, a million trillion trillion trillion carats. Nature is impossibly fertile.

"Exuberance," derived from the Latin *exuberance*—*ex*, "out of," + *uberare*, "to be fruitful, to be abundant"—is at its core a concept of fertility. Exuberance in nature is defined by lush, profuse, riotous growth; it is an overflowing, opulent, and copious abundance. Early uses of "exuberance" in English were mostly in the context of descriptions of nature, of profuse crops or of kinetic natural phenomena such as shooting stars, sulfur springs, and waterfalls. A fruitful outcome of an alchemy experiment, for instance, was characterized as "exuberated earth" in 1471, according to the *Oxford English Dictionary*, and in 1513 a particularly luxuriant plant was said to be "a pure perfyte plante . . . Marvelous by growynge . . . of grace exuberaunt." Two hundred years later, in "Spring," the Scottish poet James Thomson addressed not only "exuberant Nature," but the exuberant emotions provoked by nature:

Of mingled blossoms; where the raptur'd eye
Hurries from joy to joy . . .

. . . .

Then spring the living herbs, profusely wild
O'er all the deep-green earth, beyond the power
Of botanists to number up their tribes.

Over time, the definition of "exuberance" evolved; where it had focused on the fertility of nature, it began to center on the fertility and force of human energy and mood. Samuel Johnson, in his dictionary published in 1755, gives an example of "exuberance" taken from the writings of Joseph Addison. Its meaning is modern not only in the mood it describes but in the infectiousness of that mood: "that *exuberant* devotion, with which the whole assembly raised and animated one another." In our time, "exuberance" usually denotes a mood or temperament of joyfulness, ebullience, and high spirits, a state of overflowing energy and delight. It is more energetic than joy and enthusiasm but less intense, although of longer duration, than ecstasy. The origins of the concept of exuberance in the cyclic fertility of nature, now largely forgotten, remain critical to understanding it as a primitive life force vital to survival.

For all of nature's exuberance, there are constraints upon it. Nature proliferates, but it also kills. Predators, disease, and drought impose limits on growth, as do fluctuations in light and temperature. Precarious balances are kept and broken, and exuberancies ebb and flow in response to the changes. Underlying everything is nature itself, and it is in response to nature that we have evolved the senses and temperaments by which we variously respond to the physical world. Nature is our literal world and the genesis of our imaginative one as well. It was, as Edward Hoagland has argued, the "central theater of life for everybody's ancestors."

The experience of nature is an idiosyncratic thing. The eyes of the poet are not the same as those of the farmer: needs are different. A poet may look from a mountaintop, as Robert Crawford has, and see the beauty in the "bens and glens of stars"; a farmer may see instead the grazing promise of the lands below. A passionate emotional response may be more natural to the former, pragmatic surveillance of more use to the latter. For most in a village, a mountain range may be the border they know to be inviolable, to be kept to, but for a few the mountains will be irresistible, an enticement to risk and rapture. Some see and feel nature's presence acutely and are driven by restlessness, curiosity, or the possibilities of pleasure to explore it further. Others stay in the lands they know; they react not so passionately, seek out not quite so much.

This makes sense. It is essential that some feel so strongly that they are compelled to explore and to work out the ways of nature—it is they who discern the patterns of changing light, measure variations in rainfall, plot the movement of the stars; who monitor the sprouting of seeds or the darkening of grass, track the migration of caribou or mammoth; and who take note of the dangers of belladonna or the benefits of willow bark—for man's well-being depends upon observing the behavior of nature. To observe is to make preparation possible, to predict. "Can you bind the beautiful Pleiades?" asks God of Job. "Can you loose the cords of Orion? Can you bring forth the constellations in their seasons? Do you know the laws of the heavens?" The complexities of nature are manifold, to comprehend the laws of nature infinitely seductive.

Exuberance is not necessary for keen observation, of course. Nor is it a requirement for ascertaining the underlying patterns of nature. But, as we shall see, those who are exuberant engage, observe, and respond to the world very differently than those who are not and, what is crucial, they have an intrinsic desire to continue engaging it. This engagement is generally not a muted one; rather, it is one filled with a sense of passion, if not actual urgency: "Who publishes the sheet-music of the winds, or the written music of water written in river-lines?" asked John Muir, with the intensity that shot through his life and work. "Who reports the works and ways of the clouds . . . And what record is kept of Nature's colors?" Muir's temperament vibrated in response to all that moved around him. He *felt* nature's sounds and colors, experienced them with joy; he was compelled to keep a record of those colors and to publish the music of the winds.

There is little more important to us than to have some portion of our species respond acutely to transformations in the natural world. To hunt, plant, breed, harvest—all of these require an awareness of changes in the length of daylight and fluctuating fecundities in the physical world. Those most responsive to these changes surely had a survival advantage over those who were less alert. Many of our most intense emotional reactions tend to occur in response to precipitous changes in nature: our pulse quickens or slows; we stay in place, transfixed, or we bolt; we are filled with joy or terrified. These responses are deeply wired into our brains from our premammalian ancestors. Alertness is intrinsic to our survival. So, too, is joy. The pleasure and benefit we find in understanding a part of nature rewards our curiosity, our inquiry; expectation of pleasure makes it more probable that we will inquire yet again and act upon our curiosity in the future.

Much of our joy is in direct response to nature. "In the presence of nature," writes Emerson, "a wild delight runs through man, in spite of real sorrows. Nature says,—he is my creature, and maugre all his impertinent griefs, he shall be glad with me. Not the sun or the summer alone, but every hour and season yields its tribute of delight; for every hour and change corresponds to and authorizes a different state of the mind."

Our minds and our emotions mind nature. In her landmark study of ecstasy, Marghanita Laski found that the most common "triggers," or inducers of transcendent ecstasies, are from nature: water, for instance, and mountains, trees, and flowers; dusk and nighttime; sunrise, sunlight; dramatically bad weather; and, of course, spring. Laski alleges, and in this she concurs with the ethologist Niko Tinbergen, that these "triggers" release biologically desirable responses of attraction. They are, in essence, attracting, engaging, and bolstering forces. The joy brought about by an ecstatic reaction to spring, for instance, makes it more likely that an individual will watch for signs of its approaching, attend to subtle changes in light and temperature, and actively participate in the season's pleasures and demands once it has arrived.

We bring to spring the expectations planted by nature in the brains of our ancestors. They waited upon the sun's return and, in the while, sent out their rainmakers and firemakers to coax the day into lengthening and to beguile the winter darkness into vanishing. They believed that to recapture long days of warmth and light they needed only to drape themselves in bark and leaves and flowers; that is, they needed only to look the part of spring. Magic could blandish nature into returning her plants and sunlight.

Later, when enough seasons had passed to brand their unvarying pattern upon those who kept watch on them, magical rituals turned increasingly to an invocation of the gods. The Greek, said Sir James Frazer, "fashioned for himself a train of gods and goddesses . . . out of the shifting panorama of the seasons, and followed the annual fluctuations of their fortunes with alternate emotions of cheerfulness and dejection. . . [which] found their natural expression in alternate rites of rejoicing and lamentation." Dionysian worship—distinguished by ecstatic dance and music as well as sacrificial blood feasts—was one of the most evident of these rites. Exuberant behavior was meant to yield an exuberant harvest. An exuberant harvest, in turn, incited exuberant celebration. Man's primitive gods, as well as his moods and actions, took their cue from nature.

Our dispositions and our generativity swing with the turning of the seasons. We wait for joy to return. "Nothing is so beautiful as Spring," exclaimed Gerard Manley Hopkins. "What is all this juice and all this joy?" The contrast between the bleakness of winter and the "juice and joy" of spring; the sense of a greater distance from death; a hope for regeneration: all these go into our emotional responses to the change of seasons. Our vitalities change in response to the quickening we observe in nature. Our energy is fullest and our moods most expansive in the presence of long-lit days. They fall precipitously in the shorter, darker ones. We rejoice at the turning of winter into longer days. "The brooks sing carols and glees," Thoreau wrote of Walden Pond's transition into spring: "The sinking sound of melting snow is heard in all dells, and the ice dissolves apace in the ponds. . . . The symbol of perpetual youth, the grass-blade, like a long green ribbon, streams from the sod." Such is the contrast between winter and spring, said Thoreau. "Walden was dead and is alive again."

The festivals and celebrations to commemorate the return of light and renewed fertility are ancient, universal, and often spectacularly joyous. The Christian celebration of Easter, like pagan festivals before it, proclaims the great victory of light over the forces of darkness, of life over death. Easter, named for the Anglo-Saxon goddess whose festival was held at the spring equinox, is a movable feast: its celebration date—the first Sunday after the first full moon following the vernal equinox—is embedded in the cycles of nature, in the eternal wanderings of earth, moon, and sun. Indeed, common belief once held that the sun dances on Easter Day, a marvelous image of joy captured by the poet John Heath-Stubbs:

I am the great Sun. This hour begins
My dancing day—pirouetting in a whirl of white light
In my wide orchestral sky, a red ball bouncing
Across the eternal hills;
For now my Lord is restored; . . .
. . . .
Look, I am one of the morning stars, shouting for joy—.

The joy of the sun is ours.

Easter observations actually begin, however, in a deep darkness reminiscent of winter and death. On Good Friday, the day of

Christ's crucifixion, all that is bright or decorative in a church is removed: the altar cross is draped in black cloth; there is no singing. The Gospel reading tells of the death of a god, of betrayal and suffering. Church bells toll at three p.m., the time Christ is said to have died; a time, according to St. Matthew, when darkness covered the land. There is no remedy of light on Good Friday; only darkness obtains.

The oldest liturgy in the Christian church takes place the following night and it, too, begins in darkness. For two millennia, in open or furtive gatherings around the world, an Easter Vigil has been kept to mark the symbolic transition of darkness into light, the triumph of life over death, redemption over sin. Congregants wait for the first light from candle or fire and watch as the sun rises in the sky on Easter morning. For many, Easter Sunday is a deeply religious occasion, the celebration of the Resurrection of Christ. For others, it is different: an occasion of bright color and exuberant music, or a nod to the traditions of childhood. For most, whatever their beliefs, there is a remnant of the ancient desire to exult in the return of spring, to venerate the renewal of life.

Our response to the return of light is joy, and in that joy we recognize our beholdenness to the natural world. We believe in spring because we know we can, but we are experienced enough as a species not to take it entirely for granted. Joy recognizes these moments of uncertainty as certainly as it recognizes the glories of spring. "There was only—spring itself," wrote Willa Cather in *My Ántonia*, "the throb of it, the light restlessness, the vital essence of it everywhere; in the sky, in the swift clouds, in the pale sunshine, and in the warm, high wind—rising suddenly, sinking suddenly, impulsive. . . . If I had been tossed down blindfold on that red prairie, I should have known that it was spring." It comes in fits and starts, and we respond in kind. But we know it is spring.

Not only spring, but its passing into the long days of summer, has been cause for exuberant festivals. May Day, still celebrated in many places, albeit in a dampened way, was once one of the most riotously joyful occasions of the rural year. Festive bonfires were lit on the hills and young men and women went "a-maying" and followed the sounds of horns into the woods to cut down branches; these they decorated with flowers and hung over the windows and doors of their homes. Hoops were covered with greens and ribbons and laced with flowers, and May carols were sung. Maypoles, symbols of fertility, were cut from trees and garnished with bits of ribbons and cloth, leaves, colored eggshells, and bright flowers; villagers plaited ribbons as they danced around the maypole, celebrating the renewal of nature. "The earth / Puts forth new life again," wrote Langston Hughes. "The wonder spreads."

May Day ceremonies, and those enacted later on Midsummer Day, are rejoicings in the fullness of orchards and crops, days white with blossom and open to hope and possibility: "With the sunshine and the great bursts of leaves," wrote Scott Fitzgerald in *The Great Gatsby*, "I had that familiar conviction that life was beginning over again with the summer." The seasons of nature become the seasons and convictions of man. We celebrate spring because the sun is back, May because the trees and plants are flush, and the finish of harvest because the corn and apples are picked, the livestock ready. As days shorten and darkness dominates, we turn to midwinter fires or Christmas festivities for warmth and an assurance of the continuity of life. The joyousness of Christmas has few equals, and most of its great carols are no less exuberant than the exultant hymns of Easter—"the dark night wakes, the glory breaks," rings out the carol, "And Christmas comes once more."

From our dependence upon nature evolved senses and emotions able to respond to its danger, beauty, and opportunity. We are by our nature, by these adaptations, urgently connected to the natural world. "When we try to pick out anything by itself," observed John

Muir, "we find it hitched to everything else in the universe." The hitching is the critical thing. Exuberance, as we shall see, makes the hitching stronger and the exploration of the universe more likely: it fuels anticipation; overlooks or minimizes risks and hardships; intensifies the joy once the exploration is done; and sharply increases the desire to recapture the joy, which in turn encourages further forays into the unknown. Those most enthusiastic and energetic in their responses to nature tend to be those who most profit from it in pleasure. They are also those most likely to expand their minds to comprehend it. The physicist Richard Feynman was certainly one of these. "The vastness of the heavens stretches my imagination," he said. "Stuck on this carousel my little eye can catch one-million-year-old light. A vast pattern—of which I am a part—perhaps my stuff was belched from some forgotten star." We are part of nature; we come from the stars and we reach out to apprehend them. We are stardust in spirit and in fact; and when we delight in this, we delight in this in ourselves.

This was certainly true for Wilson "Snowflake" Bentley, a New England farmer who pursued far smaller bits of sky, the infinitely various snow crystals, with a single-minded delight. He was as exuberant in his pursuit of them as they were in their numbers. Nature was to Bentley an unbroken source of joy.

It is a rare person who remains unmoved by a first snowfall. Snow is magic: it draws us in, jostles memory, and stirs desire. It enchants. For Snowflake Bentley, snow cast a lifelong spell. Like John Muir and Theodore Roosevelt, whose contemporary he was, Bentley was incapable of indifference to the world around him. When there was a winter storm and snow was flying, he was in the fields or hills; he could not stay indoors. His delight in snow made him an astute observer of it; it then made him an infectiously enthusiastic guide. Exuberance gave him passion, stamina, and a lasting voice to speak out for small beauties.

Wilson Bentley was born on a Vermont farm in 1865, just as the Civil War was ending. He was captivated by the beauty of snow crystals even when very young, and managed to persuade his parents to buy him a camera and microscope. By the age of nineteen he had taken the first ever photomicrograph of a snow crystal. He was irretrievably smitten. "Amazed and thrilled at their matchless loveliness," Bentley wrote many years later, "the work soon became so all-absorbing that I have continued it with undiminished enthusiasm all these years. No words can convey the least idea of the intense enjoyment, the almost countless thrills, these winter studies have afforded me." Unlike John Muir, who went from the exploration of the vast wilderness lands to the apprehension of self, Bentley went from study of the infinitesimal to contemplation of the grand: "The deeper one enters into the study of Nature," he believed, "the further one ventures into and along the by-paths that, like a mystic maze, thread Nature's realm in every direction, the broader and grander becomes the vista opened up to the view."

Bentley could not remember a time when he did not love the snow. Always, from the beginning, he said, "it was the snowflakes that fascinated me most." From the first snowfall to the last, he was supremely happy. Passionate about snowflakes, he devoted his life to their study and preservation. "I found that snowflakes were miracles of beauty," he once said to an interviewer. "It seemed a shame that this beauty should not be seen and appreciated by others. Every crystal was a masterpiece of design and no one design was ever repeated." He was as stricken by their impermanence as struck by their beauty: "When a snowflake melted," he lamented, "that design was forever lost. Just that much beauty was gone, without leaving any record behind." One snowstorm brought him the most exquisite crystal he had seen to date, "a wonderful little splinter of ice, incredibly fragile," but despite his care the crystal was broken while transferring it to a slide. Even after many years had passed he was to declare the loss of the snow crystal "a tragedy," and only with effort would he be able to hold back his tears.

Bentley was insistent upon saving his "snow blossoms" for the rest of the world; he was possessed, he said, by a "great desire to show people something of this wonderful loveliness, an ambition to become, in some measure, its preserver." Just as Muir and Roosevelt could not feel as they did about the American wilderness and not do everything within their powers to save it, so too Bentley looked at snowflakes, loved them, and then did all he could do to preserve their beauty. Snow crystals existed for a reason, he was convinced: "Perhaps they come to us not only to reveal the wonderous beauty of the minute in Creation but to teach us that all earthly beauty is transient and must soon fade away. But though the beauty of the snow is evanescent . . . it fades but to come again." (Thoreau, who died only a few years before Bentley was born, also had a near-mystical response to snowflakes: "How full of the creative genius is the air in which these are generated!" he wrote in his journal. "I should hardly admire more if real stars fell and lodged on my coat. Nature is full of genius, full of the divinity; so that not a snowflake escapes its fashioning hand." Nature, he reflected with hope, had "not lost her pristine vigor yet, and why should man lose heart?")

Bentley's calling was to preserve the snow crystals and, once they were preserved, to give their loveliness an exuberant voice. He did the former with a patience that is nearly impossible to imagine, painstakingly taking photographs of more than five thousand individual crystals during his lifetime. Winter after New England winter he stood in the freezing cold as the snows fell, capturing crystals midflight, transferring them to glass plates, and photographing them before they could melt. Later, when he published their delicate images in the journals of science, his exuberance danced across the pages.

Enthusiastic descriptions of the shapes and origins of snow crystals, which bubbled up irrepressibly in his writings, were utterly out of keeping with the more circumspect language of most scientists. Indeed, Bentley's language would be stricken from any modern scientific journal; even a whiff of it would result in withering reviews and raised eyebrows from more measured colleagues. Strong emotion, more often than not, is at cross-purposes with accurate scientific description. Enthusiasm is meant to be kept on a tight rein and love itself on a short lead, although one could argue, as Cyril Connolly did, that he who is *too* much a master of his passions is reason's slave.

Bentley need not have worried about such enslavement. In one scientific paper, published in 1902, Bentley used the words "beauty" or "beautiful" nearly forty times in nine pages. The paper was about the atmospheric conditions affecting the size and form of snow crystals, as well as the classification of crystals and their occurrence and distribution in relation to various drifts and types of clouds and temperatures. But Bentley also wrote about the loveliness of the snow crystals whose photographic images he had chosen to include in the paper. They were, he said, "*marvelously beautiful* objects of nature . . . *the feast of* [their] *beauty* fills these pages." Snow crystals Nos. 716 and 718, he proclaimed, were "very choice and *beautiful*," and Nos. 722 and 723 were "charming patterns in snow architecture." He went on, enraptured by what he described as the "gems from God's own laboratory": "No. 785 is *so rarely beautiful*," he enthused, and No. 781 is "*wonderfully beautiful*," while the "*great beauty* of No. 837 will appeal to all lovers of the beautiful." Other crystals were "*exquisitely*" or "*exceptionally beautiful*." The snowstorm of February 1902, he gushed, contributed "choice examples of snow crystal architecture, as souvenirs of the skill of the Divine Artist."

Bentley was unable to contain himself, even when making scientific hypotheses. In one scientific paper, he started his speculations about the growth of crystals in a straightforward way: "I assume that the configurations of the exterior portions of the crystals surrounding the nucleus must depend largely upon the initial and subsequent movement, or the flights, downward, or horizontally, of the growing crystals within the clouds," he wrote. The objectivity of his language, to this point, is indistinguishable from that of any other scientist writing in the same journal. He continued for a while in a dispassionate vein: "We must therefore make a careful study and analysis of the interior portions of crystals. . . . These interior details reveal more or less completely the preexisting forms that the crystals assumed during their youth in cloudland."

But then Bentley's joy in the beauty of snow crystals breaks through: "Was ever life history written in more dainty or fairy-like hieroglyphics?" he asked. "How charming the task of trying to decipher them." It would be impossible, he concluded, to find the ultimate snowflake, though that would not keep him from ardent pursuit. "It is extremely improbable that anyone has as yet found, or, indeed, ever will find, the one preeminently beautiful and symmetrical snow crystal that nature has probably fashioned when in her most artistic mood."

Duncan Blanchard, an atmospheric scientist who has written the definitive biography of Snowflake Bentley, likens Bentley's search for the "preeminently beautiful snow crystal" to Sir Galahad's for the Holy Grail. This quest, believes Blanchard, "sustained and nourished Bentley with undiminished enthusiasm until his dying day. This was exuberance at its best."

Certainly, twenty-five years after writing about the "preeminently beautiful" snow crystal, Snowflake Bentley was still enthralled. And still looking. Subsequent winters provided him a wealth of new crystal photographs, and forty of the new "snow gems," he was sure, could be described as "wonderful" or "masterpieces." Individual crystals, he rhapsodized, had to be seen to be believed. "The beautiful branching one that fell December 9, 1921, No. 399, is a masterpiece of crystal architecture," he exclaimed with his usual zeal, and No. 4215 was "thrillingly beautiful." He wished that all readers of the journal in which his latest photographs appeared "could see and enjoy the snowflake masterpieces of this winter."

The images of the snow crystals reproduced in the article are indeed beautiful, and Bentley's ebullient portrayals very much make one wish one could have been there during the snowstorms as he captured the crystals falling to earth. Who would not have wanted to be there during the 1927 snowflake season as described by Bentley, especially during the "wonderfully brilliant closing" of one late February day recalled by him? On that date, he exclaimed, "the clouds for a while showered the earth with starry, fernlike gems such as thrill, amaze, and delight snowflake lovers." His delight is contagious.

Bentley is famous for his declaration that no two snowflakes are alike. He and other scientists knew that the infinite varieties of temperature and humidity conditions act together in such a way as to idiosyncratically notch crystals on their downward flight; unless collected at very high altitude before its journey is done, each snow crystal will be unique. This is true even of crystals artificially created in laboratory snow tanks. No two will be alike; each will carry the physical history of its individual travels. A single ice crystal contains some ten sextillion molecules; therefore, "considering all the ways those molecules can be arranged," argues one contemporary scientist, "the odds against any two completely identical snowflakes having fallen since the atmosphere formed some four billion years ago are enormous." Another has stated that "it could snow day and night until the sun dies before two snow crystals would be exactly, precisely alike." This is a marvelous, if unprovable, thought. Snowflake Bentley intuited such singularity and loved it.

Bentley's enthusiasm for snowflakes would be simply a footnote in the annals of enthusiasts and eccentrics were it not for the results of his sustained passion, for it was a passion which allowed him to withstand the chill of both winter and his Vermont neighbors. He endured the inevitable frustrations and failures involved in capturing and photographing a solitary snow crystal before it melted into nothingness because he felt an urgency that others did not. Bentley's temperament and sensibilities impelled him to share beauty with those less exposed to it and to proselytize those who felt less acutely than he. His exuberance brought to millions a loveliness that fell from the skies. He saw, he felt, and he captured a tiny gorgeousness for history. No bit of Nature ever had a better Boswell.

Bentley loved snowflakes above all else, but he also made important contributions to the understanding of other phenomena of nature. He observed and carefully described the date, appearance, and intensities of more than six hundred auroras and took meticulous measurements of nearly three hundred fifty collections of raindrops. He was a pioneer photographer of clouds, frost, and dew, and his work on cloud physics, in the assessment of Blanchard and other atmospheric scientists, was forty years ahead of its time.

The people in his New England village, however, regarded him as a little cracked. Being Vermont farmers and less than transfixed by snow, they found Bentley's intoxication odd. Why take pictures of snowflakes, they asked, when "you can't sell them and you can't eat them." Fortunately, the American Meteorological Society disagreed and awarded the self-educated dairy farmer its first research grant. His photomicrographs of snow crystals made their way into scientific journals—sixty were published in *Nature* alone—as well as into popular newspapers and magazines, and they influenced naturalists, photographers, scientists, and jewelry designers at Tiffany. There is no equivalent of his photographic collection, nor is it likely that there will ever be one.

When Bentley died in 1931, even his Vermont neighbors had a sense of the importance of his life and passing. "John Ruskin declared that genius is only a superior power of seeing," wrote his hometown newspaper. "Wilson Bentley was a living example of this type of genius. He saw something in the snowflakes which other men failed to see, not because they could not see, but because they had not the patience and the understanding to look."

Nor had they his capacity for joy or exuberant pursuit. "So long as eyes shall see and kindle at the beautiful in Nature," Bentley said, his camera and pen would be there. It was this capacity to be kindled, of course, that set Bentley apart. His urgency and passion ensured that his message would be both seen and heard. The physicist W. J. Humphreys, one of the many eminent scientists who were deeply impressed by his work, wrote the text to accompany Bentley's photographic masterpiece, *Snow Crystals.* In it, he observed that Bentley had pursued his life's work with the "insistent ardor of the lover and the tireless patience of the scientist," that he had "made it possible for others to share at leisure, and by the comfortable fireside, the joys that hour after hour bound him to his microscope and his camera in an ice cold shed." Bentley brought indoors an otherwise invisible beauty from the skies.

Bentley's was a magnificent obsession, plumb-line true and enduring. Just days before he died, he wrote in his weather notebook for the last time. "Cold west wind afternoon," the entry reads. "Snow flying."

CHAPTER THREE

"Playing Fields of the Mind"

For most mammals, including ourselves, early exploration of the world is enhanced, indeed often made possible, through the exuberant play of youth. Such play, it has been said, is the business of childhood, but play is more than that: it is a deadly serious business. Much learning must get done in not much time, for youth is, indeed, a stuff which will not endure. The time is short when a young animal, still protected and provided for by its parents and not yet bound to the waiting demands of hunting, mating, and procuring shelter, can run flat out, gambol, and improvise with impunity. "In the sun that is young once only," wrote Dylan Thomas, "Time let me play and be / Golden in the mercy of his means."

We play because we have an exuberance of spirits and energy, but we are also exuberant because we play. We seek to play not only because it is a part of our evolutionary history, but because we know that more often than not it will bring pleasure. That pleasure, in turn, makes us more likely to act in ways that increase our chances of survival and sway. Long before our species adapted to the seasons and terrains dealt us by nature, other animals had learned how best to capture food and reach water, how to outflank predators, to be aware of their own and wary of strangers. They had evolved ways to fashion strong bonds with kin and learned the particulars of their home territories through exploration and risks taken. A coupling of instinct with learning, of pleasurable play with group bonding, of joy with curiosity and invention meant that animals with such capacities were more likely to respond with facility to changes in their environment. The varieties and combinations of behaviors tried out in play were among those that increased the odds of behavioral flexibility. Play exists, in significant measure, to promote plasticity and to teach an animal to take advantage of opportunity.

Play is a vital facilitator, shaper, and motivator: it allows the pleasurable practice of improbable twists and turns in instinctive behaviors which, in turn, creates for the animal a wider range of possibilities for future actions. It shapes the developing brain in potentially lifesaving ways. "Natural selection," wrote the philosopher Karl Groos in 1898, "will favour individuals in whom instinct appears only in an imperfect form, manifesting itself in early youth in activity purely for exercise and practice—that is to say, *in animals which play.*" It is not unlikely, Groos went on to say, that "*the very existence of youth is due in part to the necessity for play;* the animal does not play because he is young, he has a period of youth because he must play."

By its nature, play is rather diffusely defined; the origins of the word remain surprisingly obscure. *Plein,* meaning, in Middle

Dutch, "to dance about, jump for joy," is thought to be a linguistic ancestor of the English word *play;* so, too, is the Old German word *Spilan,* denoting a "light, floating movement." "Play" takes up seventeen long columns and accounts for more than one hundred individual definitions in the *Oxford English Dictionary.* To these, modern science has added its own numbingly precise ones. Most definitions center on a feeling of free or unimpeded movement; activities involving fun and amusement and characterized by swift, exuberant, irregular, or capricious motions; and a springing, flying, or darting to and fro, a joyous gamboling and frolicking about.

Play, as Stephen Miller of Harvard has put it, is a "soup of behavior," something we recognize when we see it, but find hard to pin down in language. Miller studied zoo-goers watching animals at play and found that they were quite consistent in what they labeled as play, noting, for example, that "it didn't look like it was for real" or that the animals "looked like they were enjoying themselves." Scientists, however, found it far more difficult to label the diverse behaviors of play. Miller believes that the zoo-goers were almost certainly responding to a wide variety of cues which, because of their very subtlety (and, one would also guess, their effects on ancient, preverbal portions of the brain), scientists felt unable to measure objectively.

There are several typical features of play. Physical movements are often exaggerated; they are much slower or faster than usual, or much larger or smaller. Animals at play, Miller observes, move in ways that display "much flailing, bobbing, exaggeration, and indirect, ineffective action. In short . . . a 'galumphing' appearance." "Galumphing," borrowed from its inventor, Lewis Carroll, is a nearly perfect word. The *O.E.D.* notes that it is reminiscent of "gallop" and "triumphant," conveying a sense of "marching on exultingly with irregular bounding movements." Carroll, a master of linking play with pleasure and wit, used "galumphing" in *The Hunting of the Snark* to describe the actions of the seagoing, lacemaking Beaver, who, having hunted the Snark, "went simply galumphing about." The slaying of the Jabberwock, in *Through the Looking Glass*, also entails galumphing:

One, two! One, two! And through and through
The vorpal blade went snicker-snack!
He left it dead, and with its head
He went galumphing back.

. . . .

"O frabjous day! Callooh! Callay!"
He chortled in his joy.

"Galumphing" completely captures the joy and bounce of play: it is a word that sounds as play is; is in itself inventive, as play is; and is as open to meaning and possibility as the freewheelingness of play.

The exaggerated, cavorting quality of play acts in part to signal to other animals that it, and not other, more aggressive behavior, is intended. Rhesus monkeys running to play, for example, are distinguished by their bouncing locomotion and rotating heads or torsos. These physical movements, like the "play face" exhibited by other primates, or like laughter in humans, invite other members of the species to play. The relaxed facial expressions of black bear cubs convey playful intent, as do the highly specific positionings of their large and mobile ears. There is also a repetitive characteristic to playful encounters, as well as a "mock" quality, that is, a pretending or "as if" stance toward reality. This so-called nonliterality is one of the most reliable indicators of play in children. Play, said Karl Groos, "gives the whole world of appearance a special colouring, distinguishing it from everything that is real, and rendering it impossible that even in our utmost absorption we should ever confuse the make-believe with the real."

Play, unlike hunting or defense or most other behaviors,

appears to provide little immediate benefit to the individual animal or its species. The importance of the activity seems to lie for the most part in future benefits that accrue from the process of play itself, rather than in achieving any particular short-term goal. But, as we shall see, the process is vital. It is also pleasurable, and that pleasure, to the extent it encourages social or inventive behaviors beneficial to the individual and ultimately the species, is likewise vital.

Pleasure is a largely subjective state, of course, and, until recently, animal behaviorists have been reluctant to acknowledge its presence in nonhuman species. Few, however, deny the exuberant playfulness of young mammals. George Schaller, a biologist with the New York Zoological Society, describes the ebullient behavior of a two-year-old panda after it was released from a dark cell into the outdoors: "It exploded with joy. Exuberantly it trotted up an incline with a high-stepping, lively gait, bashing down any bamboo in its path, then turned and somersaulted down, an ecstatic black and white ball rolling over and over; then it raced back up to repeat the descent, and again."

In her classic study of Australian wombats, Barbara Triggs writes: "Wombat play is made up of several characteristic movements and attitudes performed in no particular order but with tremendous enthusiasm and exuberance. Typically, a young wombat signals the beginning of playtime by standing absolutely still with its front legs stiff and straight. Then it jerks its head and shoulders up, sometimes lifting its front feet right off the ground. Then, but not necessarily in this order, it tosses its head from side to side; jumps in the air with all four feet off the ground; rolls over on to its side; races off at a rocking gallop before coming to a sudden stop, reversing through 180 degrees 'on the spot' and racing back to its mother, stopping or veering sideways just before the expected collision; lies flat on its stomach, head thrown back and swinging from side to side, lips drawn back in a wombat 'grin.' If it is playing on or near a slope, it will sometimes run up the slope and roll down, tumbling over and over on its side." Triggs, acknowledging the infectious quality of the marsupial's exuberance, concludes by saying "I defy anyone to watch a wombat at play without laughing aloud."

Some species—the primates, the marine mammals (especially dolphins, sea lions, and seals), the *Mustelidae* (which includes weasels, otters, and ferrets), and the dog and cat families—are more exuberantly playful than others. The sheer zippiness of the animals can be breathtaking. This is particularly true of weasels, which have been wonderfully described as "hair-trigger mousetraps with teeth." One owner described his weasel's pelting-about: "From whichever retreat hid him for the moment, a wedge-shaped head and wicked pair of eyes would appear. Then out he'd roll, turning cartwheel after cartwheel like an acrobat going round the circus ring. He moved so fast that it was impossible to distinguish where his head began and his tail finished. He was like a tiny inflated rubber tyre bowling round the room."

Irrepressible playfulness has been observed in river otters as well. Gavin Maxwell, who wrote *Ring of Bright Water* and for whom one race of otters is named, described the zestfulness of the otters living with him in the Scottish highlands. The principal otter characteristic, he came to believe, was perpetual play. Otters, whom he depicted as "extremely bad at doing nothing," were mesmerizing, if often exhausting companions. Maxwell recounted one otter playing for hours at a time with "what soon became an established selection of toys, ping-pong balls, marbles, india-rubber fruit, and a terrapin shell that I had brought back from his native marshes. The smaller among these objects he became adept at throwing right across the room with a flick of his head, and with a ping-pong ball he invented a game of his own which would keep him engrossed for up to half an hour at a time. An expanding suitcase that I had taken to Iraq had become damaged on the journey home, so that the lid, when closed, remained at a slope from one end to the other. Mij discovered that if he placed the ball on the high end it would run down the length of the suitcase unaided. He would dash round to the other end to ambush its arrival, hide from it, crouching, to spring up and take it by surprise as it reached the drop to the floor, grab it and trot off with it to the high end once more." At other times, Maxwell wrote, the otter "would set out from the house carrying a ping-pong ball, purposeful and self engrossed, and he would still be at the waterfall with it an hour later, pulling it under water and letting it shoot up again, rearing up and pouncing on it, playing his own form of water polo, with a goal at which the human onlooker could but guess."

Many physical aspects of play seem to create pleasure for its own sake, although this is necessarily a subjective call. Marc Bekoff, a biologist at the University of Colorado in Boulder, is one of the leaders of a group of animal behaviorists that is giving new credence to the surprisingly understudied field of animal emotions. Like an increasing number of scientists, Bekoff argues that to assume that only humans experience an emotion such as joy is arbitrary; it makes no sense at all from an evolutionary point of view. Animals engaged in exuberant play seem a particularly striking example of this. Although it may appear subjective to label an animal's experience as pleasurable or joyful, it is unscientific to assume out of hand that a pleasurable emotion does not exist, or to ignore it simply because it is difficult to measure. There are limits always on what we can know about ourselves, about others of our kind, and certainly about those of another species. (We will never know, for example, what a weasel thinks about. Annie Dillard has put this beautifully: "He won't say. His journal is tracks in clay, a spray of feathers, mouse blood and bone: uncollected, unconnected, loose-leaf, and blown." Each species keeps its own emotional and mental ways. We scarcely understand our own, much less those of other animals.) But the limits on our knowledge should not lead us to dismiss the existence in animals of intense and enjoyable emotions. Evolution would be odd indeed to endow only humans with zest and joy.

Animals seem to take particular pleasure, as we do, in movements that are enjoyable to the senses—great leaps, fast forward motions, floating or rocking movements—as well as in actions that are particularly energetic in their own right. Speed is intoxicating, and the quickened heartbeat and fast respiration of vigorous play are often highly enjoyable states. Katy Payne, the acoustic biologist who was the first to establish that elephants communicate through infrasound, gives a marvelous account of young elephants taking delight in the chase: "Young bulls love to chase things; they relish the exhilarated, chin-up, feet-splayed rush and the sight of other animals in flight, and they magnify the impact of their assaults by chasing in pairs or small gangs. Once I saw a rush of which the object was a butterfly. Eyes wide, a gang of young male elephants collectively weighing some twenty or maybe forty tons thundered to a stop as the small fairy, white and weightless, rose up out of their midst. Then each turned on his heels and fled."

Gliding and coasting motions, which give rise to an often exhilarating sense of freedom, seem to delight both animals and man. Young sea lions will ride ocean waves again and again. Although this wave surfing may serve some adaptive function later in the animal's life—adult sea lions will occasionally catch waves in order to make it to the high rocks during storms—it not uncommonly seems to be done simply for the joy of it. Harbor seals of all ages play exuberantly: they pirouette through the air in 360-degree spins; they "hovercraft" just over the surface of the water, splashing as they go; they "whoosh" along the surface on the incoming surf; and they "bubble chase," pelting after their expelled air bubbles underwater and then following them along as they rise to the surface. Biologists studying harbor seals in Nova Scotia believe that ebullient play is pleasurable to the seals and intrinsically reinforcing.

Play is most common in fine weather, but snow also seems to bring out playfulness in many species. Sliding across the snow and ice is a rapid and efficient means of getting from place to place—river otters traveling over mountain passes or making their way down from mountain lakes often leave uninterrupted slide trails that are several hundred meters long—but many animals will repeatedly, and seemingly unnecessarily, slide down the ice or snow on their stomachs or, in the case of human children, on their sleds, and then gallop up to the top of the slide and plunge back down again. Tobogganing has been observed in the giant panda, as Schaller and his Chinese colleagues observed in Wolong: "One animal, probably a subadult, walked down a steep slope. Whenever it came to a forest opening where snow lay deeply, it slid downhill on chest and belly for 4 to 7 m[eters], leaving a deep furrow. There were at least 6 such sites; at one, the panda apparently walked back uphill to repeat the slide." One of the pandas tobogganed when it was clear he could more safely have walked.

Bears of other species have also been observed sliding down snowbanks on their chests and stomachs, and have been seen making snowballs and playing with them. (Japanese macaque monkeys, too, have been sighted constructing and playing with large snowballs.) John Fentress, an adviser to the Wolf Release Project in Yellowstone National Park, relates the apparent joy of a young wolf's first experience of snow as it ran through snowfields, "having fun." Bernd Heinrich, in *Mind of the Raven,* describes ravens pushing themselves forward in the snow on their bellies and then rolling sideways down a bank of snow.

At some point, exuberant play veers upward into near dance. "Dancing" associated with mating behaviors is well known in many species, of course. But occasionally an animal engages in rhythmic, apparently joyous movements in a seemingly purposeless way. Katy Payne describes dancelike behaviors in mother elephants watching their young chase wildebeest and suggests that dance is one of many communal activities that brings elephant families closer to one another.

Ronald Rood, who raised an orphan baby porcupine on his property in Vermont, discovered that the young rodent was inordinately fond of solitary dancing on the porch. "He'd whirl in a circle, stiff-legged, with quills raised," writes Rood, and then, "slashing his tail back and forth, he'd run forward a few steps, spin several times, and then run backward. Sometimes he'd grunt a challenge at an unseen foe. More often he'd dance in silence. Occasionally we'd turn on the porch light long after dark, and find our porcupine dancing in the center of the floor."

Lest this be seen only as a sentimental interpretation of aberrant captive behavior, it should be noted that scientists in New York have documented a similar "exercise dance" in members of their porcupine colony. This dance, carried out with widely varying degrees of playfulness, depending upon the individual porcupine, was virtually indistinguishable from that shown by the hand-reared animal. The porcupines, in solitary play, would rock from side to side, "alternately raising and stamping with the hind feet, like a marcher marking time." The movement of the hips and hind feet was accompanied by a swinging of the arms, which eventually became a "rhythmic swing." The porcupines engaged in their "dances" on most days.

Not all animals play. With unusual exceptions, it is the more intricately brained animals, those with the longest and most parentally protected youth, that engage in sustained stretches of high-energy play. These animals, few or singly born, sheltered and fed by their parents for extended periods of time and shielded from predators and the extremes of weather, have the time and energy to engage in exuberant play and exploration. They are given safe harbor by their parents and a long tether on which they can canvass, delight, and learn. As they grow older, the survival requirements initially taken on by their parents become their own, and the food and reproductive obligations of their species take on increasing moment. Play declines necessarily and precipitously as youth gives way to full growth. Young weasels and stoats, for example, play outside their nests when they are approximately four to six weeks old. This time of intense play overlaps the period when they are still nursing and immediately precedes the acquisition of permanent teeth and subsequent killing of prey on their own. Most weasel families break up when the kits are three or four months old and able to hunt and survive independently. Play recedes to the background of the weasel's life, replaced by tasks more immediately necessary for survival. But the early days of weasel play, like those of most young mammals, are full of energy and zip.

"The random high spirits of youth are as necessary and inevitable as the serious and restrained pertinacity of maturity," maintained the zoologist Chalmers Mitchell in his 1911 Christmas lectures at the Royal Institution in London. But high-spiritedness, he observed, is not present in all young animals. "Caterpillars, young cockroaches or grasshoppers, lobsters or crabs or snails are not to be distinguished from their seniors by any excessive gaiety," he noted dryly. "The exuberance of youth begins with the higher animals and increases as we ascend the scale of vertebrate life." Chalmers, in an argument not unlike that put forward a decade earlier by Karl Groos, proposed that play was not simply the result of the excess energies available during a highly evolved and protected youth. Rather, he believed, young mammals, unlike reptiles or insects, are fed and protected "in order that they may have surplus energy, and they require the surplus energy for the experimental business of their youth. . . . Limbs, claws, nose, teeth, tail, all their senses are exercised in every possible way, are applied in every possible direction, on everything that comes within their reach."

Play certainly seems to facilitate an animal's ability to move in a more deftly coordinated and responsive way. Chasing, tumbling, and scrambling about are a requisite part of strengthening muscles, honing eye-limb coordination, and laying down essential synaptic pathways in the brain. Limb and eye-limb movements are coordinated in the cerebellum, and the number of cerebellar synapses is significantly influenced by behavior. Not surprisingly, there is a relationship between play and synaptic growth. Young mice, for example, begin playing when they are about fifteen days old and play most vigorously between days nineteen and twenty-five, which is also the period of maximum synaptic formation.

Physical skills, such as those involved in hunting or defense, stem at least in part from the complex behaviors learned and practiced during play. Some of these survival benefits accrue early. Very young cheetah cubs, for instance, show a high rate of locomotor play, which researchers suspect helps the cubs in escaping from their predators. Most young mammals play with objects, an activity which, among other things, teaches them how to catch prey and how to explore their physical world. Many species—including our own, nonhuman primates, rats, and those in the mustelid family—show a preference for complex and novel play rather than simple object manipulation. George Schaller found that Serengeti lion cubs play with nearly everything that comes their way: pieces of bark, tufts of grass, elephant droppings, tortoises, and the twitching tassels on the ends of the tails of other lions.

Moose have been observed repeatedly pushing sticks under water, watching them pop back up again, and then resubmerging them, and even domestic cattle will play with objects they encounter in the field. Sea lions and seals play with driftwood and kelp and, on occasion, with animals of other species such as marine iguanas. Dolphins have been seen to carry debris and seaweed on their pectoral fins, flip the "toy" off, and then catch it on their tail flukes. River otters, like those described by Gavin Maxwell, play with a wide variety of objects for extended periods of time. The naturalist Hope Ryden described the antics of otters she encountered. When curious, she says, they would "stretch their necks above the surface and rotate their heads from side to side like submarine periscopes scanning for enemy ships. Not infrequently, one would flip over and swim on its back or pick up a floating bottle or some other object and play with it. Once I watched a young otter capture a painted turtle he had no intention of harming. He simply held it in his front paws while he did the backstroke around the pond." On another occasion she saw "two daredevil otters make sport of an enormous snapping turtle, repeatedly try to make the predacious creature stand on end." Otters, she agrees with Maxwell, are "irrepressible."

Occasionally, play with objects occurs between animals of different species. Most often, of course, this happens between dog and man, but the biologist Joyce Poole describes a wonderful example of play between a young elephant and herself. Poole threw her rubber flip-flop at Joshua, an adolescent male she was observing in Kenya, and watched his response. He smelled, touched, and tasted it and then tossed it up in the air. "Finally, he tossed my shoe, undertrunk, back to me. I picked it up . . . and threw it back to him. We did this a couple of times until something else caught my attention, and I looked away for a minute. The next thing I knew something hard landed on my head and fell to the ground with a thud. Joshua had found a small piece of wildebeest bone and had thrown it at me with surprising accuracy. It seemed clear to me that morning that Joshua understood and was amused by our game: There we were, two species out on the plains playing catch."

Birds also play. A one-year-old raven at the Zoological Gardens in Copenhagen "quickly learned to throw pebbles, snail shells and a rubber ball vertically in the air, catching them again with great dexterity," according to one observer. The bird would also "often lie down on its back and shift its playthings from the beak to the claw and back again." Bernd Heinrich observed ravens drop objects in flight and then swoop down to catch them; the birds also played furiously in the air, with many unnecessary but apparently pleasurable spins and bankings.

Play is conspicuously rare in reptiles, who have little youth to speak of. Usually a reptile gnaws or claws its way out of the egg to find itself only one in a clutch of hundreds, neither sheltered nor protected by its parents and left to its own instinctive devices. For animals subject to the strict metabolic constraints of reptilian life, play is a high-energy, risky, and inessential activity, a luxury to a young animal with limited vitality and restricted thermal resources, vulnerable from birth to tides and predators. Now and again, the predator is one of its own parents. Paul MacLean, a neuroanatomist at the National Institutes of Health, makes explicit an essential difference in the rearing of young reptiles and mammals: "The young of the Komodo dragon," he writes, "must take to the trees for the first year of life to avoid being cannibalized, while immature rainbow lizards must hide in the underbrush in order to prevent a similar fate. With the evolution from reptiles to true mammals there appears to have come into being the primal commandment, 'Thou shalt not eat thy young or other flesh of thine own kind.' "

Mammals, as they evolved, spent proportionately more time caring for fewer and fewer offspring. Social behavior and communication took on a commensurate importance. Play, which MacLean and other biologists believe acted originally to promote harmony in the communal nest, was a critical part of the development of social bonds, communication, and ways of learning how to learn. Through exploration and exchange of sensory information, play introduced an animal to the touch, smell, sound, and sight of others within its own family or species. Such sensory information helped to lay down the tracks of memory and to establish social affinities. These affinities created by play often extend beyond the period of youth: as Marc Bekoff observes, animals that play together tend to stay together. In many instances, play also begins the process of learning how to hunt or forage collectively. Through play, animals learn how to size up the physical attributes of their peers as well, and begin to establish social hierarchies and assess the suitability of other animals for mating purposes. (Once the dominance structure is established in the communal den of spotted hyenas, for example, aggressive behavior tends to turn more playful.)

Animals also learn through play how to curb their aggressive instincts toward others of their own kind. The pleasure of play reduces the chance of inflicting serious injury on those who have shared in it. Beavers, for example, are highly social animals that form intense family bonds. Hope Ryden, who for four years studied a family of beavers living in a lily pond in New York State, was initially incredulous that large families of beavers, sometimes as many as fourteen, could survive throughout a winter sharing cramped living quarters and depending upon food from a common stockpile. To do so, she reasoned, the animals had to have highly complex social strategies allowing them to communicate in subtle ways, extensive sources of mutual and communal pleasure, and, above all, a high threshold for the release of aggressive behavior. Throughout much of the year, Ryden noted, beavers did in fact build up strong social bonds through pleasurable nibbling and grooming of one another, close physical proximity, and extended periods of ebullient play.

"How do beavers keep from getting on one another's nerves?" she asks in *Lily Pond*. "Not a single night had passed but I had watched my beavers swim alongside one another, or touch noses, or 'speak' to one another. And how often had I seen two or more of them seek one another out for no other reason than to feed in company? Yet the lilies they consumed side by side were available all over the pond. And what of the precision diving and porpoising bouts I had witnessed? Were these not expressions of exuberant play? [The beavers] engaged in nonstop aquabatics, plunging under and over each other, swimming together and down. Suddenly, up again. First one, then the other, rolling, porpoising, somersaulting. This was exuberance." Nature, Ryden concluded, had given the beavers strong social glue; fast-held pleasures kept their aggression at bay. Exuberance bound beaver to beaver into a close and united group.

Exuberant play appears to be particularly important in nourishing social affinities in very young animals that later become members of a cohesive social unit. Wolves, who form close packs as adult animals, show more playful behavior when young than do coyotes and red foxes, who grow up to be more solitary. Common seals engage in highly exuberant play when young; as adults they form tightly cohesive social groups and display little evidence of fighting or aggressive competition for mates. Grey seals, on the other hand, are much more likely as adults to disperse along a shore than live together in closely bound groups. They are also less likely to vigorously play together as pups. It is not just the amount of contact that is important, however, but the nature of that contact. The ethologist Desmond Morris has suggested that exuberant movements during play may have a "catalytic" effect on the formation of social affinities and that these bonds would not be so strong if the young seals "merely nuzzled each other in a tranquil manner."

Play is critical in diffusing social tensions as well. The naturalist

Benjamin Kilham, who rears orphaned wild black bear cubs, recounts an occasion when, after a week had elapsed during which he could not spend time playing with the cubs, one of them "slow-bit" his hand with a canine tooth. "Something was obviously amiss between us, so . . . I took them on a walk. But even that was abnormal and strained. Reluctant even to start out, the cubs moved slowly along behind me, feeding on beechnuts for a while before following me up the hill toward a bear tree. Halfway up the hill they sat down, so I joined them. Then they decided to play . . . a twenty-minute roughhouse session ensued. Afterward I took a half-hour nap entwined in an ursine mass. Everything was back to normal." Play was restorative to the temporarily disrupted social bonds.

Elephants are particularly known for their tight social bonds and for an extended period of maturation. The young suckle for four years, and births are few and far between. Each young calf represents an enormous investment of time and energy by its mother and others in the close-knit community. Cynthia Moss, who directs the Amboseli Elephant Research Project in Kenya, describes the family reaction to the birth of an elephant calf as "almost delirious with excitement." There is, she says, much screaming and trumpeting and communal expression of joy. These intense emotions "are part of what they are doing right. They are the glue that keeps the families together."

Joyce Poole, like Cynthia Moss, has studied African elephants for decades. Expressions of joy are frequent in elephant families, she observes, especially when they greet one another after having been apart. The "greeting ceremony" may involve as many as fifty elephants and occurs after elephants have been separated for as short a period as a few hours or as long a time as several weeks. The greeting, Poole writes, is "pandemonium." The elephants "rush together, heads high, ears raised, folded, and flapping loudly, as they spin around urinating and defecating, and secreting profusely from their temporal glands. During all this activity they call in unison with a powerful sequence of low-frequency rumbles and higher frequency screams, roars, and trumpets." Poole believes, as other elephant researchers do, that the joy female elephants feel when they reunite is part of a response essential to their survival. Calves born into large and closely united families are more likely to survive, and strongly shared positive emotional responses reinforce the social bonds within those families. Elephants, Poole observes, are raised in "an incredibly positive and loving environment." She says she has never seen a calf disciplined: "Protected, comforted, cooed over, reassured, and rescued, yes, but punished, no." As Cynthia Moss has put it, "elephantine joy plays a very important role in their social lives."

Play and other exuberant social behaviors also have a contagious effect. Moods are by nature infectious, and joyous moods tend to spread rapidly throughout a group or herd of animals, heralding as they go that it is safe to enjoy, rest, hunt, explore, or play. Adelie penguins in the Antarctic, for example, exhibit an "ecstatic display" when they return to their colony: "The bird suddenly stretches its head and bill upward," writes one observer, "and then, with rhythmic beats of its flippers and its head still pointing to the sky, slowly emits a [sound] not unlike the slow roll of a drum." The behavior, which is repeated over and over again, is highly contagious, spreading from bird to bird throughout the community. As many as a hundred thousand birds have been observed taking part in this display.

Sometimes contagious exhilaration is preparatory to a group activity that requires both social cohesiveness and physical risk or exertion. This is certainly true for many human activities, as we shall see, but it also occurs in other mammals. Prior to a hunt, African wild dogs will gather together, sniffing one another and bounding about in energetic play. As time goes by, the playing gets wilder and rougher, finally reaching a climax when the whole pack masses together and then sets off after a gazelle or wildebeest. Field biologists who have observed this play-then-hunt behavior contend that the progressive buildup of excitement before hunting looks like "nothing so much as a 'pep rally,' that serve[s] to bring the whole pack to hunting pitch."

Social play in rats, which occurs after weaning but before sexual maturation, is critical to the development of their social and cognitive skills. Play behavior in these juvenile rodents, not surprisingly, is highly rewarding to them and has been shown to be regulated by the powerfully reinforcing opioid systems in the brain. Drugs that block these opioid systems reduce the urge to play. Behavior patterns laid down during the early weeks or months of intensive play are of lasting importance not only to the individual animal but to other animals in its social network. Play, which the neuroscientist Jaak Panksepp has described as the "brain source of joy," is thought to be tied to a variety of other beneficial physiological responses as well, including strengthening the immune system and increasing resilience under stress.

Panksepp believes that play probably increases gene expression in the frontal lobe for a protein involved in brain development. Experimenters have also shown that mice raised in cages filled with tunnels and toys play more, explore more, and ultimately generate more new neurons than mice raised in standard cages. Extensive psychological research in humans, to which we shall return, shows a highly beneficial effect of positive mood on learning and flexibility in thinking, in addition to its significant influence on social behavior. Play is the headwater of this elated, shaping mood.

We tend, as a thinking species, to emphasize the beholdenness of our emotions to our thoughts, rather than to trace our thinking to the ancient powers of our emotions. Yet our emotions were laid down far earlier than language or imaginative thought, sculpted by the realities of survival that we share with all other animals: to explore and to know our territories, to stay out of the grasp of our predators, to scavenge food and mate, and to set up safe havens. Each requires the complex yoking of swift and intense emotion to physical agility and, increasingly, with the evolution of higher animals, to mental acumen. Survival depends on comprehending the elements of the environment and then acting effectively upon that comprehension. Trout raised in hatcheries, for example, have smaller brains than trout born in the wild, who must learn to recognize and evade predators and to spot, chase, and capture prey. Charles Darwin had observed this more than a century earlier. "I have shewn," he wrote in *The Descent of Man*, "that the brains of domestic rabbits are considerably reduced in bulk, in comparison with those of the wild rabbit or hare; and this may be attributed to their having been closely confined during many generations, so that they have exerted their intellect, instincts, senses, and voluntary movements but little." To explore further is to learn more, and to learn more is to acquire the means of dealing with an unpredictable and changing world. Play promotes and encourages this.

Play helps the animal acquire knowledge about both the potential and the dangers of its world; it sets and becomes the physical arena for exploring new objects and for combining physical activities with sensory experiences in ways that might otherwise remain untried. Play increases the scope of the animal's experience and the range of its skills, generates a greater sense of control, and allows the animal to test its competence. Jane Goodall has emphasized the central role of play in making young chimpanzees familiar with their environment. The young ape, she writes, "learns during play which type of branch is safe to jump onto and which will break, and he practices gymnastic skills, such as leaping down from one branch and catching another far below, which when he is older will serve him in good stead—during an aggressive encounter with a higher-ranking individual in the treetops, for instance." The young chimps, in short, learn to go out on a limb.

Behavior that expends such energy, that is potentially dangerous yet intensely reinforcing, and that is nearly universal in the more cognitively complex animals must be of consequence. Play is unscripted. In being so, it introduces and rewards flexibility, prepares the animal for the unpredictable, and makes enjoyable the animal's testing of the boundaries of what it knows and what it has yet to know. Play is about learning how to learn. It is a kind of controlled adventure, an exploration of both new and familiar worlds. Play and curiosity are inevitably linked.

"Inasmuch as new objects may always be advantageous," wrote William James, "it is better that an animal should not *absolutely* fear them." It is important to explore new objects, he went on, and to ascertain "what they may be likely to bring forth." In that light, James suggested, some "susceptibility for being excited [by novelty] must form the instinctive basis of all human curiosity." Both play and exploration are intrinsically motivated behaviors, and they have in common many elements of curiosity and inquisitive behavior toward new objects and situations. "It's cat and monkey spirit," said Eugene Walter in *Milking the Moon*. "Let's see what's over there. Let's just have a look."

But there are differences between play and exploration. The systems in the brain that govern them appear to be relatively distinct, and exploration tends to occur prior to play, and in new environments. Play is more likely only after objects or circumstances have been explored and become at least somewhat familiar. Mood is also different during play and exploration—more joyful during the former, more serious during the latter. The exuberance of play is, in some respects, a joyous improvisation on the knowledge newly acquired through exploration.

Humans, who spend at least one fifth of their lives in childhood and adolescence, are uniquely playful and exploratory animals. When young, we run out into the world, regard and grapple with what we find, and absorb into our lives that which we have newly seen or felt. The freshness of these experiences lingers and insinuates: "There was a child went forth every day," wrote Walt Whitman in *Leaves of Grass:* "And the first object he look'd upon, that object he became, / And that object became part of him for the day or a certain part of the day, / Or for many years or stretching cycles of years." Everything the child sees or feels, said Whitman, becomes part of him: early lilacs, water plants, the horizon's edge, the fragrance of a salt marsh. Everything he touches or plays with or explores becomes a part of who he will be.

A child is impressionable by nature, and made more lastingly so through play. Studies of children find that memory is sharper as a result of playing and that play increases performance on a variety of measures of intelligence. In many respects, as Bernd Heinrich suggests in his study of ravens, play is quite similar to the workings of intelligence. It is, Heinrich writes, "an acting out of options, among which the best can then be chosen, strengthened, or facilitated in the future." But whereas the maneuverings of intellect are abstract, in play the options are played out overtly. Play, a substantial body of research has shown, promotes flexibility in children's thinking and behavior, much as it appears to increase the behavioral options available to other young animals. The playwright James Barrie expressed a similar idea in a letter he wrote to the boys who had been his inspiration for *Peter Pan*. "One by one," he said, "as you swung monkey-wise from branch to branch in the wood of make-believe, you reached the tree of knowledge."

The more playful the child, psychologists find, the more creativity he or she is likely to demonstrate. Highly creative children and adolescents are far more playful than their highly intelligent but less creative peers. Play appears to exert a particularly strong effect on children's ability to produce flexible and original associations when they are shown an object or placed in a new setting. The level of elation affects the imaginativeness of play. The more joyful and exuberant the child is while playing, the more creative the structure and content of the play itself.

Psychologists who study children have discovered two dimensions of play that are directly relevant to the concept of exuberant play. "Physical spontaneity" refers to the high-energy physical activity level shown in running, skipping, hopping, and jumping about. Ellen Winner, a professor of psychology at Boston College and the author of *Gifted Children,* observes that exuberance, especially of the sort found in highly gifted children, is often first and most strikingly apparent as high energy during infancy. Parents of gifted children, Winner finds, report that their children, even when very young, were unusually active, slept far less than other children, and were exceptionally alert and curious. She believes this drive is a stable one, a characteristic that carries on into and characterizes the adult lives of those who are exuberant and creative.

"Manifest joy," on the other hand, refers to a child's expression of enjoyment during play; that is, the child's level of enthusiasm and exuberance. Manifest joy, like physical spontaneity, is a persistent trait; if a child scores high on this measure in kindergarten he or she is very likely to score high on it in high school (and almost certainly beyond high school as well, although this has not been studied in a systematic way). The centrality of elated mood to playfulness has been demonstrated in many investigations.

Children need the freedom and time to play. Play is not a luxury; the time spent engaged in it is not time that could be better spent in more formal educational pursuits. Play is a necessity. This is a lesson too often lost on competitive parents and educators. The average school-age child in the United States, it has been estimated, now has 40 percent less free time than twenty years ago. Recess has been entirely eliminated in many elementary schools, and lawsuits have brought a "safe" sterility to the equipment on most playgrounds. Chemistry kits explode less often, but they are also a bit less magical. Long lazy days of just "messing about" are now filled with lessons, and games so structured as to teach little of what could be more interestingly and originally learned in wide-open roughhousing and aimless exploration.

It is essential to explore wild places, to expect that hazard will exist in most interesting places and circumstances, and to not be so fearful of injury or germs as to make childhood a shell of what it should be. "It is better to have a broken bone than a broken spirit," as one opponent of "safe" playgrounds has put it. Children need to be given a long lead to explore and the encouragement to play heedlessly and exuberantly with other children; to make painful mistakes; to fall down, lollop, get lost in the woods, run madly about. They need to galumph.

We all need both wild woods and sheltered riverbanks. Complex environments allow for complex play and probing. Like the brains of the wild trout and caged mice, our brains will acquire the complexity to which we are exposed. Exuberant play thrusts us into more elaborated and wonderful worlds, and the delight we take in such worlds kindles, in turn, a desire to persist in play and exploration.

Exuberant playfulness ends for many with childhood or adolescence. For others it remains. Margaret Mead, for one, was intrigued by this. "I am interested," she wrote, "in what happens to people who find the whole of life so rewarding that they are able to move through it with the same kind of delight in which a child moves through a game." Or, as T. H. Huxley said, the ability to perceive the world anew, the temperament "to face nature like a child." This exuberance, this freshness and playfulness of mind and mood, the capacity to galumph, are surely things to try to hold on to. It was said of John Muir that he never grew old, that he retained throughout his life a "child heart." And Theodore Roosevelt, at the age of fifty-five, was still a wildly enthusiastic man and utterly determined to help map the unexplored River of Doubt, which flows through the Amazonian rain forest. "I have to go," he said. "It's my last chance to be a boy!" (He did go. It was a trip of Rooseveltian proportions, fraught with poisonous snakes, equatorial fevers, violent rapids, insanity, and drownings. The expedition was high drama from beginning to end, and in the end, perhaps not surprisingly, the river was renamed Rio Roosevelt.) Nothing, in fact, was able to kill Roosevelt's passionate enthusiasm until the death of his son Quentin in World War I. Then, a friend of his noted, "the boy in him had died."

To keep alive the exuberance of youth is to keep alive the possibilities of imagination and play. "Each morning I am something new," sings the young girl in Delmore Schwartz's poem:

> *"I am cherry alive," the little girl sang,*
> *"Each morning I am something new:*
> *I am apple, I am plum, I am just as excited*
> *As the boys who made the Hallowe'en bang:*
> *I am tree, I am cat, I am blossom too:*
> *When I like, if I like, I can be someone new,*
> *Someone very old, a witch in a zoo:*
> *I can be someone else whenever I think who,*
> *And I want to be everything sometimes too:*
>
> . . . .
>
> *But I don't tell the grown-ups: because it is sad,*
> *And I want them to laugh just like I do*
> *Because they grew up and forgot what they knew*
> *And they are sure I will forget it some day too.*

They are wrong. They are wrong. When I sang my song, I knew, I knew!
I am red, I am gold, I am green, I am blue,
I will always be me, I will always be new!"

The young girl instinctively shields her joy from the adults who have forgotten or left behind that which she knows so clearly to be true. Childhood is not the only province of discovery and exuberance, but it is perhaps the most natural.

CHAPTER FOUR

"The Glowing Hours"

Always childhood ends. Nearly always, the unrestrained exuberance of youth ends with it. The kite is wound in, wonder shades into familiarity, and the skipping stops. Restraint accrues slowly, giving way to greater sophistication and savoir-faire. Childhood enthusiasm forfeits a portion of its charm: more and more it is to be dampened or subtly honed, kept to oneself, remolded into more worldly intimations of pleasure. The rising expectations of life exact a toll from the young as they are obliged to face them.

It is fortunate that the muting of exuberance is neither rapid nor absolute. Youth is, after all, a time to fly and fall on enthusiasm, to act with audacity. The world of the young is meant for scuttling about and, as we have seen, nature gives the time and means for this. Evolution invests heavily in the child's long days of eager adventure, reaping its returns in the adult's more informed sallyings forth of mind and body. But youth does it first and with greater abandon. This is the time, as Robert Louis Stevenson has it, to go "flashing from one end of the world to the other." It is a rash and full and delighting time.

Yet youth is a confusing, disturbing time as well. Emotions are inconstant and hard to sort through. Exuberant moods swing into darker ones, often ferociously and without apparent reason. Acquaintances differ greatly in character and temperament, and it is not clear how to respond to the differences or whom to trust. Play and other activities of childhood provide some experience for navigating these straits, but more is needed. Someone, a Virgil or a Merlin, is required to help guide the young through the emotional terrain they will encounter; someone to enchant and enthrall; someone to lay out the geography of the child's imagination, to provide an introduction to the bewildering array of personalities and situations likely to be met; someone to make sense of the discord and to help put conflicting emotions into meaningful perspective; someone to sort out friend from foe, valuable enthusiasm from folly. Experience will do this to some extent, of course, and so will watching and learning from the actions of those who are older. But writers, too, chart the minds and feelings of children; they draw up imaginary worlds for the young and fill those lands with hopes and dangers, discord and adventure, so that children might explore them under the protection of imagination before having to take them on in reality.

These inventors of worlds are the great writers of children's stories—Robert Louis Stevenson, J. M. Barrie, Kenneth Grahame, E. B. White, A. A. Milne, Walter R. Brooks, Pamela Lyndon Travers, Louisa May Alcott, and L. Frank Baum, among others—and they lay out brilliantly the anxieties and possibilities of youth, able to do so at least partly because they themselves had such ready access to their own experiences as children. Max Beerbohm wrote that James Barrie, the creator of *Peter Pan,* "stripped off himself the last remnants of a pretense of maturity. . . . Mr. Barrie is not that rare creature, a man of genius. He is something even more rare—a child who, by some divine grace, can express through an artistic medium the childishness that is in him." Barrie himself said it somewhat differently: "I think one remains the same person throughout [life], merely passing, as it were, in these lapses of time from one room to another, but all in the same house. If we unlock the rooms of the far past we can peer in and see ourselves, busily occupied in beginning to become you and me." And: "Perhaps we do change; except a little something in us which is no larger than a mote in the eye, and that, like it, dances in front of us beguiling us all our days. I cannot cut the hair by which it hangs." Kenneth Grahame, who wrote *The Wind in the Willows,* remarked to a friend, "I can still remember everything I felt then; the part of my brain I used from four till about seven can never have altered."

Writers with such genius for tapping into the emotions of childhood nearly always find ample room for exuberance and joy in the worlds and characters they create. They give a second-to-none view of what it is like to be exuberant—by nature or transiently so—and what it is like to be in the often delightful, occasionally annoying company of someone who is infectiously, boundingly ebullient.

There can be no more unforgettable examples of exuberance than Toad in *The Wind in the Willows* and Tigger in *The House at Pooh Corner.* They are the *grands mousseaux* of the temperament: bubbling and exhausting; exasperating, irrepressible, and unavoidable. Both are irritating, charming, and faintly if not overtly ridiculous. Their enthusiasms are urgent but fickle.

Tigger, a less nuanced exuberant than Toad, bounces into and out of the lives of Winnie-the-Pooh and the other inhabitants of A. A. Milne's Hundred Acre Wood. Nearly everyone is unsettled by Tigger's high-springing, discombobulating ways, but his presence is particularly ungluing to the dyspeptic Eeyore and timorous Piglet. "*Could* you ask your friend to do his exercises somewhere else?" Eeyore peevishly implores Pooh. "I shall be having lunch directly, and don't want it bounced on just before I begin. A trifling matter and fussy of me, but we all have our little ways." Piglet, a Very Small Animal, finds the high-voltage Tigger a Very Bouncy Animal, and unnerving, "with a way of saying How-do-you-do, which always left your ears full of sand." Despite Tigger's warmth and friendliness, his energy and impulsiveness overwhelm the anxious Piglet. To Rabbit, who finds Tigger more annoying than intimidating, Tigger is the sort "who was always in front when you were showing him anywhere, and was generally out of sight when at last you came to the place and said proudly 'Here we are!' "

Pooh, who is less disconcerted by Tigger (even though Tigger hides behind trees and jumps out on Pooh's shadow when he isn't looking), and more predictably focused on food and figure, takes his measure of Tigger into verse:

But whatever his weight in pounds,
shillings, and ounces,
He always seems bigger
because of his bounces.

Pooh's point is a good one: exuberance tends to leave the impression that its possessor is larger than life. (Piglet, less interested in the fine points of temperament, objects to the shillings: "I don't think they ought to be there." Pooh explains, "They wanted to come in after the pounds so I let them.")

The lively and gregarious Tigger bounds about the forest, leaping from one short-lived enthusiasm to the next: honey to haycorns, haycorns to thistles, thistles to Extract of Malt, which he then has for breakfast, dinner, and tea. The certainty with which Tigger holds his enthusiasms is met only by the quickness with which he abandons them. He is sure beyond reckoning, until forced to reckon. Tiggers are very good flyers, he exults, "Stornry good flyers." And excellent jumpers. And swimmers. Or, at least, until they are not. "Can they climb trees better than Pooh?" asks Roo, himself no piker in the bounce-and-joy division. "Climbing trees is what they do best," declares Tigger without equivocation. And thus begins their catastrophic scramble up the tree and subsequent plummet through the branches. It is considerably easier to be propelled up the tree by one's exuberance, Tigger finds, than to get back down. Tiggers cannot climb downward—their tails get in the way—and only by crashing to the ground can Tigger finally get himself untreed. Tigger is irrepressible, however; despite the ignominy, he springs upward and onward, unencumbered by the prudence that might attach to anyone else less helium-borne. Experience slows him not at all.

Nearly twenty years earlier, Kenneth Grahame had described in *The Wind in the Willows* a similarly irrepressible animal. The "gay and irresponsible" Mr. Toad of Toad Hall is a whirligig of energy and contradictions: self-absorbed, yet generous; self-satisfied, yet quick to contrition; restless, yet oddly content. Most of all, Toad is a caricature of exuberance: carefree, expansive, impulsive, and hopelessly given to short-lived enthusiasms. He is, in all things and at all times, utterly over the top. Toad chases after one horizon only to find that he really seeks another. He is dazzled by new fads and smitten by delusions of his own cleverness.

Mole, an enthusiastic but not exuberant animal, is eager to meet Toad but warned of his excesses by the Rat, who knows Toad all too well: "Once, it was nothing but sailing," said the Rat. "Then he tired of that and took to painting. . . . Last year it was house-boating. . . . It's all the same, whatever he takes up; he gets tired of it, and starts on something fresh." Toad's current enchantment, Mole learns, is a canary-yellow, horse-drawn Gypsy caravan with bright red wheels. Soon Toad's infectious enthusiasm becomes Mole's:

> "There you are!" cried the Toad, straddling and expanding himself. "There's real life for you, embodied in that little cart. The open road, the dusty highway, the heath, the common, the hedgerows, the rolling downs! Camps, villages, towns, cities! Here to-day, up and off to somewhere else to-morrow! Travel, change, interest, excitement! The whole world before you, and a horizon that's always changing! And mind, this is the very finest cart of its sort that was ever built, without any exception." . . . The Toad simply let himself go. Disregarding the Rat, he proceeded to play upon the inexperienced Mole as on a harp. Naturally a voluble animal, and always mastered by his imagination, he painted the prospects of the trip and the joys of the open life and the roadside in such glowing colours that the Mole could hardly sit in his chair for excitement.

Mole is no sooner swept up by Toad's ardor for life on the road than Toad is fanatically into his next obsession, a magnificent motorcar, which blasts down the highway stirring up dust and

Toad's combustible passions with it. The canary cart is forgotten; it no longer has any hold on him. Toad is capsized by rapture: a new world is in front of him: "The poetry of motion! The *real* way to travel! The *only* way to travel! Here to-day—in next week tomorrow! Villages skipped, towns and cities jumped—always somebody's else's horizon! O bliss! O poop-poop! O my! O my!" Toad's world is the road ahead.

> "What are we going to do with him?" asked the Mole of the Water Rat.
>
> "Nothing at all," replied the Rat firmly. "Because there is really nothing to be done. You see, I know him from of old. He is now possessed. He has got a new craze, and it always takes him that way, in its first stage. He'll continue like that for days now, like an animal walking in a happy dream, quite useless for all practical purposes."

Toad's exuberance, like that of Tigger, is put into relief by the countervailing temperaments of the other animals. In Tigger's world there are, in addition to the bouncier characters, cautious, melancholic, and fearful animals as well: Piglet, for example, is afraid of All Things Fierce, and Eeyore is a moper who tries to convince Pooh that "We can't all, and some of us don't . . . [do] Gaiety [or] Song-and-dance." Toad, too, is surrounded by characters very different from himself: the wise and gruff and balanced Badger; the practical, then poetic and dreamy-minded, and then again practical Rat; and Mole, "an animal of tilled field and hedgerow," of quiet but determining enthusiasms, an animal needful of the anchorage of his old home, yet passionately open to the wider world of sun and air and the River. Their adventures are far different from the flamboyant ones of Toad.

It is Rat and Mole, not Toad, who hear the Piper at the Gates of

Dawn, the strangely elusive and beautiful music of Pan: "'O, Mole!' cries the Rat, 'the beauty of it! The merry bubble and joy, the thin, clear happy call of the distant piping!'" It is they, the more sensitive and reflective of the animals, not the swift-to-act and slow-to-think-it-through Toad, who propel themselves into action when faced with Otter's anguish that his young son, who has swum away, might drown or perish in a trap. It is Rat and Mole who set out on a midsummer night to search river and field in order to track down the baby otter, and it is they who find—in contrast to the dust and distractions of Toad's motorcars and the open road—a night-time miracle on an island fringed with willow and silver birch and alder, crabapple and wild cherry and sloe. Their music is Purcell and Mozart, not Sousa.

The reactions of other animals to the exuberant Toad and Tigger reflect the complexity of children's attitudes toward those more energetically enthusiastic than themselves, and prefigure some of the benefits and liabilities of exuberance seen in adults: the joie de vivre and infectious, expansive (often imaginative) qualities on the one hand, and the intimidating, interfering, rash, and impulsive characteristics on the other. Milne and Grahame overdraw both sides of this ambivalence, of course, and few of the exuberant are in reality as extreme in behavior, or as impervious to reflection, as Tigger and Toad. Still, the other animals' reactions to their bubbly friends are illuminating.

Most are attracted to the sheer life force of Tigger and Toad, to the ebullience, adventure, and excitement they create in the wake of their enthusiasms. But the less exuberant animals are wary as well, mindful of the Right Way to do things and feeling a need for the social order. Exuberance is not entirely to be trusted or admired: it may be delightful, for a while, but it is potentially reckless and disorderly; it may lead to new places in the mind and heart, but it is not always to be taken seriously. Tigger and Toad are lively, but they are buffoons. They are enthralled with the possibilities and pleasures of life but, disconcertingly to the more restrained animals, they also tend to have a glorious time in the midst of their self-made maelstroms. Things are fabulous, until they are catastrophic. The two exuberants are intensely independent actors upon their worlds until disaster hits. Then the other animals, who are more usually overshadowed by the ebullient Tigger and Toad, gather power from the need to reestablish order and to exert moral authority. The ballasting animals act out of concern, outrage, and often a trace of envy as well. When necessary, they band together to take the erring animal in hand.

Rabbit, for one, in the wake of suspicions that Tigger has bounced Eeyore into the river, determines that Tigger is "too bouncy." He goes further: "It's time we taught him a lesson." The problem with Tigger is that "there's too much of him, that's what it comes to." Eeyore, the aggrieved, is indeed offended: "Taking people by surprise. Very unpleasant habit. I don't mind Tigger being in the Forest," he says, "because it's a large Forest, and there's plenty of room to bounce in it. But I don't see why he should come into *my* little corner of it, and bounce there."

Piglet, who is inclined to defend the affable Tigger, protests—"He just *is* bouncy . . . and he can't help it"—but gradually he, too, is brought around to Rabbit's plan for Tigger's redemption: "Piglet settled it all by saying that what they were trying to do was, they were just trying to think of a way to get the bounces out of Tigger."

Rabbit concocts a plan for Piglet, Pooh, and Rabbit to take Tigger to a place he has never been before, to lose him, and then find him again the next morning. He will be, Rabbit assures Piglet and Pooh, "a different Tigger altogether . . . he'll be a Humble Tigger . . . a Sad Tigger, a Melancholy Tigger, a Small and Sorry Tigger, an Oh-Rabbit-I-*am*-glad-to-see-you Tigger." Tigger will be deflated, unbounced, newly appreciative, and cut down to size:

"If we can make Tigger feel Small and Sad just for five minutes," explains Rabbit, "we shall have done a good deed."

Far from losing Tigger in the Forest, of course, Pooh, Piglet, and Rabbit themselves become hopelessly lost in the mist. Tigger effortlessly finds his way out. Pooh and Piglet, after much aimless and anxious wandering about, eventually make their way to the clearing, but Rabbit remains stranded, unable to navigate back to safety. The maligned and still very much bounced Tigger bounces to Rabbit's rescue, and into a different perspective:

> Tigger was tearing around the Forest making loud yapping noises for Rabbit. And at last a very Small and Sorry Rabbit heard him. And the Small and Sorry Rabbit rushed through the mist at the noise, and it suddenly turned into Tigger; a Friendly Tigger, a Grand Tigger, a Large and Helpful Tigger, a Tigger who bounced, if he bounced at all, in just the beautiful way a Tigger ought to bounce. "Oh, Tigger, I *am* glad to see you," cried Rabbit.

Toad is similarly taken to task by his fellow animals after exhibiting a level of rashness staggering even to those who know him well. Having obtained his fabulous motorcar, Toad speeds his way into disaster. He smashes up his car, not once but many times, recklessly forces others motorists off the road, and is put into hospital on three separate occasions. Badger has had it: "We'll take Toad seriously in hand. We'll stand no nonsense whatever. We'll bring him back to reason, by force if need be. We'll *make* him a sensible Toad." As with Tigger, this is easier said than done. Certainly, verbal entreaties go only so far. Badger at first believes he has persuaded the Toad of the error of his overly exuberant ways, but Toad soon sets him straight: "I'm *not* sorry. And it wasn't folly at all! It was simply glorious!"

Badger and the others dig in, strip the backsliding Toad of his freedom and finery, and lock him up in his bedroom. His ill-advised enthusiasm for motorcars is treated as a fever that wants breaking: "It's for your own good, Toady, you know," says the Rat. "Think what fun we shall all have together, just as we used to, when you've quite got over this—this painful attack of yours!" Mole assures Toad that his money will be well looked after, not wasted as it had been: "We'll take great care of everything for you till you're well, Toad." The animals settle in for a nursing siege not altogether dissimilar to the type used in dealing with manic patients on slightly more conventional psychiatric wards:

> They descended the stair, Toad shouting abuse at them through the keyhole; and the three friends then met in conference on the situation.
>
> "It's going to be a tedious business," said the Badger, sighing. "I've never seen Toad so determined. However, we will see it out. He must never be left an instant unguarded. We shall have to take it in turns to be with him, till the poison has worked itself out of his system."
>
> They arranged watches accordingly. Each animal took it in turns to sleep in Toad's room at night, and they divided the day up between them. At first Toad was undoubtedly very trying to his careful guardians. When his violent paroxysms possessed him he would arrange bedroom chairs in rude resemblance of a motor-car and would crouch on the foremost of them, bent forward and staring fixedly ahead, making uncouth and ghastly noises, till the climax was reached, when, turning a complete somersault, he would lie prostrate amidst the ruins of the chairs, apparently completely satisfied for the moment. As time passed, however, these painful seizures grew gradually less frequent, and his

friends strove to divert his mind into fresh channels. But his interest in other matters did not seem to revive, and he grew apparently languid and depressed.

Toad's dark mood does not linger long, but he uses its approximation to gull his keeper, the Rat, into abandoning guard duty. Rat is no sooner on his way to seek a doctor for his friend than the malingering and unrepentant Toad is out of bed in a hop. Straightaway knotting his bedsheets together, he slips out of his window, lowers himself to the ground, and trots briskly down the high road into his worst disaster yet. The "poop-poop" of a passing car sets him off and, before he can muster an ascertainable trace of restraint, he is caught up irremediably in the throes of his old passion and headlong ways. He is, he chants to himself, "Toad once more, Toad at his best and highest, Toad the terror, the traffic-queller, the Lord of the lone trail, before whom all must give way or be smitten into nothingness and everlasting night."

Toad is also, however, a thief—having stolen the car under whose spell he had fallen—and is sentenced to prison. Alternately abject and defiant, Toad's mood rises and falls upon the changing circumstances of his existence. Contrition is short-lived in the presence of the possibility of escape. A successful breakout from jail sends his mood and ego airborne: "It was too late," exclaims Rat after he catches Toad exulting over his escape, "Toad was puffing and swelling already." Puffing and swelling build to reckless ecstasy, then to open-throated Toad Whoops and verses of self-puffery. And more disaster. By book's end, however, after relentless deflatings by his friends, Toad is proclaimed by the other animals to be at last a Reformed, Modest, and Altered Toad.

Perhaps. Yet one cannot help but think that someday, somewhere, the bubbles will rise again and Toad, giddied by some curious and marvelous enthusiasm, will pelt off toward a new horizon and hop afresh into utter mishap. Exuberance is nothing if not irrepressible.

Most characters in children's stories are not like Tigger and Toad. They are not exuberant by temperament; rather, they respond with exuberance to certain external circumstances: an event may trigger it, or it may be sparked by the infectious joy of others, or some magical circumstance will set it off. Children's authors often invoke such magic or events in order to conjure exuberance; the mood is deceptively difficult to convey through words.

P. L. Travers, like many other writers, used flight as a vehicle for provoking and portraying exuberance. Her major character, the astringent nanny Mary Poppins, flies to London's Number Seventeen Cherry Tree Lane on a parrot-handled umbrella and dances down through the sky on the "heaven-end" of a kite string. When she leaves for uncertain places, to return at an unknown time, she sweeps off on the gusts of the west wind or she flies away, to a blast of trumpets, on a spinning merry-go-round that whirls upward to the stars. She flies, and with her fly Michael and Jane, the young children who are in her charge. She sweeps them off to extraordinary adventures: bobbing balloons carry the children and masses of other Londoners high up into the air, "rainbowy" and joyous over the park; they step into chalk pictures and jump peppermint-stick horses over lilac bushes and ride them home to dinner, or, perhaps, to the uttermost ends of earth. But they fly, in mood if not in fact.

Mary Poppins gives the children magical hours filled to the brim with a glimmering, evanescent, infectious, but ultimately vulnerable exuberance. The mood is delightful, but it is wide open and liable to puncture. The children and Mary Poppins visit her uncle, Mr. Wiggs, for instance, and enter a world of contagious, propulsive gaiety. When they arrive, Mr. Wiggs is sitting on air, close to the ceiling: "I'm a cheerful sort of man," he tells them in greeting, "and very disposed to laughter." If he laughs when his birthday falls on a Friday, he explains, "I become so filled with Laughing Gas that I simply can't keep on the ground. . . . The first funny thought, and I'm up like a balloon." Mr. Wiggs's bouncing and laughing and bobbing are completely infectious; presently, the children are "rolling over and over on the floor, squealing and shrieking with laughter."

Jane rises on her newly caught mood. As she laughs, she feels herself "growing lighter and lighter, just as though she were being pumped full of air. It was a curious and delicious feeling and it made her want to laugh all the more." Soon everyone is floating on air, clutching their sides and gasping with laughter, rolling and bobbing about, until Mary Poppins, conjurer first, and then realist, announces to the children that It Is Time to Go Home. Her statement abruptly shatters the mood, as reality will, and sends all of them bumping to the floor: "The thought that they would have to go home was the first sad thought of the afternoon, and the moment it was in their minds the Laughing Gas went out of them." Exuberance begets exuberance and exuberance, flight. For a while. But reality keeps enough lead in its pockets to pull those in flight back down to earth.

P. L. Travers creates a string of enchanted escapades that loop through the children's minds like Chinese lanterns, but they usually end with the acerbic Mary Poppins sniffing indignantly at the children's suggestion that anything out of the ordinary has happened. She casts her dazzling spells only to break them with her acid tongue or a precipitous return to home, but nearly always she leaves behind a memento of what took place—a light and bright balloon, now deflated but still suggestive of a magical flight over the park; a snakeskin belt from an astounding evening at the zoo; a small pink starfish that twinkles like diamonds, from a trip to the ocean floor—

and these tokens of magic times keep the children dotted with a "still wondering" quality: What is real? What is not? What actually happened? Will it happen again?

Mary Poppins leaves enough to the children for them to know that there are things that can be explained and things that cannot. The gift of wonder—and the joy of it—is the legacy she leaves Michael and Jane as she soars over Cherry Tree Lane for the last time: "Where and How and When and Why—had nothing to do with them. They knew that as far as she was concerned those questions had no answers . . . but the gifts she had brought would remain for always." Children need to hold on to the bits and dreams, the joys of childhood. So do adults.

It is hard to imagine anyone more brilliantly able to capture the joys of childhood and to bridge the worlds of adult and child than cartoonist Charles Schulz. Umberto Eco describes Schulz's comic strips as "interrupted poetry"; Art Spiegelman puts it somewhat differently: Schulz's work, he says, has the "simplicity and depth charge of a haiku." Garry Trudeau states that for himself and his fellow cartoonists Schulz is the "gold standard." (The public's enthusiasm is reflected in numbers; before Schulz's death in 2000, *Peanuts* was syndicated to nearly three thousand newspapers, published in seventy-five countries, and translated into more than twenty-five languages; it reached an estimated audience of 350 million people.) These tributes are importantly true, but Schulz also had an almost unerring genius for portraying the anxieties and delights of childhood; he felt, then drew, the elemental association between the emotions of children and those of adults.

There is, in *Peanuts*, an underlying and profound sadness which reflects not only Schulz's own struggles with depression but his sensitivity to the quiet terrors of human loneliness. "The most terrifying loneliness is not experienced by everyone and can be understood by only a few," Schulz said. "I compare the panic in this kind of loneliness to the dog we see running frantically down the road pursuing the family car. He is not really being left behind, for the family knows it is to return, but for that moment in his limited understanding, he is being left alone forever, and he has to run and run to survive." It is this heart-stopping poignancy which gives indisputable credibility to Schulz's work. The great artists, wrote the poet Edward Thomas, have seen what they have imagined. Surely this is true of Schulz.

But there is an essential exuberance in Schulz's work as well, a combative hopefulness that wins the day over sadness. Nowhere is this joyfulness more fully manifest than in the mind of Snoopy, the imaginative beagle born at Daisy Hill Puppy Farm and introduced to the world in October 1950. (The character of Snoopy was based on a dog Schulz had had as a boy. "He was the wildest and smartest dog I've ever encountered. Smart? Why, he had a vocabulary of at least 50 words. I mean it. I'd tell him to go down to the basement and bring up a potato and he'd do it.")

Snoopy is a seriously exuberant animal. He is also independent, quirky, debonair, keenly intelligent, selfish, mischievous, and an incurable romantic. His observations on life reflect the wide-ranging interests of his creator, who loved Tolstoy, Scott Fitzgerald, Dostoevsky, and Flannery O'Connor. Like them, Snoopy is subject to a certain world-weariness and now and again needful of a newly imagined life.

Whimsy and an irrepressible joyousness serve Snoopy well in his Walter Mitty imaginings. He spins his fantasies with the inventive energies of a child and elaborates them with the delicacy and detail of a Venetian glassblower. He appears to have inherited from his creator an infinitely playful mind. "I wonder why Snoopy is willing most of the time to simply lie on the top of that doghouse," Schulz asked once, apropos of seemingly nothing. "Why doesn't he roll off? I remember a veterinarian telling me once that when birds fall asleep sitting on a limb of a tree, their brain sends a message down to their claws telling the claws to stay clamped on the limb so the bird doesn't fall off and land on his head. So, I justify this by saying that perhaps it is the same with Snoopy's ears. I think when he falls asleep, his ears clamp onto the top of the doghouse." Schulz asks the kind of question a child would ask, and answers it with the delightful absurdity that a child understands and loves.

Snoopy, like Delmore Schwartz's little girl who exclaims, "Each morning I am something new. . . . I will always be me, I will always be new!" embraces and becomes the open-ended possibilities of life. His exuberance and inventiveness not only entertain him, they stave off a tendency toward ennui. Snoopy, according to Schulz, "has to retreat into his fanciful world in order to survive. Otherwise, he leads kind of a dull, miserable life. I don't envy dogs the lives they have to live. They're trapped living with families that they never knew anything about."

Snoopy is far from trapped. He lives in a doghouse, seen only in side view, that contains a pool table and a television set, stained-glass windows, carpeting, a cedar closet and a potted philodendron, paintings by van Gogh and Andrew Wyeth, and a picture of Tiny Tim. He rows his canoe in the birdbath, occasionally into coastal fog, and sleds down snow slopes in his food bowl. He poofs dandelions. He dines by candlelight on top of his doghouse and sleeps there with the moon as his night light (although now and again he panics that the moon will fall on his head). He loves eggs Benedict, doughnuts, chocolate chip cookies, root beer, and silver-dollar pancakes. He blows bubbles, and he skates better than Peggy Fleming. He served not just honorably in World War I but as a flying ace in his Sopwith Camel, taking on the Red Baron from the top of his doghouse.

He is an astronaut who goes to the moon, a yet-to-be published novelist, and a sometime truffle hunter. He is everything his imagination can create: an anteater and a partridge in a pear tree, a Scott Fitzgerald hero, a piranha, Hucklebeagle Finn, Mickey Mouse, a fierce snow snake, a Riverboat Gambler, Dr. Beagle and Mr. Hyde, an authority on dragonflies. He gives kisses sweeter than wine and travels to the Sahara and France. He is a devotee of the absurd but he is unjaded. He believes that as long as you can see the moon you can never be lost in the woods. You will be facing west, he explains. The moon "is always over Hollywood."

"Life for Snoopy is just such a fantasy," says Judy Sladky, a five-time U.S. national skating champion and the person Charles Schulz chose to portray Snoopy off the ice as well as on. "If reality isn't what he likes, he changes it around to suit him. He thinks of things and does them." When Sladky puts on her forty-pound Snoopy costume and assumes his personality, she says, "Snoopy just takes over. I have done things I have never done before. Backflips, for example. He wears me out. He exhausts me. He can do anything." Sladky believes that "exuberance is where Snoopy starts from. I have to get into the mood, then the action follows." Acting exuberant, she says, creates exuberance, although she believes that the energy and joy come at a cost. "After the exuberance is over then there is a cliff, exhaustion, depression." Snoopy "lives in a place to go somewhere else. At the end of every wonderful thing Snoopy does, he falls down, gets shot down." But then, of course, he always gets back up again.

Snoopy has an extraordinary capacity for celebration. He leaps for joy; he dances with delight. "To those of us with real understanding," he says, "dancing is the only pure art form." A frame or so later, he continues, "To live is to dance, to dance is to live." Snoopy's mind is a different dance of life, but a dance all the same. He is his own Balanchine. No occasion need pass unmarked by dance: There is a First Day of May dance, which differs only slightly from the First Day of Fall dance, which differs also only slightly from the First Day of Spring dance. In fact, he admits, "even I have a hard time telling them apart." There is a Suppertime dance, which doesn't always work, and a Rain dance, about whose success we know very little. The Be Kind to Animals dance, Snoopy is quick to point out, "symbolizes the last days of 'Be Kind to Animals Week.' " There is no Third of May dance, although there is a Second, and he has a special dance he performs for Lucy, the "Haha you have to go to school and I don't dance." The critical difference between the Second Day of Spring dance and all others, he explains, is in the action of the toes.

Snoopy's glee as he dances is infectious: it is impossible to look at his outstretched arms, twirling ears and spinning feet, and not feel delight in his energy and pleasure. "Snoopy has the freedom to express uncontrolled joy," observes Jeannie Schulz, the cartoonist's widow, who is wonderfully exuberant in her own right. "Snoopy doesn't have to have any controls. . . . Sparky [Charles Schulz] always said that when Snoopy began to walk on two feet, his whole personality opened up, that after that the character took on a life of his own . . . the exuberance grew out of Sparky's pen."

In the television classic *It's the Easter Beagle, Charlie Brown*, Snoopy, who has become separated from his friends in a department store, comes upon a stack of brightly colored Easter eggs. He picks up an elaborately decorated one and with one eye closed peers into an opening at the end. Inside the egg there is a diorama of bunnies in a colorless landscape. (These bunnies are the famed Bunnie-Wunnies, stars of Snoopy's favorite books. Snoopy, who has a bunny coloring book as well, "loves bunnies" more than almost anything else. They are among the most affectionately held of his passions.) Suddenly the dull landscape changes to vibrant greens and daisies and colored Easter eggs. The bunnies link paws and break into a Matisse-like dance. Snoopy leaps into the diorama, the world within the egg, and joins the bunnies in their dance. Ears sailing, they all twirl together for a while and then Snoopy breaks into a wildly exuberant solo—ears extended and flying out from his head, he pirouettes and kicks out his feet in a Russian folk dance. It is a moment of complete magic, one that draws the viewer into a private world of unrestrained rejoicing. "On with the dance!" Byron had written: "Let joy be unconfined; / No sleep till morn, when Youth and Pleasure meet / To chase the glowing Hours with flying feet." Snoopy's feet are not perhaps those imagined by Byron, but no one's dancing and joy could be more delightful or unconfined.

Snoopy's bliss comes not just from dance. He revels in re-creating himself, delights in the absurd, and takes endless pleasure in playing with ideas and seeing where they go. He is well-read, although not as much as he might have been, perhaps slowed down by his tendency to move his lips when he reads. He is enchanted by the esoteric. "Dragonflies sew up your lips so you can't eat, and you starve to death," Snoopy tells Woodstock at one point. "If you chew wintergreen candy in the dark, it makes sparks!" he exclaims elsewhere. He exchanges philosophical musings with a fruit fly, who has a lifespan of twenty-four hours, and learns that the fly has only one regret: "I wish I knew at nine o'clock what I know now," he confides to Snoopy.

Snoopy is a sensitive observer of nature, and his joy in life is reflected in his sorrow at its passing. He finds the falling of leaves almost unbearably sad. He watches one leaf as it falls to the ground and says, "Well! The first falling leaf of the season . . . The first leaf to make the courageous leap! The first leaf to depart from home! The first leaf to plunge into the unknown . . . The first leaf to die!" Later, he looks up as a leaf falls from the tree and says to it, "Don't stay here . . . They'll come and get you with a rake." "Nobody ever tells them about the guy with a rake," he adds. Snoopy has a child's capacity to wonder and exult, but his exultation is tempered by an adult awareness of the inherent sadnesses of life. His imagination is part child, part adult. It improvises, spins, and revels in its fancies.

"The man's true life, for which he consents to live," wrote Robert Louis Stevenson, "lies altogether in the field of fancy. The clergyman, in his spare hours, may be winning battles, the farmer sailing ships, the banker reaping triumph in the arts: all leading another life, plying another trade from that they chose. . . . For no man lives in the external truth, among salts and acids, but in the warm, phantasmagoric chamber of his brain, with the painted windows and the storied walls." Snoopy, dining by candlelight on the top of his doghouse, with his stained-glass window and van Gogh below, would agree.

"All children, except one, grow up," wrote James Barrie in *Peter Pan*. "They soon know that they will grow up, and the way Wendy knew was this. One day when she was two years old she was playing in a garden, and she plucked another flower and ran with it to her mother. . . . Mrs. Darling put her hand to her heart and cried, 'Oh, why can't you remain like this for ever!' This was all that passed between them on the subject, but henceforth Wendy knew that she must grow up. You always know that after you are two. Two is the beginning of the end."

The one child who never grows up is, of course, Peter Pan: "I ran away the day I was born," he explains, "because I heard father and mother talking about what I was to be when I became a man. . . . I want always to be a little boy and to have fun. So I ran away to Kensington Gardens and lived a long long time among the fairies." That Peter cannot and will not grow up is at the heart of Barrie's play: Peter is guide and master of the magical Neverland, but he can never be a part of the more completely human world from which the children he enchants come, to which they return, and where they grow up. Peter is almost always at play and joyful, but his play and joy go nowhere, cannot move beyond childlike pleasure into the complexities of human relationships and life. Peter's world, that of childhood imagination, is unbounded, yet it is also hemmed in by an inability to keep up with the expectations of life. Peter's joys are real, but ultimately unsustaining. The children he beguiles with his exuberance and with the magic of Neverland move on, but Peter does not.

Peter leaps buoyantly from adventure to adventure, from the Mermaids' Lagoon to the Pirate Ship, with little or no memory of where he has been and what he has done. He is captivating, but thoughtless and capricious. He teaches Wendy, John, and Michael how to jump on the wind's back and to fly—"You just think lovely wonderful thoughts and they lift you up in the air"—and he leads them, "Second to the right, and straight on till morning," to Neverland. But, tellingly, Peter forgets to teach the children how to stop, and while he goes off to have adventures they are left to struggle with their new powers: "He would come back laughing over something fearfully funny he had been saying to a star, but he had already forgotten what it was, or he would come up with mermaid scales still sticking to him, and yet not be able to say for certain what had been happening. It was really rather irritating." Wendy raises the disquieting possibility: "If he forgets them so quickly, how can we expect that he will go on remembering us?"

The children's adventure is only an imaginary one, a glorious way station on their journey to the rest of their lives, but Neverland is Peter's past, present, and future. Because of this, his life is tantalizing to the children, but his own emotions are arrested. He learns from life even less than he remembers of it, and Barrie fates him to experience over and over again the same events, each time anew, with the generations of children he seduces away to Neverland, where time is very odd and "all the four seasons may pass while you are filling a jug at the well."

Neverland as imagined by Barrie is an unforgettable, forgotten place of personal adventure and memory, and one of the most wonderfully construed ideas in children's literature:

> There are zigzag lines on it, just like your temperature on a card, and these are probably roads in the island; for the Neverland is always more or less an island, with astonishing splashes of colour here and there, and coral reefs and rakish-looking craft in the offing, and savages and lonely lairs, and gnomes who are mostly tailors, and caves through which a river runs, and princes with six elder brothers, and a hut fast going to decay, and one very small old lady with a hooked nose. It would be an easy map if that were all; but there is also first day at school, religion, fathers, the round pond, needlework, murders, hangings, verbs that take the dative, chocolate pudding day, getting into braces, say ninety-nine, three pence for pulling out your tooth yourself, and so on; and either these are part of the island or they are another map showing through, and it is all rather confusing, especially as nothing will stand still. . . .
>
> On these magic shores children at play are for ever beaching their coracles. We too have been there; we can still hear the sound of the surf, though we shall land no more.

Like Toad in *The Wind in the Willows,* Peter moves heedlessly from engagement to engagement: "He was fond of variety," writes Barrie, "and the sport that engrossed him one moment would suddenly cease to engage him." He is a genius at make-believe but unlike the other boys, who know the distinction between truth and invention, Peter does not. Nor does Peter learn from friendship or adversity in the same way that most others do. "Peter," his creator tells us, "had seen many tragedies, but he had forgotten them all."

And Wendy remarks, "Fancy your forgetting the lost boys, and even Captain Hook!" "I suppose," she says on reflection, "it is because you have so many adventures." Fun leads to more fun, but nowhere else. There is much joy but no learning, much exuberance but no wisdom.

When Peter duels Captain Hook—who, the playwright reminds us, has eyes as blue as the forget-me-not and long curls "which look like black candles about to melt," speaks with elegant Etonian diction, and "is not wholly evil; he has a *Thesaurus* in his cabin, and is no mean performer on the flute"—most who have seen the play remember the verbal exchange that takes place when Hook asks Peter: "Pan, who and what art thou?" and Peter crows back exultantly, "I'm youth, I'm joy." But a different side of Peter's nature is revealed in a fight he has with Hook, for it gives an unimpeded look into the limitations of never growing up: "[Peter] saw that he was higher up the rock than his foe. It would not have been fighting fair. He gave the pirate a hand to help him up. It was then that Hook bit him. Not the pain of this but its unfairness was what dazed Peter. . . . No one ever gets over the first unfairness; no one except Peter. He often met it, but he always forgot it."

Peter moves neither forward nor backward in his dealings with himself and the world. He repeats his mistakes as he repeats his adventures and, accordingly, advances not at all in his knowledge of himself or others. The promise of joy compels him, but the joy he finds is fleeting, unremembered, and put neither to good nor particular use. Not surprisingly, James Barrie confided many years after he had written *Peter Pan* that "its true meaning came to me—Desperate attempt to grow up but can't."

The imagination of children has its limitations, as does the exuberance that accompanies it. The nature of both must change if a child is to live resourcefully in a world that changes. The characters in children's stories stay as they are, of course, caught in time—

Christopher Robin and Pooh, for instance, go off together, and "wherever they go, and whatever happens to them on the way, in that enchanted place on the top of the Forest, a little boy and his Bear will always be playing"—and they remain there, waiting for new generations of children: unforgettable for the worlds they open up, and irreplaceable for the facets of human nature they reflect. They serve us well, and in different ways, as we negotiate youth and then take our leave of it.

Youth is moved away from only to be missed, of course. Answerability to the tasks of life fosters a more circumspect kind of exuberance, which inevitably is accepted but rued. As the years pass, greater pains are taken to recapture youth's vehement joys. "I remember my youth and the feeling that will never come back any more," wrote Joseph Conrad, "the feeling that I could last forever, outlast the sea, the earth, and all men; the deceitful feeling that lures us on to joys, to perils, to love, to vain effort—to death; the triumphant conviction of strength, the heat of life in the handful of dust, the glow in the heart that with every year grows dim."

Yet Robert Louis Stevenson, better able than nearly anyone to capture the lost lands of childhood and adolescence, spoke as forcefully of the easements of getting older: "The regret we have for our childhood is not wholly justifiable," he wrote. "What we lose in generous impulse, we more than gain in the habit of generously watching others; and the capacity to enjoy Shakespeare may balance a lost aptitude for playing at soldiers. . . . We take our pleasure differently."

It is possible for some, but not for most, to hold on to the heat of life and, although they take their pleasures differently from when they were young, they continue to take them with a full measure of joy.

CHAPTER FIVE

"The Champagne of Moods"

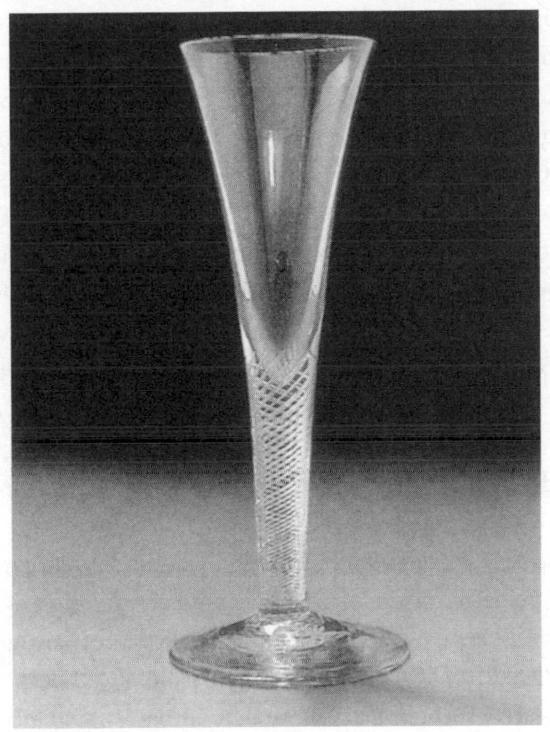

Improbably, the English invented Champagne. Decades before French winegrowers produced their first bottle of sparkling wine, Christopher Merret described to the Royal Society in London the methods being used by English wine-coopers to make brisk and sparkling wines. The addition of vast quantities of sugar and molasses to a finished wine, he reported in 1662, provoked a second fermentation, which created bubbles. This process, according to the Champagne historian Tom Stevenson, made the English wine not just lively but "unequivocally sparkling." Merret's contemporary the great French wine master Dom Pérignon, far from cultivating bubbles in his wines, spent a great deal of time and energy attempting to annihilate them.

The bubbles won out. Every second around the world seven bottles of Champagne are uncorked. This, at 250 million bubbles in an average bottle, is a gloriously unimaginable amount of bubble and fizz. Champagne launches ships and marriages, marks the race won, the examinations finished. It is uniquely the wine of celebration, of joy, and of elegance. When Scott Fitzgerald wrote that in Gatsby's gardens "men and girls came and went like moths among the whisperings and the champagne and the stars," only "champagne" could evoke the mood he wanted. Champagne *is* a mood, an austerely beautiful signifier and creator of moment and emotion. Dom Pérignon, despite his initial misgivings, is said to have exclaimed when he first tasted it, "Come quickly! I am tasting stars."

Champagne is coolly and joyously incandescent. Evenings gain from it in vivacity, and its pleasures spread among those in its presence. Desire sharpens. "Hardly did it appear," wrote an eighteenth-century drinker of Champagne, "than from my mouth it passed into my heart." Its bubbles generate an intoxicating gaiety; indeed, one wine authority believes that "Champagne should laugh at you." The bubbles, he contends, should be "extremely animated and persistent: When the glass is held to the light, it should be possible to spot them forming right down near the stem and watch them rocketing upwards like balls in a juggler's hands." Champagne, in short, is exuberant.

Human temperaments, like Champagnes, come in different degrees of effervescence. Some are *grands mousseaux*, fully sparkling, unstoppable, bubbling, and relentlessly high-spirited. Others are *pétillants,* only faintly sparkling. Most are *crémants,* somewhere in between. The *grands mousseaux* infect others with their liveliness and bring to life's delights and setbacks a seemingly inexhaustible energy and resilience. They carry into adulthood that full measure of joy which so many others leave behind with youth. It is written into the wild expansiveness of Whitman—"O the joy of my spirit—it is uncaged—it darts like lightning! / It is not enough to have this globe or a certain time, / I will have thousands of globes and all time"—and we feel it in Churchill's uninhibited passion for the brilliant colors on his palette: "I must say I like bright colours. I cannot pretend to feel impartial about the colours. I rejoice with the brilliant ones, and am genuinely sorry for the poor browns. When I get to heaven I mean to spend a considerable portion of my first million years in painting, and so get to the bottom of the subject. But then I shall require a still gayer palette than I get here below. I expect orange and vermillion will be the darkest, dullest colours upon it, and beyond them will be a whole range of wonderful new colours which will delight the celestial eye."

For those less exuberant or not at all, one globe may be more than enough, and bright colors, while pleasing, will not transport. We vary in our capacities for enthusiasm because a diversity of temperaments serves our collective good. We know intuitively that some will be quick and passionate in their responses, as we know that others, less urgently moved, will wait and be more deliberative. For each the world has space and reason.

In their musical *Gigi,* based on the novella by Colette, Alan Jay Lerner and Frederick Loewe put human differences in the capacity for joy, and the universal desire for it, at the heart of their words and music. Honoré, played by the ebullient Maurice Chevalier, loves women, wine, love, Paris, everything. He exults in life and is as enchanted by it as his nephew, Gaston (played by the mannerly, devastatingly handsome Louis Jourdan), is jaded and bored. Honoré experiences the world as a source of wonder and thrill; for Gaston it provokes nothing but indifference and malaise. The French, it is said, touch Champagne to the lips of newborn babies. Clearly some remain under its spell longer than others.

Only the young schoolgirl Gigi (played by Leslie Caron), exuberant by youth and nature, can make Gaston laugh and relieve his ennui. But Gaston's worldliness and imperturbability have, in turn, charm for Gigi: he is ballast to her scattered effervescence, a challenge to her unschooled gaieties. He can introduce her to the world of Maxim's, white tie, and tails; to Champagne and dancing; to love, desire, and restraint. He has the pleasure of being disarmed, she of disarming. Yet it is Gigi's exuberance and joie de vivre that linger long after the film is over. To "fly to the sky on Champagne / And shout to everyone in sight" is, as Gigi sings in "The Night They Invented Champagne," the essence of exuberance. Uninhibited joy such as hers is captivating, contagious, and a powerful psychological force. It is, however, a psychological force that has been of more interest to songwriters than to most psychologists.

Exuberance, it is safe to say, has not been a mainstay of psychological research. Until recently, psychology textbooks have devoted more than twice as much space to "negative" emotions like depression and anxiety as they do to "positive" ones like joy and happiness; the most dynamic of the positive emotions, exuberance, is scarcely studied at all. For every hundred journal articles on sadness or depression, calculates psychologist Martin Seligman, only one is published about happiness. Cross-cultural analyses of language find that in virtually every society there are many more concepts for negative emotional states than for positive ones. This, in the context of the richness of human experience, seems on the face of it hard to comprehend. Yet it makes sense, at least up to a point, to focus on psychopathology and potentially destructive emotions or dangerous circumstances. They, not joy or happiness, raise awareness of immediate threats to the individual and to society. Pathological behavior can incapacitate or kill, and it can create dangerous instability within a group. Survival is made more likely by a biological and emotional system that is highly focused, alert to peril, and ready to handle threat with dispatch. Positive emotions, in this context, could be viewed as an evolutionary luxury.

Indeed, brain imaging studies conducted at the University of Iowa demonstrate that when subjects are shown emotionally pleasant pictures (for instance, landscape scenes, fireworks, and dolphins playing with a ball); unpleasant pictures (a bird covered in oil, a dead soldier with part of his face missing, a rotting carcass of a dog); or pictures that are neutral (an open umbrella, a woven basket, leaves on a tree), the unpleasant pictures provoke activation in the primitive, subcortical parts of the brain conceptualized by scientists as an ancient danger-recognition system. The pleasant pictures, on the other hand, activate a phylogenetically much younger part of the brain, the prefrontal cortex. Danger, the researchers conclude, requires a quick and relatively simple response system; the ability to appreciate the positive in situations requires, on the other hand, a more sophisticated level of processing in the brain.

But, as we have seen, nature has taken care to create a capacity for delight and prolonged enthusiasm. These exuberant and otherwise affirmative states can generate alternative solutions to menace and hazard, foster resilience and social bonds, and reward successful behavior with an infusion of pleasure. The brain systems necessary to appreciate the auspicious and to fashion a fitting response are recent but vital. The study of the pursuit and harnessing of the auspicious, however, has been skipped over in favor of understanding danger, distress, and disease.

My colleagues and I, for example, published a paper more than twenty years ago about positive experiences associated with mania.

In our review of the medical and psychological literature on mania—an admittedly destructive state, but one that in its milder forms is often characterized by many advantageous qualities such as high energy, exuberance, increased sexual desire, and rapid, creative, and expansive thinking—we were stunned to see how disproportionately psychological research had focused on negative emotions and how slight was the mention of temperamental strengths. Our field was more interested in the depressed and anxious brain than in what Coleridge so marvelously described as the "Bright Bubbles of the . . . ebullient brain." As clinicians, we of course knew that psychologists and psychiatrists are obligated to ameliorate suffering, not to root around for benefits that might derive from it. We are asked to find remedies. Suffering demands action in a way that pleasure and success do not.

Still, as a clinical psychologist interested not only in psychosis and suicide but in creativity and the arts as well, I was disconcerted to see how far our field had moved away from the wide-ranging and profound interests of David Hume and William James, how far behind we had left our earlier attempts to understand passion, imagination, and the nature of human greatness. I was far from alone in these concerns.

Psychology has begun to catch up with its earlier, broader interests. In the last two decades, psychologists have brought new life and better science to the study of positive psychological traits. The number of articles published about "positive psychology" and "positive emotions" has quadrupled over the last two decades. In January 2000 an entire issue of *American Psychologist* was dedicated to the topic of "Positive Psychology: Happiness, Excellence, and Optimal Human Functioning." The issue's coeditors, Martin Seligman of the University of Pennsylvania and Mihaly Csikszentmihalyi of the Claremont Graduate University in California, are well-respected researchers and writers who have pioneered the study of optimism and optimal experiences. They argued powerfully for an emphasis on those aspects of human nature which enhance life and productivity, capacities that might one day prevent mental illness, not simply contend with it once it occurs. "Our message," they wrote in *American Psychologist,* "is to remind our field that psychology is not just the study of pathology, weakness, and damage; it is also the study of strength and virtue. Treatment is not just fixing what is broken; it is nurturing what is best. Psychology is not just a branch of medicine concerned with illness or health; it is much larger. It is about work, education, insight, love, growth, and play." Their statement was an eloquent call to the field.

Psychology has always, if insufficiently, concerned itself with individual differences in personality and temperament. Basic emotions such as joy, anger, and fear are universal, but individuals vary enormously in the nature, quickness, and intensity of their emotional responses. Emotions are innate, although susceptible to alteration through experience and the environment, and they exist to alert us to specific and significant situations such as danger or opportunity, so that we respond in a visceral way to gain advantage or to increase our chances of survival. These emotional responses—such as increased heart rate, a surge of adrenaline, or physical shutdown—tend to be of rapid onset and short duration. Most consistently, they are characterized by two principal psychological dimensions, "pleasantness" and "activation."

Emotions can be placed along a continuum of subjective enjoyment that ranges from pleasant to unpleasant, and another continuum of alertness and energy that ranges from high activation to low. This dimensionality approach to emotions to some extent circumvents William James's apt observation about the futility of rigid categories: "The trouble with the emotions in psychology," he wrote more than a hundred years ago, "is that they are regarded too much as absolutely individual things. So long as they are set down as so many eternal and sacred psychic entities, like the old immutable species in natural history, so long all that *can* be done with them is reverently to catalogue their separate characters, points, and effects." Modern psychological science views emotions in far more complex ways than observers did in James's time.

Exuberance, under this model, can be conceptualized as high on the pleasantness (or "positive affect") dimension, as well as high on activation. This combination of positive emotion and high energy is far more likely to result in an active engagement with the world than depression or anxiety, which are psychological states hallmarked by avoidant and fearful behaviors and lacking in drive. Exuberance is also more likely than happiness alone to lead to new and energetic pursuits. Happiness is a less activated emotional state and one that is, by definition, more content with the way things are than eagerly gauging possibilities of how things might be in the future. Joy, on the other hand, lures the individual forward with further promise of pleasure for, as C. S. Lewis has observed, anyone who experiences joy will want it again.

Exuberance encourages exploration and rewards it with the possibility of joy and greater opportunity for food, territory, and mates. But exploration also increases the likelihood of danger to the individual. The active, exploring animal is more vulnerable to predators and the elements than the timid one who is likely to remain sheltered and camouflaged, more protected by remaining within the group than if it ventures out on its own. Captive foxes who are fearful of new situations and strangers, for example, are far less likely than their bolder littermates to be killed by predators and automobiles once they are released into the wild. Guppies, even, show a range of intrepidness. Most, sensibly enough, will keep their distance when placed near larger fish. A fearless and curious few males, however, will swim toward a potential predator. Not surprisingly, they are more likely to be eaten, but those who are not prove to be more attractive mates to the surviving female guppies. Trepidation cuts both ways.

Antonio Damasio, a neurologist and writer at the University of Iowa, uses the sea anemone to draw in a simple and elegant way the differences between an active, exposed animal and a closed and withdrawn one: "This fundamental duality is apparent in a creature as simple and presumably as nonconscious as a sea anemone. Its organism, devoid of brain and equipped only with a simple nervous system, is little more than a gut with two openings, animated by two sets of muscles, some circular, the others lengthwise. The circumstances surrounding the sea anemone determine what its entire organism does: open up to the world like a blossoming flower—at which point water and nutrients enter its body and supply it with energy—or close itself in a contracted flat pack, small, withdrawn, and nearly imperceptible to others. The essence of joy and sadness, of approach and avoidance, of vulnerability and safety, are as apparent in this simple dichotomy of brainless behavior as they are in the mercurial emotional changes of a child at play."

The exuberant person, far from simply responding to the environment in which he finds himself, acts vigorously upon it or seeks out new ones. Whether through play, through exploration, or through engagements of the imagination, those who are exuberant *act*. Spirited play, as we have seen, rewards exploration with pleasure, and propels young animals and children into more intimate and varied contact with their physical environment; play sees to it that necessary skills are acquired and a diversity of experiences is tried. Joyous states do other critically important things as well. They strengthen the bonds between members of a group and make more likely the group's participation in shared activities that will benefit the group as a whole; they fortify the ties between parent and child, teacher and student, leader and follower, lover and lover.

The energy, enthusiasm, and optimism of those who are exuberant tend to make them more socially outgoing, as well, and more likely to take risks; this, in turn, almost certainly increases their attractiveness to the opposite sex and, accordingly, their chance of reproductive success.

We have some sense of what exuberance does, but what actually is meant by it, and how is exuberance measured? What elements combine to make an exuberant temperament? How often, and with what steadiness, does the temperament persist over time? To what extent is exuberance determined by genes and to what extent is it influenced by the environment? We know more than we did in William James's time, but not nearly as much as one would like.

Hippocrates, Aristotle, and Galen described individual differences in temperament, as did others over the centuries to follow, but it was not until the twentieth century that scientists and clinicians studied and classified temperament in a more empirical way. Temperament, which can be broadly defined as the relatively stable pattern of moods and behaviors first manifest early in life, has been more fully described by the psychologist Gordon Allport as the "class of 'raw material' from which personality is fashioned." "Personality" generally denotes the unique or most distinctive aspects of an individual, characteristics shaped by innate forces operating under the influence and constraints of upbringing and environment. "Temperament," according to Allport, is the "internal weather" in which personality evolves. "The more anchored a disposition is in native constitutional soil the more likely it is to be spoken of as temperament. . . . [It comprises] the characteristic phenomena of an individual's emotional nature, including his susceptibility to emotional stimulation, his customary strength and speed of response, the quality of his prevailing mood, and all peculiarities of fluctuation and intensity in mood." These phenomena, Allport assumed, were largely inherited. In practice, the terms "temperament" and "personality" are often used interchangeably, although temperament is assumed to be more genetically determined.

One of the most reliably measured differences in temperament is between those individuals who are highly enthusiastic and active, who reach out to new people and new experiences, and those who are less energetic and outgoing, more likely to avoid the unfamiliar. The German psychiatrist and philosopher Karl Jaspers, in his classic 1913 textbook *General Psychopathology*, described a continuum of active, ebullient temperaments. The "euphoric" temperament, as set out by Jaspers, was distinguished by the "abnormally cheerful" individual who "bubbles over happily . . . is blissfully light hearted about everything that happens to him and is contented and confident. The happy mood brings a certain excitement with it including motor excitement." The "sanguine" temperament, he believed, was abnormally excitable: "It reacts quickly and in lively fashion to every kind of influence, it lights up immediately but excitement dies down equally fast. The individual leads a restless life, and likes extremes. We get a picture of vivacious exuberance or of an irritable, troubled hastiness."

More recently, Hagop Akiskal, a psychiatrist at the University of California at San Diego, has developed the concept of "hyperthymia" to describe the cheerful, overly optimistic individual, more often male than female, who is talkative, extraverted, self-assured, and filled with plans and ideas. He or she needs little sleep and possesses the kind of energy which leaves others gasping. Akiskal, who estimates that at least one person in a hundred meets the research criteria for hyperthymia, cautions that while there are many advantages to this type of temperament—gregariousness, indefatigability, and the ability to handle highly stressful situations with relative ease—there is, as well, an instability in mood that can lead to intemperate behavior.

Extraversion, of all of the traits examined by psychologists to date, is the one most clearly and directly related to exuberance. The extravert, as defined by a variety of extensively and well-validated personality assessment measures, is energetic, outgoing, lively, cheerful, enthusiastic, forceful, active, and talkative, and tends to seek excitement. Extraverts have low levels of social anxiety, high levels of self-esteem, and are exquisitely alert and sensitive to reward signals. In experimental situations, for example, they react far more intensely than introverts when shown photographs of people with happy faces. (This is consistent with findings from neuropsychological studies of patients with manic-depressive illness. When manic, patients attend and respond far more to positive words presented to them in an experimental task; when depressed, they are much more likely to pay attention and respond to negative words. The state of one's mood also clearly affects the content of the material that is remembered. Neuropsychologists repeatedly find that depressed patients disproportionately recall words with depressive content and that nondepressed subjects do the opposite. Depressed patients are also more likely to remember failures and other negative experiences in their lives, as well as to underestimate their performance on a variety of psychological and intellectual tests.) Mood state influences what is noticed, how it is remembered, and how it is retrieved from the memory. Those whose temperaments afford them extended periods of positive mood, or periods of particularly intense and exuberant moods, experience the world in a very different way from those who are more even-tempered or dyspeptic.

Differences in temperament exist in many other species. In a cross-species review of temperament in nonhuman animals, "extraversion" was the most universal. Seventeen out of nineteen studies identified a factor related to extraversion, such as sociability in pigs and dogs, "vivacity" in donkeys, or a "bold approach versus avoidance" dimension in octopuses. The reviewers point out that the manner in which temperament shows itself depends upon the species: "Whereas the human scoring low on Extraversion stays at home on Saturday night, or tries to blend into a corner at a large party, the octopus scoring low on Boldness stays in its protective den during feedings and attempts to hide itself by changing color or releasing ink into the water." (Scientists find significant differences between octopuses on several dimensions of behavior, most consistently in levels of activity, reactivity, and avoidance. These differences show themselves early, by the third week of life, and researchers believe that the diversity of temperaments increases the likelihood that the octopuses will better adapt to their highly variable habitats, predators, and prey. Octopuses often live in exceptionally fluctuating near-shore environments that are subject to violent storms, extreme changes in tide and season, and pollution. Variability in temperament is also seen in animals that rely upon learning in order to adapt to changing circumstances; octopuses, it has been known for a long time, are good learners.)

Some species—rhesus and vervet monkeys, for instance, as well as hyenas and pigs—score particularly high on measures of curiosity, playfulness, and exploration. Primates vary a great deal from one species to another. Chimpanzees are more outgoing and impulsive than gorillas, who are shyer; rhesus monkeys are more exploratory and socially active than pigtail macaques. Primatologists speculate that species with the most diverse diets, those who must forage farther afield and actively defend against predators, need to be bolder, more curious, more energetic, and more exploratory than, for example, the less active fruit-eating monkeys.

Cats, dogs, and monkeys vary enormously within their own species in how fearful or bold they are in approaching new situations or unfamiliar individuals. Perhaps one house cat in seven, for instance, will avoid new situations and strangers and only unusually attack a rat. One in five young rhesus monkeys is easily distressed and fearful, a figure comparable to that found in human infants. Similar differences may also be seen in some species of birds and fish. Among zebra finches and pumpkinseed sunfish, for example, those animals most likely to explore novel objects are less likely to fly or swim in close proximity to other animals of their species; they are also more likely to be leaders, to forage independently and over a greater range, and to be better able to adapt to novel environments. They, by dint of their behavior, are also more likely to put themselves in danger's way. The anthropologist Melvin Konner has discussed at length the importance of such risk-taking behavior in his excellent book *Why the Reckless Survive.* If an animal is designed for survival and reproduction, he points out, it is not designed for perfect safety.

Joyce Poole studied a family of twenty-four elephants in Kenya and found significant personality differences among them. Some elephants, she says, were "just plain boring," others were timid, and yet others were "full of games and mischief." She believes that some elephants are inherently "mercurial," while others "just plod along." One of the elephants she observed, for instance, a female named Ebony, was "always up for some mischief and full of life and exuberance most of the time." Others were more phlegmatic and only intermittently ebullient.

Few animal researchers have looked at exuberance per se, although one intriguing study of brown bears living in an open area of sedge and flats on Admiralty Island in Alaska did assess "sparkliness," "liveliness," and "spiritedness." Robert and Johanna Fagen, zoologists at the University of Alaska, collected data on seven adult bears during three years of summer salmon runs. "Sparkly" bears were defined as "bubbly, cheerful and full of sprightly movements." "Spirited bears," in like vein, were defined as "vivacious, animated and energetic, [approaching] life with abundant physical and mental energy." The Fagens repeatedly rated the bears on a wide variety of measures and concluded that there are four general dimensions—liveliness, irritability, confidence with other bears, and fishing behavior—that best describe the personality of brown bears. They noted that although liveliness is an essential characteristic in bears, it is very little studied: "We find it interesting that a dimension (Lively) that includes lively curiosity and spirited movements, animation, flamboyance, sparkle, and a tendency to show off should emerge as a prominent feature of bears' individual personalities. . . . It seems to measure qualities of behaviour and personality rather than absolute levels of activity or amount of movement. Previous studies of individuality in nonhumans and in humans did not report lively curiosity, spirited movements, or the other items included in our Lively dimension." Rosemary Bolig and her colleagues at Ohio State University did find a related personality trait in rhesus monkeys, however, which consisted of opposing behavior patterns of "exuberance" and "nurturance" (the researchers referred to the monkeys with these qualities as "party animals" and "homebodies"). The exuberant animals were assessed by the researchers as highly active, curious, and playful; the nurturant animals, on the other hand, scored high on maternal and protective behavioral dimensions.

Extraversion in humans, extensive research shows, predisposes the individual to experience and display more positive emotions such as enthusiasm, interest, excitement, and joy. The relationship between positive emotions, or pleasant affect, and extraversion is one of the most consistent findings in studies of personality; it is also one of the strongest (the correlation between the two traits approaches 0.80 in several studies). Indeed, some psychologists have argued that positive emotionality is the glue that holds together the component parts of extraversion. Extraversion is also closely related to the number of close friends an individual has and to how likely he or she is to be selected for leadership roles. Extraverts as a group tend to be happier than their more introverted peers. The psychologists Ed Diener and Martin Seligman studied more than two hundred undergraduates, for example, and found that those students who described themselves as "consistently very happy" were much more extraverted than those who described themselves as less happy. It is to some extent a self-perpetuating phenomenon. Extraverts are gregarious and enthusiastic; such characteristics tend to be attractive to others and to create more opportunities for pleasure through greater contact with other people and the surrounding world. Enthusiasm itself leads to a more fervent engagement with ideas and an impassioned pursuit of interests.

Extraverts are not only more likely than introverts to experience positive moods, they also feel a greater intensity in such moods. In responding to questionnaires, they tend to agree with items like "When I feel happy it is a strong kind of exuberance," "When something good happens, I am usually much more jubilant than others," and "When I'm happy I bubble over with energy." Introverts do not. In addition to experiencing more intensely positive moods, extraverts also feel and perform better in stressful and challenging circumstances.

People who are extraverted are more likely to act, to move, to engage. They lope, not amble; they fizz. They are like the infectiously exuberant writer Eugene Walter, who says, "We are the ones who gallop ahead two hundred miles and then stop and say, 'What country is this?' If we could organize, we would have taken over the world way back, but we are interested in so many things that when we head for California, we end up in Florida. You know. Our emblem is the centaur: half animal, half man. And shooting that arrow at the moon. Centaurs have all four feet on the ground, but that arrow is whizzing off to a distant planet."

Why are some so vital and others not? Why do some people gallop full throttle into adventure while others fall back, fearful, intent on avoiding misadventure? There are many reasons, but the most notable differences in temperament are rooted in genetics and in the architecture and chemistry of the brain. Heredity unequivocably plays a critical role in temperament. Some individuals are simply more biologically predisposed to respond with fear when confronted with a new or uncertain situation; others are inclined to enjoy or investigate the unfamiliar. The genetic contribution to temperament, especially to extraversion, is strong. An analysis of 24,000 twins found that if one identical twin is an extravert, the other is very likely to be one as well. This is not nearly as true for fraternal (nonidentical) twins. (The correlation between identical twins is 0.5 or more, suggesting that at least 50 percent of extraversion is due to genetic factors. The correlation between fraternal twins is 0.2.)

Thomas Bouchard and his colleagues at the University of Minnesota examined identical twins who had been raised apart and found that if one twin was extraverted, the other twin, who had identical genetic material but a different environment of upbringing, was very likely to be extraverted as well (the correlation was high: 0.6). This gave strong support to the argument that there is a powerful genetic influence on temperament and that family upbringing has less of a determining role. "Joy, good cheer, and bubbliness," the Minnesota researchers reported, were particularly heritable in their study. Studies of young children who are identical twins and who have been classified as having either an inhibited or an uninhibited temperament, analogous to extraverted and introverted temperaments in adults, also show correlations of between 0.5 and 0.6 for these types of temperament. If only those twins classified as extremely inhibited or extremely disinhibited are examined, the correlation is even higher, ranging between 0.7 and 0.9.

We are not so different from domestic dogs in our heritability and differences in temperament. All breeds of dog show a powerful genetic influence on behavior (again, generally in the range of 0.5). As with humans, there are significant differences in temperament between individual dogs within a breed, but there are also striking differences across breeds. The behavioral traits that most reliably differentiate breeds of dogs are excitability and general activity level, traits related to extraversion in humans. Basset hounds, for example, who would be on few judges' lists for Most Exuberant in Show, are among the lowest scorers on both activity and excitability; terriers, on the other hand, are highly excitable and propulsively active. Breeds of dogs also vary enormously in the extent to which they demonstrate inhibited or excitable social behaviors, such as body and tail posture, tremor, exploratory and escape behaviors, and changes in heart rate. Some breeds are timid, others more curious or fierce. Swedish researchers studied behavior in more than 15,000 dogs of 164 species and found a major "personality" factor which consisted of playfulness, the tendency to explore, an interest in the chase, and sociability. This trait, which pervasively influenced the dogs' behavior, is also observed in seven- to nine-week-old wolf pups. Confronted with a novel situation, the most fearless pups are also the most playful and exploratory. The shyer dogs, on the other hand, are uninterested in playing, more cautious, and less curious. The most exploratory and fearless pups, when tested a year later, are the most "dominant" within their pack.

There is little difference in playfulness between male and female dogs; studies of human children, on the other hand, find boys are more likely than girls to be physically spontaneous, to explore more and to explore larger areas, to engage in "rough-and-tumble" play, and to be less socially inhibited. Studies of adult humans also find that men are more likely than women to be energetic and to be perceived by others as enthusiastic and unrestrained.

Our species, like others, shows a wide diversity of temperaments: some of us rush toward the new and assume that it will bring pleasure, not wretchedness. As many at least step back from life, stay within a sprint of our foxholes, and watch as others make the forward moves. The disposition to advance or to retreat, to be enthusiastic or to be fearful, shows itself early in life. In a landmark series of studies, Jerome Kagan and his colleagues at Harvard identified temperament profiles in infants and young children that are strikingly similar to those seen in adults. Behaviorally inhibited infants and children, like introverted adults, will, when confronted with new people or new situations, actively avoid engagement with the unfamiliar. If they do approach, they do so slowly and reluctantly. They cling to their parents, are quieter, and move less energetically than their more extraverted peers.

Uninhibited children, on the other hand, readily initiate contacts with others, laugh and smile a lot, and are unusually talkative. They are extraverts in the making. A subgroup, perhaps one in ten of those studied by Kagan and his coworkers, displays an unusually high level of energy, smiles frequently, and laughs with "zeal." This quality, the researchers conclude, is "difficult to name, for it is not captured simply by activity level. Other children run a lot but do not possess the enthusiasm and vibrancy that is distinctive of these children. The term *vitality* comes closest." The Harvard scientists found that this characteristic manifests itself at a remarkably young age. At four months, the infants who show a vibrant and positive mood also babble and smile a lot, exhibit little or no anxiety, and are "utterly fearless." When examined more than ten years later, these high-mood and high-energy infants— whom Kagan calls "Ethel Merman types"—are very likely to have remained extraverted and energetic.

Uninhibited children, in addition to being more exuberant, are far more likely to take risks than those who are inhibited. In a study of five-and-a-half-year-olds, for example, children were asked to choose the distance from which they preferred to throw a ball into a basket. The inhibited children opted to stand very close to the basket, only one or two feet away. The uninhibited children, on the other hand, more often chose to stand four or five feet back. The investigators noted that these children looked as though they enjoyed the greater challenge.

Nathan Fox, of the University of Maryland, describes a group of highly energetic and enthusiastic infants whom he and his colleagues have studied. Characterized by the researchers as "exuberant," these infants are highly and eagerly reactive to novelty; they smile, coo, and gurgle a great deal, are sociable and enthusiastic, and seem exceptionally eager to explore the world around them. They show little fear in unfamiliar situations. "From the youngest age," the researchers observe, "these infants appeared to exhibit exuberance for novelty and social interaction that was unique." Their description of the moods and behaviors of these exuberant infants was strongly confirmed by the observations made by the children's parents. Fox and his colleagues found, as Kagan did, that the "exuberant" infants made up about 10 percent of those they studied.

The psychologist Ellen Winner, who studies artistically and intellectually gifted children, states that even in infancy exuberant children show very high energy levels and are unusually alert and curious. "From an early age," she writes, "these children find things that interest them and they throw themselves into these domains." One eight-year-old child she studied "created hundreds of soldiers, each wearing a uniform of a particular country and rank (which he learned about by reading). He made these soldiers out of paper. When he was finished with this project, he created hundreds (literally) of zoo animals, each one very realistic, and designed cages for them." She describes his behavior as exuberant, she says, because "he became completely immersed in this activity for months, it required high energy, and it clearly gave him pleasure. He lived in his own private world."

Many studies find a strong positive relationship between curiosity, a preference for novelty, and subsequent measures of intelligence and academic performance. Infants who gaze more at novel objects, for example, score higher on later tests of cognitive ability. (Likewise in our primate cousins. Curiosity predicts successful problem-solving in small-eared bushbabies.) Shy and anxious children, who tend to avoid novel objects or situations, perform less well on tasks measuring creativity. Seeking out novelty shows itself early. Researchers at the University of Southern California tested the tendency to seek out stimulation in 1,795 three-year-olds and then assessed their cognitive abilities when they were eleven. Children who scored high on stimulation-seeking when young—defined by the researchers as greater exploratory behavior away from their mothers, more friendliness and talking to strangers, and more active social play—scored significantly higher on IQ tests (12 points). They also had higher scholastic and reading ability than the children who had scored lower on stimulation-seeking.

There are several possible explanations for this unusually strong correlation (0.5 to 0.9) between an active, curious temperament and cognitive ability. Young children who are shy or anxious may so fear criticism from others and be so desirous of pleasing that they do not take the risks necessary to enhance their lives of play and imagination. These children, often more afraid of failing than excited by the chance of winning, put such a premium on "getting it right" that they limit their exploration of the field of possibilities. Young children who seek out novelty or actively explore the world around them, on the other hand, create a very much enriched environment in which to learn. They are probably innately more highly energetic and motivated, as well. Research indicates that curious, enthusiastic, and cheerful children also have a more positive effect on their parents, as well as on other children, teachers, and other adults. This in turn positively affects the child's overall social and learning environment. Exuberant children create more complex and rewarding environments for themselves than do shyer, more timid children.

The interaction between environment and temperament is inordinately complex, of course, and it only gets more complex the more it is studied. All animals are shaped not only by their genes, but by the circumstances under which they develop; as the science writer Matt Ridley has put it, "Nature versus nurture is dead. Long live nature via nurture." Mice and rats, we know, grow more neurons in the hippocampus when they are brought up in an enriched environment rather than in a standard laboratory cage. Rat pups removed from their nests and given more freedom to explore for a few minutes every day for the first three weeks of life show less anxiety and more exploratory behavior later. Rhesus monkey infants raised in an enriched environment, with a terrycloth-covered movable surrogate monkey and a terrycloth diaper covering a "water bed"—rather than in a standard environment, with a stationary surrogate and only a terrycloth diaper—do much better when they are tested later on a variety of problem-solving and motor tests. There is, as well, an interaction between an animal's innate temperament and how it is raised. Those monkeys who do the best on subsequent testing are those who rate low on fearfulness during their first month of life and who are also reared in an enhanced environment. Those who do worst, on the other hand, are those who display a fearful temperament when very young and have no exposure to an enhanced environment.

Strong temperamental differences in young nonhuman as well as human infants, together with findings from studies of identical and nonidentical twins demonstrating a strong genetic influence on temperament, make it clear that biological factors are importantly involved in exuberant behavior. One of the most consistently implicated of these biological factors is dopamine, a neurotransmitter of ancient origins. Dopamine, one of many neurotransmitters in the brain responsible for the regulation of moods and motivated behaviors (others include, importantly, serotonin and norepinephrine), long predates our earliest mammalian ancestors; indeed, its existence can be traced back to mollusks and other invertebrates that lived more than half a billion years ago. Dopamine does many things: it heightens attention to unexpected happenings in the environment; it regulates movement; and, most critically here, dopamine is central to reward. Dopamine is released in the brain in response to behaviors of obvious evolutionary value such as obtaining food, drink, and sex. But the plunge downward on a roller coaster also increases dopamine, as does a win at gambling or a line of cocaine. Music, too, as we shall see, has a powerful effect. Brain imaging studies conducted while a person is listening to music show that there are increases in cerebral blood flow in the same reward areas of the brain that are active when food, sex, or highly addictive drugs are involved. (Music may also, like other inducers of positive mood, decrease activity in those regions of the brain associated with negative emotions, such as anxiety or revulsion.)

In the 1950s, James Olds and Peter Milner at McGill University in Canada carried out a famous series of experiments on the brain's "pleasure center." They placed an electrode deep into the brains of rats and demonstrated that rats will press a bar connected to the electrode in order to stimulate a particular area within the hypothalamus, a part of the brain crucial to, among other things, regulating reward and punishment. Indeed, rats find the pleasure so extraordinarily reinforcing they will press the bar thousands of times an hour and actually starve to death rather than stop. A major nerve pathway, which runs through this pleasure area of the brain, pours out dopamine when stimulated. Rats injected with drugs that block the effects of dopamine press the bar far less frequently.

Dopamine, which is concentrated in the frontal lobes of the brain and in the limbic system, or "emotional brain," strongly influences the emotional system that motivates both exploratory and anticipatory behaviors. The anticipation of sex or food, of meeting up with a close friend, or of discovering or experiencing something new brings with it an expectation or hope that pleasure will follow. Expectation, in turn, motivates behaviors likely to lead to the desired outcome. The release of dopamine that accompanies both the anticipation and the consummation of these activities makes it more likely that the behaviors will be repeated.

Scientists believe that the brain's sensitivity to dopamine is correlated with extraversion. If, for example, a drug that increases dopamine transmission is injected into the brain of a mouse, the animal will become more "outgoing" and exploratory. A mouse born without the genes necessary to make dopamine, on the other hand, becomes essentially catatonic, unable to take action. The dopamine-deprived mouse is as pervasively uninterested in its environment as the dopamine-enhanced mouse is actively curious.

Psychologists find that extraverts are exquisitely sensitive to rewards. Researchers at the University of Illinois at Urbana–Champaign, for example, studied individuals from thirty-nine countries and found that only those facets of Extraversion that are most linked to reward sensitivity—Affiliation (friendliness, enjoyment in the company of others); Ascendance (leadership and social dominance); Venturesomeness (seeking out of exciting, stimulating situations); and Social Interaction (preference for being with others)—cluster together on a single higher-order Extraversion factor. This factor, in turn, correlates strongly with positive emotion. It may well be that individuals who seek out novelty and adventure are biologically more likely to feel an intense rush from having done so and, therefore, are more likely to seek out novelty yet again.

Dopamine is strongly associated with positive affect and importantly implicated in its most pathological manifestation, acute mania. The staggeringly high level of physical energy and mental activation that, together with an abnormally excited mood, is so characteristic of mania creates an extreme state of mental and physical exuberance. Amphetamines promote the release of dopamine and inhibit its uptake; they usually produce hypomania if given to patients who have a genetic vulnerability to manic-depressive illness. So does the dopamine precursor L-dopa, which is used in the treatment of Parkinson's disease. Likewise, drugs that increase dopamine levels tend to have an antidepressant effect. Antipsychotic medications, in turn, exert much of their therapeutic effect against mania by selectively blocking dopamine receptors in the brain.

There are specific areas in the brain particularly linked to pleasure and high mood, as well as to exploratory behavior. Greater activation in the left frontal area of the brain is associated with joyful emotions, for example, and is more common in highly uninhibited children and in adults who are unusually enthusiastic and energetic. The left frontal portion of the brain is associated with physical and cognitive behaviors involved in novelty-seeking and reward, and the left prefrontal cortex with the anticipation of pleasure. Neuroimaging studies show that photographs with highly interesting or positive content activate the left amygdala but not the right (the amygdala is involved in emotional processing and the formation of emotional memories). An individual with damage in the left frontal areas of the brain is more likely to be depressed and apathetic; damage or abnormal growths in the right frontal region of the brain, on the other hand, not uncommonly result in inappropriate laughter or actual mania. Recent brain imaging studies show that patients with manic-depressive illness are more likely to have a reduction in gray-matter volume in areas of the right prefrontal cortex than in the left.

In a 2003 study reported in *Science,* Carl Schwartz, Jerome Kagan, and their colleagues reported that at least one aspect of brain functioning associated with temperament is relatively consistent from infancy through early adulthood. They used functional magnetic resonance imaging (fMRI) techniques to measure the response of the amygdala to neutral expressions on either a novel face or a familiar one. Their subjects were young adults who, in the second year of their lives, had been categorized as having either an inhibited or uninhibited temperament. When shown novel faces, those who had been evaluated as inhibited at two years of age showed greater responses in both the left and right amygdala than those who had been categorized as uninhibited. There was no difference between the two groups when they were shown familiar faces. Kagan suggests that the amygdala is primarily responsive to novelty, rather than to a fear-inducing stimulus, and that exuberant individuals seek and enjoy new experiences in part because of their amygdalar chemistry (not because they are not fearful).

Exuberance is not only greatly influenced by the biological activity of the brain, it in turn exerts its own sway over mind and body. That effect is a salutary one. "A merry heart doeth good like a medicine," says Proverbs, and mirth, wrote Robert Burton four hundred years ago in *The Anatomy of Melancholy*, is one of the true nepenthes: it "purgeth the blood, confirms health, causeth a fresh, pleasing, and fine colour . . . whets the wit, makes the body young, lively, and fit for any manner of employment." The merrier the heart, alleged Burton, the longer the life. Modern science tends to support his contention: positive emotions such as joy act as breathers from stress and in doing so they help to restore physical and psychological health after draining or stressful times.

In one test of this idea, Barbara Fredrickson and her colleagues at the University of Michigan conducted an experiment in which they induced a high-arousal, negative emotional state in their research subjects. The experimenters measured the cardiovascular effects of the negative induction; they then followed up the negative induction with one that produced either a neutral mood, mild joy, sadness, or contentment. They found that subjects who received positive inductions, of joy or contentment, took far less time to return to their normal level of cardiovascular functioning than those who had not.

Shelley Taylor of the University of California at Los Angeles, along with others, suggests that positive attitudes such as optimism and the propensity to find benefit in difficult experiences—such as chronic or terminal illness, natural disaster, or being the parent of an acutely ill newborn—may improve the course and outcome of the illness or distress. Psychologists increasingly believe that positive emotions and expectations may improve the body's immune functioning, make it more likely that an individual will act in healthy ways, and increase the chances of having better and more sustaining personal relationships. (A study of 180 nuns who had been asked when they were in their twenties to write brief autobiographical statements found that 24 percent of those who had expressed the most positive emotions had died by the age of eighty; in contrast, 54 percent of those who expressed the least positive emotions had died.) William Hazlitt wrote that our attachment to life depends upon our interest in it, that "passion, imagination, self-will, the sense of power, the very consciousness of our existence, bind us to life, and hold us fast in its chains, as by a magic spell." Those who are exuberant or positive by nature hold on to life and move forward when many others cannot.

Positive emotions affect not only physical health but thinking and behavior as well. "Lively passions commonly attend a lively imagination," wrote David Hume over 250 years ago, and his observation finds support in recent psychological research. Alice Isen at Cornell University has, with others, developed a variety of ways to temporarily induce elevated moods in experimental subjects in order to compare changes in their behavior and thinking with those of subjects whose moods have not been elevated. Methods used to evoke mood changes include music, film comedies, cartoons, and unexpected gifts or praise. Most of these inducers are relatively mild but the results are nonetheless striking. (It would perhaps require drugs, dance, or ecstatic music to provoke actual exuberance.) Those who have had their mood experimentally elevated are more likely to make decisions quickly and efficiently, to help others when given the opportunity, to speak more and faster, to be more sociable, and to take greater risks. They also more actively explore their surroundings and engage in a greater variety of activities. It is as if a dollop of galumphing had been injected into their brains.

Originality and fluency of thinking are particularly affected by changes in mood. This has been shown not only in mood induction studies but also in research that has looked at cognitive changes during mild manias. (Mild mania is almost always accompanied by a vivid elevation in mood; exuberance, as we shall see later, is a frequent feature of both mild and more severe manias.) Mood induction studies repeatedly show that individuals whose moods have been experimentally elevated give a larger number of responses, as well as a larger number of unusual responses, to neutral words presented during a word association task, a measure that is linked to creativity. (If, for example, a subject is asked to offer as many words as possible in response to the word "tulip," both the number of words, and the number of unusual words, can be compared with the responses of thousands of others who have been given the same task.) They are also more likely to classify visual forms and verbal concepts in a global way than in a specific way. An individual in a positive mood tends to see the forest and the pattern among the trees; an individual who is in neither a positive nor a depressed mood picks out the trees. Someone who is depressed focuses in on the bark (and then notes, as well, where it is peeling). "Not by constraint or severity shall you have access to true wisdom, but by abandonment and childlike mirthfulness," wrote Thoreau in his journal. "If you would know aught, be gay before it."

In a typical study of positive mood induction, twenty-two eighth-grade students were given tests designed to assess creativity and problem-solving ability. One task required the students to come up with as many words as possible in response to the words "fruit" and "bird." In the other, each child was given a box, some tacks, a candle, and a book of matches and asked to set up the candle vertically so that it could act as a lamp without dripping wax. Positive mood was induced in half of the students and a neutral mood in the other half. Those students in the positive mood group generated significantly more words than those in the neutral group (twenty-nine and sixteen words, respectively), and the words they generated were much more likely to be unusual. Seven out of eleven in the positive mood group were able to solve the candle problem, whereas only two of the eleven in the neutral mood group were able to do so.

Positive mood, in this and in other studies, increased creativity and flexibility in thinking. No one really understands how these changes in thinking are brought about. To some extent, positive mood may work simply because it is incompatible with anxiety and other negative emotions that hinder productive thought. But it is of course more complicated than that. Isen and her colleagues suggest that positive mood influences the way in which cognitive material is organized and retrieved from the memory, a sensible hypothesis that has yet to be adequately tested. Because positive mood is associated with increased dopamine levels in the brain, they also speculate that the increase may account for the facilitation of cognitive processing during elevated mood states. Surges in dopamine appear to draw attention to unexpected events or conditions, including those that predict reward; this, in turn, may make more likely the active pursuit, exploration, and understanding of the circumstances surrounding such events.

The effect of positive mood on thinking and behavior is by no means straightforward, nor is it consistent across studies. Scientists at Washington University in St. Louis found that college-age subjects in whom a positive mood had been evoked did better on verbal tasks but worse on visual ones. Subjects in whom an anxious mood had been generated, however, did precisely the opposite: they performed better on visual tasks and less well on verbal ones. Perhaps because the detection of physical danger is based largely on visual cues, it makes sense that anxiety and fear, which are highly primitive and instinctive emotions, focus and improve visual perception. Language, a far more recent addition to the brain's repertoire, may be more obviously linked to younger systems in the brain that generate pleasure.

It is not obvious to what extent mental activation, and not just mood, is important in increasing cognitive fluency. Too-great activation may overwhelm and ultimately undermine originality and productivity. But some activation is crucial to attentiveness, motivation, and the capacity to translate thought into action. Intellectual and creative advantages prompted in a laboratory situation may not be important in the real world, unless the individual is also physically alert and compelled to act.

Manic-depressive illness, because it is strongly heritable and often characterized by exuberance (as well as being related to a spectrum of exuberant temperaments), is a particularly important, naturally occurring psychological laboratory for looking at the effects of mental activation and euphoric and expansive mood states on thinking and behavior. Manic-depression, also known as bipolar disorder, is unique in its importance to understanding these effects. It is possible to see in mania and depression the impact of powerful mood changes on an individual and to compare in the same individual the effects of positive mood changes (for example, those of mild mania) with those that are negative (for example, those of depression).

Manic-depression is a well-studied illness that has been carefully observed and described since the time of Hippocrates, five hundred years before the birth of Christ. It is relatively common, occurring in its most severe forms in 1 percent of the population and, in its very mild and briefer variations, in at least another 5 percent. Bipolar illness encompasses a wide range of temperaments and psychopathologies. In the classic form of the disease, both mania and depression are severe, potentially life-threatening clinical conditions characterized by extreme changes in mood, thinking, energy levels, sleep patterns, and behavior. Mania—a state during which, as the composer Hugo Wolf put it, "the blood becomes changed into streams of fire"—is distinguished by an often wildly exuberant mood (albeit one joined by irritability and, not infrequently, depression), expansive and grandiose thinking, cascading speech, phenomenally high levels of energy, little need for sleep, a frenzied tendency to seek out others, terrible judgment, and rank impulsiveness. Mania, an inherently unstable state, is often accompanied by florid psychosis, that is, by delusions and hallucinations. Depression, in stark contrast, is characterized by a flat, irritable, and apathetic mood, sluggish and morbid thinking, profoundly disturbed sleep, reclusive behavior, and an almost unimaginable lethargy and mental pain. Psychic life is dull and colorless, and virtually every aspect of living is laborious. Depression conserves energy; mania expends it. If exuberance is the Champagne of moods, mania is its cocaine. Mania is exuberance gone amok.

Mood is not just "positive" during mania, it is, as the nineteenth-century German psychiatrist Emil Kraepelin put it, "unrestrained, merry . . . exalted and cheerful . . . overflowingly so." Aretaeus of Cappadocia, who is often called the clinician of mania, wrote two thousand years ago that those who are manic are gay, active, and expansive. They are, he said, naturally passionate and joyous. They laugh, they joke, "they show off in public with crowned heads as if they were returning victorious from the games; sometimes they laugh and dance all day and all night." They become "wholly mad," he wrote. "Some run along unrestrainedly, and, not knowing how, return again to the same spot. . . . Some flee the haunts of men, and going to the wilderness, live by themselves." Aretaeus also observed—accurately, as it turns out—that mania, which is strikingly dependent upon the length of daylight, "is connected with temperature of the season," and that "in those periods of life with which much heat and blood are associated, persons are most given to mania, namely, those about puberty, young men, and such as possess general vigour."

Soranus of Ephesus, writing in the second century A.D., emphasized both the anger and merriment so evident during mania; he also noted the "continual wakefulness" of the manic state. The nineteenth-century French alienist Jean-Etiènne Dominique Esquirol agreed that manic patients were lively and irritable and enthusiastic, but he emphasized that they were, as well, volatile and takers of risk. Although susceptible to rapid switching into irritability and anger, mood and energy generally soar during mania. And with them soars thought. (The artist Benjamin Haydon once said of himself that when manic he was "like a man with air balloons under his armpits and ether in his soul.")

People think, feel, and act very differently when they are manic. Manic speech differs unmistakably from depressed speech: not only is it faster and far more abundant, it employs more colorful language, more action verbs, and more adjectives. Rhyming and sound associations dramatically increase. (Vincent van Gogh, for one, wrote that his exuberant mood propelled not just his art but his speech: "There are moments," he said, "when I am twisted by enthusiasm or madness or prophecy, like a Greek oracle on the tripod. And then I have great readiness of speech.") Depressive speech, in contrast, is slowed, vague, and punctuated by qualifying adverbs. Artwork produced during mania is distinguished by its vividness, expansiveness, and bright colors as well as by a euphoric, positive, and excited emotional tone. Depressive artwork, on the other hand, is characterized by cold and dark colors, and a poverty of ideas.

Observers from ancient times and scientists from modern ones have emphasized that quick speech and thought almost invariably accompany the elated mood of mania. Indeed, rapid or pressured speech (which is more objectively measured than rapid thought) is present in virtually everyone who is manic. Unusual talkativeness and "flight of ideas," or the subjective experience that one's thoughts are racing, are among the formal diagnostic criteria for mania. The opposite is also true: depressed mood is associated with greatly slowed thinking in more than 90 percent of patients with bipolar depression; a paucity of ideas is one of the defining features of the clinical condition.

A Royal Navy officer named John Custance wrote a remarkable account of mania and depression in his 1952 memoir, *Wisdom, Madness, and Folly: The Philosophy of a Lunatic*. The exquisite sense of well-being he felt during early stages of his manic attacks permeated every aspect of his being: "First and foremost comes a general sense of intense well-being," he wrote. "I know of course that this sense is illusory and transient . . . [but] the general sense of well-being, the pleasurable and sometimes ecstatic feeling-tone, remains as a sort of permanent background of all experience during the manic period."

Custance's thoughts took flight with his mood. "As I sit here looking out of the windows of the ward," he wrote,

> I see flocks of seagulls who have been driven inland by the extreme cold. The mere sight of these seagulls sets up immediately and virtually simultaneously in my mind the following trains of thoughts:—
>
> 1. A pond called Seagull's Spring near my home.
> 2. Mermaids, i.e. "Sea girls," sirens, Lorelei, Mother Seager's syrup, syrup of figs, the blasted fig-tree in the Gospels, Professor Joad who could not accept Jesus as the supremely perfect Man owing to particular incident . . .
> 4. Gulls equal girls, lovely girls, lovelies, film-stars, countless stars in the infinite wastes of space, query: is space really infinite? According to Einstein it is not. . . .

Manic thought initially gallops along in a straight line, but chaos ensues as the mania progresses. Thoughts proliferate malignantly and race mindlessly about in the increasingly overloaded and cluttered brain; then they collide and splinter. All governance is lost. Madness settles in and swiftly obliterates whatever advantages mood may have given the mind. The wide-flung ideas of mania and the rush of their flight lead to a terror not imaginable to those who have not lost their minds in this particular way. One patient, writing a hundred years ago, described the initially compelling, then ultimately horrifying path she took to insanity: "All the problems of the universe came crowding into my mind, demanding instant discussion and solution—mental telepathy, hypnotism, wireless telegraphy, Christian Science, women's rights, and all the problems of medical science, religion, and politics. I even devised means of discovering the weight of a human soul, and had an apparatus constructed in my room for the purpose of weighing my own soul the minute it departed from my body. . . . Thoughts chased one another through my mind with lightning rapidity. I felt like a person driving a wild horse with a weak rein, who dares not use force, but lets him run his course, following the line of least resistance. Mad impulses would rush through my brain, carrying me first in one direction and then another. To destroy myself or to escape often occurred to me, but my mind could not hold on to one subject long enough to formulate any definite plan." Like Custance's mind, this patient's was on high alert and lit up like a pinball machine: everything seemed to connect and ideas jumped. But then, as her mood ratcheted up, her thinking pelted utterly out of control.

To understand the connection between exuberant mood and thought is to understand the striking changes in both the quantity and the quality of thinking that occur during mania. John Ruskin described the pell-mell traffic of his thoughts while manic: "I roll on like a ball, with this exception, that contrary to the usual laws of motion I have no friction to contend with in my mind, and of course have some difficulty in stopping myself when there is nothing else to stop me. . . . I am almost sick and giddy with the quantity of things in my head—trains of thought beginning and branching to infinity, crossing each other, and all tempting and wanting to be worked out." Samuel Taylor Coleridge, in a different but quite wonderful way, described his times of mental exuberance: "My thoughts," he said, "bustle along like a Surinam toad, with little toads sprouting out of back, side, and belly." The abundance of thoughts during manic exuberance is matched only by their bounce and velocity.

Manic thought, like exuberance, is usually infused with a sense of immediacy and high significance. It demands an audience or a record: notice must be taken. Individuals when manic are inclined to act—to write, paint, compose, petition—rather than to remain passive. (Robert Lowell astutely observed that "mania is a sickness for one's friends, depression for oneself." Little remains unstirred anywhere within hailing distance of a manic.) The writer Morag Coate captured this sense of fervid grandiosity in describing one of her psychiatric hospitalizations: "I must record everything and later I would write a book on mental hospitals. I would write books on psychiatric theory too, and on theology. I would write novels. I had the libretto of an opera in mind. Nothing was beyond me. . . . I wrote a fairy tale; I wrote the diary of a white witch. . . . It was all vitally important."

There is a surprisingly large body of empirical evidence linking creative thinking to manic-depressive illness. More than twenty studies show that artists, writers, and other creative individuals are far more likely than the general public to suffer from mood disorders, especially manic-depression. Clearly, most people who are creative do not have a mood disorder, and most people who have a mood disorder are not unusually creative. But, as a group, creative individuals have a disproportionately high rate of depression and bipolar illness. Many explanations have been put forward to explain this finding—high levels of energy and enthusiasm, a tendency to take risks, an underlying restlessness and discontent, more finely tuned senses, a need to impose order on chaos, and a range and intensity of emotional experiences common to the artistic and manic-depressive temperaments—but the most commonly and persuasively suggested are the types of changes in mood and thinking that take place in both manic and creative thought.

Creative and manic thinking are both distinguished by fluidity and by the capacity to combine ideas in ways that form new and original connections. Thinking in both tends to be divergent in nature, less goal-bound, and more likely to wander about or leap off in a variety of directions. Diffuse, diverse, and leapfrogging ideas were first noted thousands of years ago as one of the hallmarks of manic thought. More recently, the Swiss psychiatrist Eugen Bleuler observed: "The *thinking* of the manic is flighty. He jumps by by-paths from one subject to another. . . . With this the ideas run along very easily. . . . Because of the more rapid flow of ideas, and especially because of the falling off of inhibitions, artistic activities are facilitated even though something worthwhile is produced only in very mild cases and when the patient is otherwise talented in this direction." The expansiveness of thought so characteristic of mania can open up a wider range of cognitive options and broaden the field of observation.

Both individuals who are manic and those who are writers, when evaluated with neuropsychological tests, tend to combine ideas or images in a way that "blurs, broadens, or shifts conceptual boundaries," a type of thinking known as conceptual overinclusiveness. They vary in this from normal subjects and from patients with schizophrenia. Researchers at the University of Iowa, for example, have shown that "both writers and manics tend to sort in large groups, change dimensions while in the process of sorting, arbitrarily change starting points, or use vague distantly related concepts as categorizing principles." The writers are better able than the manics to maintain control over their patterns of thinking, however, and to use "controlled flights of fancy" rather than the more bizarre sorting systems used by the patients.

Manic thought is often wildly combinatory, humorous, and playful; it galumphs and in so doing, especially in its milder forms, contributes significantly to performance on tests measuring creativity. Kraepelin and other nineteenth-century physicians tested their patients for fluency of word associations during mania and found that it increased dramatically. Later, in the 1940s, researchers at the Payne Whitney Psychiatric Clinic in New York demonstrated that, when elated, their manic-depressive patients showed greatly increased verbal and associative ability. They also found that normal subjects (or, more accurately, medical students), when given a mild dose of the stimulant dexedrine sulfate, improved on tests of associative fluency. The performance of the normal subjects while on dexedrine remained significantly lower than that of the manics, however.

Verbal associations increase in different ways during mania. Researchers find, for example, that the number of unusual responses to word-association tasks (similar to those used in studies looking at the effect of positive mood induction on associational fluency in normal subjects) increases threefold. The number of statistically common responses drops dramatically. The increase in word associations is generally proportionate to the severity of manic symptoms.

The changes in mood and thinking that accompany mania are far more intense than those induced by psychologists during experimental studies. This suggests several things. The relationship between elated mood and fluency of thinking is, up to a point, linear: the more elevated the mood, the more fluent and diverse the thinking. Too much elevation, however, results in fragmented thinking and even psychosis. Likewise, the level of enthusiasm with which an idea is held has an impact on the likelihood that it will be put into action. (People who have manic-depressive illness are, when manic, more likely to act, to be utterly certain of their convictions, and to put their ideas into action without the brakes or judgment provided by rational thought.) In other words, mood affects not only thought, but the uses, if any, to which thought is put. This is particularly important for those individuals whose work is strongly tied to mood.

Several years ago I conducted a study of eminent writers and artists and found, like most researchers before and after me, that they were much more likely than the general population to have been treated for mania or depression. One of my major interests was to look not just at psychopathology, however, but to try to better understand the role of moods in the creative process itself. Virtually all the writers and artists had experienced extended periods of intense creative activity, which were characerized by striking increases in enthusiasm, energy, rapid and fluent thought, and self-confidence. Most reported that a sharp rise in their mood *preceded* the onset of their creative work. Ruth Richards and Dennis Kinney, in a Harvard study of manic-depression and creativity, found that the overwhelming majority of their subjects experienced at least mildly elevated moods during the periods when they were most creative. When their moods were most elevated, so were the ease and expansiveness of their thinking and the quickness of their mental associations. Eugene Fodor, of Clarkson University in New York, has shown that students vulnerable to manic-depressive illness are particularly creative when their mood has been experimentally elevated.

The act of being creative, it is assumed, occasions elation. No doubt this is true, but studies of mania and creativity, along with results from studies of experimentally induced positive mood, suggest that the opposite may be at least as important: that is, elevated or expansive moods come first; creative thinking follows. Creativity may then, in its own right, elevate mood; this can lead to extended periods of reverberating moods, energies, and imagination. It may lead to decreased sleep as well, which can further elevate mood.

The links between manic-depression and creativity suggest some ways of understanding the connections between positive mood, thinking, and behavior. But mania is a pathological condition and represents extremes in mood and behavior. More illustrative of positive emotions and high energy—of "normal" exuberance—are the temperamental qualities closely allied with manic-depressive illness. Emil Kraepelin, who remains for most of us who study the illness the preeminent authority on the subject, describes a temperamental variant, which he called a manic predisposition. "The slightest forms of the disorder," wrote Kraepelin, "lead us to certain personal predispositions still in the domain of the normal. It concerns here brilliant, but unevenly gifted personalities with artistic inclinations. They charm us by their intellectual mobility, their versatility, their wealth of ideas, their ready accessibility and their delight in adventure, their artistic capability, their good nature, their cheery, sunny mood." But, Kraepelin goes on to say, there are liabilities to this temperament: "they put us in an uncomfortable state of surprise by a certain restlessness, talkativeness, desultoriness in conversation, excessive need for social life, capricious temper and suggestibility, lack of reliability, steadiness, and perseverance in work, a tendency to building castles in the air." People who have this temperament, he observed, may also be inclined to depression, although it is usually their relatives who show the most severe pathology. This tendency toward depression, together with a typically strong family history of manic-depression, led Kraepelin to his belief that the manic predisposition (or the sanguine temperament) is to be regarded as "a link in the long chain of manic-depressive predispositions."

Modern researchers tend to support Kraepelin's belief in the existence of "manic-depressive predispositions." Hagop Akiskal and his colleagues, as we have seen, find that certain temperaments, including hyperthymia—characterized by optimism, high energy, enthusiasm, and extraversion—are part of a continuum of traits in the general population. Individuals who score high on hyperthymia are more likely to switch into mania than those who score very low. This is also true for college students who obtain high scores on the Hypomanic Personality Scale, a test developed by psychologists to identify people at high risk for developing bipolar disorder. A thirteen-year follow-up study of students who scored high on the test found that they were more likely to experience subsequent attacks of mania and depression. (Typical items on the scale are: "I am frequently so 'hyper' that my friends kiddingly ask me what drug I'm taking"; "I often have moods where I feel so energetic and optimistic that I feel I could outperform almost anyone at anything"; "I often get so happy and energetic that I am almost giddy"; "I often get into excited moods where it's almost impossible for me to stop talking.") The Hypomanic Personality Scale correlates with personality dimensions such as extraversion and openness to new experiences. Six percent of college students tested meet the criteria for hypomanic tendencies, a figure reasonably close to the 8 percent who were categorized as hyperthymic in a study of American and Italian students. These estimates are comparable to the 10 percent of children described as exuberant by Jerome Kagan at Harvard, and the 10 percent of infants characterized as exuberant by Nathan Fox at the University of Maryland.

It is clear from clinical observation and decades of scientific research that exuberance is an important phase of acute mania for most of those who become ill. A significant percentage of people who have manic-depressive illness also have an underlying exuberant temperament. Clearly, however, most people who are exuberant do not have manic-depressive illness. There are overlapping characteristics—high mood and expansive energy, among others—but there are critical distinctions as well. Exuberance is far from a pathological state for most who have it. It is, instead, a highly valued and integral part of who they are and how they meet the world. Understanding the role of exuberance in manic-depression can provide one perspective on exuberance—extremes in behavior will almost always illuminate more normal behavior—but there are limits to the comparisons that can be made. Still, the shading of normal exuberance into "pathological enthusiasm," as

Robert Lowell once described a manic attack, is an important as well as a cautionary phenomenon, and it is one to which we will return.

Enthusiasm is intoxicating: it goes to the head. Just as it is hard to remember while under the spell of Champagne that beneath the fizz lies a dangerous undertow, so it is with exuberance. But if dangerous on occasion, far more often it is a delight, a lift, and a boon.

CHAPTER SIX

"Throwing Up Sky-Rockets"

People *like* to be humbugged, declared P. T. Barnum. They need sizzle and flair to lighten their otherwise "drudging practicalness": they long to catch fire, be bedazzled, be a part of something that delights them, excites them, and binds them to others. They want someone to splash their world with color; someone, as Barnum put it, to "throw up sky-rockets." Barnum was more than glad to oblige; most people, as he knew to his advantage, shared neither his genius for invention nor his irrepressible exuberance.

He was, he wrote, "blessed with a vigor and buoyancy of spirits vouchsafed to but few"; it would be, he continued, "utterly fruitless to chain down the energies peculiar to my nature." As we shall see, he never tried. Far from chaining down his energies and spirits, Barnum set them loose to infect others.

Emotions are contagious. We survive because we apprehend quickly in others, and then speed on their way, those emotions that alert us to risk or prospect. Emotions are part of the social glue of our immensely social species, elemental to the reverberating emotional circuitry that compels us at times to pull together and at others to disperse. Nothing, it would seem, is quite so wildly contagious as exuberance, and yet it remains curiously absent from psychological explorations of group behavior. Certainly it is easier to find studies of despair and anxiety than of infectious enthusiasm; perhaps the bent of the human condition dictates this, spelled as it is by uncertainty, suffering, and death. An undercurrent of darkness runs throughout our philosophical beliefs, and melancholy is woven into our great literature and music: "We can hear it in all acuteness in Schubert and Schumann," wrote Leon Edel. "It sounded for us in the cosmic cadences of Beethoven; it comes at us from almost every page of poetry." Words for desolation come apace, those for exuberance less so.

Anyone who teaches about moods knows this to be true: it is far easier to convey the essence of depression to young doctors and graduate students than it is to depict mania or other elated moods. In part this is because the language for melancholy is such a rich and nuanced one, but it is also because those who are being taught about moods are more likely to have experienced depression than either mild or full-blown mania. This discrepancy was brought home to me when I was director of the UCLA Affective Disorders Clinic and responsible for teaching psychiatric residents and clinical psychology interns about depression and manic-depressive illness. There was a glaring and disproportionate emphasis on clinical teaching about depression, even after taking into consideration the fact that depression is more common than mania.

Elated mood states were given short shrift in the medical and psychological research literature as well; their seductiveness to the individuals who experienced them was seldom mentioned and there was little or no discussion of their highly infectious nature. Anyone who treats the early stages of mania knows the exhilarating maelstrom they create, but few clinicians and scientists were writing or talking about this. As someone who had experienced the fleeting glories of mania firsthand, I found the oversight grating and difficult to understand. Yet I was as incapable of describing mania as the rest of my colleagues were.

At the height of my frustration in teaching about exultant and manic states I happened to see Jim Dale's Tony Award–winning performance in *Barnum,* a musical shot through with exuberance. In a high gust of enthusiasm I wrote to Mr. Dale, explained my teaching conundrum, and asked him if he would be willing to meet with me to talk about how actors portray moods. He generously agreed to do so and turned out to be a highly intelligent observer of human behavior, as well as exceptionally thoughtful on the subject of how to depict exuberant moods and how to ignite them in others.

When words are neither the only thing nor the most important, Dale emphasized, then action is. And music. For it is in action, in dance and in music, in the kinetic thrusting upward of arms and legs and the throwing up and back of the head, that great joy finds its highest expression. Everyone on the *Barnum* set, he pointed out, is in near-constant, rollicking motion. The music is fast, loud, brassy, and exhilarating. Barnum talks fast and moves faster; others who are onstage are either singing or dancing, juggling, bouncing, or leaping. Or doing them all simultaneously. Everything—music, lyrics, balloons, streamers—blasts out in primary, audacious color. Indeed, it is in the exuberance of color that the musical's librettist, Michael Stewart, has Barnum sum up his life: "The colors of my life," Barnum sings,

Are bountiful and bold
. . . .
The dazzle of a flame
The glory of a rainbow
I put them all to shame.
No quiet browns and grays
I'll take my days instead
And fill them til they overflow
With rose and cherry reds.
And should this sunlit world
Grow dark one day
The colors of my life
Will lead a shining light
To show the way.

Jim Dale's dazzling energy as Barnum electrified the rest of the cast and the audience. His actions and moods were, simply put, contagious.

The real P. T. Barnum showed little inclination toward muted colors or a quiet existence: "I have lived so long on excitement, pepper, & mustard," he wrote to a friend, "that plain bread & milk don't agree with me." He saw little reason to tone down his natural flamboyance; on the contrary, he delighted in it and used it to infuse fizz and spark into the world around him. Everything Barnum created was larger than life; his was a world of razzle-dazzle and poster colors. Neither time nor setbacks muted his zeal. When he was nearly eighty years old, the inexhaustible Barnum took his circus to

London. The bewilderingly complex production required three ships to transport the twelve hundred performers and more than four hundred horses and other circus animals. Billed as a "triple 100 act circus," it featured a "Roman Hippodrome," one hundred chariots, and a reenactment of the destruction of Rome. There were "wondrous mid-air feats" and "Mirthful and Astounding Visions." The "united enchantments, delusions and displays of all ancient and modern magicians of every clime," proclaimed one circus poster, were "commonplace and puerile, compared with the Supernatural Illusions [which are] for the first time exhibited, without extra charge, in the Great Show's Electric-Lighted Wizard's Temple." Barnum never met an excess he didn't like, never saw a rainbow whose colors he couldn't improve.

Action and distraction, he knew, will trump the still. In the musical based upon his life, Barnum is portrayed as keeping at bay his own and the world's ennui by spinning off energy and joy. "Through a night as dark as space / And cold as the sea," he sings, "Someone's got to make it bright / Shoot a rocket, shine a light." Someone, in short, has to build a fire, distract, and amuse. Exuberance, Barnum knew, is complicated: it may exist on its own or it may keep darker company; it is something people not only want but need.

Following my talk with Jim Dale, and after having returned to the theater to once again watch him captivate his audience with his electric irrepressibility, I added two songs from *Barnum* to my clinic lectures about mania and other elated states. Even the doctors who had been on call the night before—usually a study in the ability, shared with horses, to sleep upright, with their eyes giving only the illusion of being open—were tapping their feet. Exuberance, the real thing, *is* contagious. But why?

There are many reasons. The rapid and accurate communication of emotion among members of a group is essential if they are to survive. Split-second transfer of fear alerts the group to potential danger and compels a swift, coordinated response. Likewise vital, if less obviously so, the quick dispersal of exuberant or triumphant emotion accelerates the spread of the news of victory, opportunity in the environment, or a new idea. It sends the message that it is time to explore, to gather as a group, to celebrate, to have fun. When there is cause for celebration, or collective enthusiasm and energy are required, infectious fervor will further a swift dissemination. Malcolm Gladwell argues in *The Tipping Point* that behaviors spread as viruses do. Exuberance is a potent vector; it shoots emotion and opportunity into a group just as brisk, high winds carry pollen and seeds into new fields and habitats. Exuberance is a propitious thermal, which first lifts and then ferries energy, enthusiasm, and hope.

Exuberance draws people together and primes them to act boldly; it warrants that the immediate world is safe for exploration and enjoyment and creates a vivifying climate in which a group can rekindle its collective mental and physical energies if depleted by setback, stress, or aggression. It answers despair with hope: "How I long for a little ordinary human enthusiasm," wrote John Osborne in *Look Back in Anger*. "Just enthusiasm—that's all. I want to hear a warm, thrilling voice cry out Hallelujah! Hallelujah! I'm alive." By capturing many in its far-flung web, exuberance overrides the inhibition that blocks action or innovation; like other positive emotions, it also enhances learning and fosters communal generosity. Infectious joy pumps life into social bonds and creates new ones through collective celebration and lively exchange. Shared joys rather than shared sufferings make a friend, Nietzsche believed, and there is much truth in this. High spirits beget high spirits; the memory of delight is laid down, the expectation of joy seeded.

C. S. Lewis laid stress upon our need for close proximity to intense experience: "Good things as well as bad," he wrote in *Mere*

Christianity, "are caught by a kind of infection. If you want to get warm you must stand near the fire: if you want to get wet you must get into the water. If you want joy, power, peace, eternal life, you must get close to, or even into, the thing that has them. . . . They are a great fountain of energy and beauty spurting up at the very centre of reality. If you are close to it, the spray will wet you: if you are not, you will remain dry."

Joy infects. Katharine Graham once said of an editor that "he had fun and he gave it to others." But how? How does an emotion spread from one person to another? Are some people better able to transmit emotions and, if so, is that because they themselves are more emotional? Psychologists have asked these questions for years. In an early study, conducted in the 1970s, researchers filmed individuals as they viewed slides whose content was highly emotional in nature (photographs of burned bodies, for example, or of laughing children). People whose facial reactions to the pictures were particularly expressive and easy to read were labeled "powerful senders"; those whose faces displayed scant or ambiguous emotional cues, on the other hand, were designated "weak senders." Further investigation revealed that the powerful senders, those who displayed a rich nonverbal language, scored high on measures of extraversion. The weak senders, in contrast, were far more introverted; their nonspoken language of emotions was barren and tightly held.

Carl Jung had observed this four decades earlier. When the extravert expresses emotion, Jung wrote, he makes a "visible and convincing appearance" before his public. Although both the extravert and introvert possess enthusiasm, "that which fills the extravert's heart overflows from his mouth; the introvert's lips are sealed by the enthusiasm that moves him within." The introvert, Jung continued, "kindles no flame of enthusiasm in the world around him. . . . [H]is laconic expression and the mystified lack of comprehension it produces in his public" lead others to doubt that he has anything "extraordinary to say." The extravert, on the other hand, immediately appears intriguing; his manifest success in life is a "vitalizing and invigorating factor."

The psychologist Howard Friedman, of the University of California at Riverside, devised a test to measure individual differences in nonverbal emotional expressiveness. The Affective Communication Test, a thirteen-item self-report scale, contains such items as "I show that I like someone by hugging or touching that person," "When I hear good dance music, I can hardly keep still," and "At small parties I am the center of attention." People who score high on this test tend to be colorful and charismatic, playful, more attractive to others, outgoing, dominant, and able to inspire others to act. (Highly expressive physicians, for example, have more patients than doctors who are less obviously emotional; people who are more expressive are also more likely to be attracted to lives in politics, lecturing, or acting.)

Expressive individuals strongly influence the moods of those who are unexpressive, but the reverse is not true: unexpressive people have little impact on the emotions of those who are expressive. Psychologists find that more emotional information is conveyed by expressive individuals and that their emotional responses attract greater attention from those around them. Women, although they in general score somewhat lower on measures of extraversion than men, tend to score higher on emotional expressiveness. (It may be that men score higher on characteristics of extraversion, such as impulsiveness, which are not as directly related to expressiveness.)

The transmission of emotions is rapid. Viewing faces with "happy" or "sad" expressions, for example, quickly evokes those feelings in the viewer. Barbara Wild and her colleagues at the University of Tübingen in Germany found that communication of most facially expressed emotions takes place within half a second;

the time frame is particularly short if people are looking at "happy" faces. That "happy" faces are registered so swiftly may be partly because happiness, at least as measured by spontaneous facial expression, appears to be the most accurately communicated of the emotions. Psychologists who study the relative communicability of emotions report that happiness is correctly communicated from one individual to another 48 percent of the time. Fear, in contrast, is correctly registered only 10 percent of the time, anger 13 percent, sadness 17 percent, and disgust 23 percent.

Negative emotions, although less accurately transmitted than positive ones (a strange finding, given the importance of the swift communication of fear), are more contagious; that is, expressive individuals are better able to infect others with negative emotions than with positive ones. This is consistent with recent psychological research, which finds that many types of negative stimuli, such as negative words, are detected faster than positive ones. Experimental subjects more quickly locate an angry face among happy ones than a happy face among angry ones. This is almost certainly because the *immediate* survival of an individual or a group is dependent upon the threat of danger being quickly recognized and spread among its members.

Happiness, which is what psychologists have studied rather than joy or exuberance, is a dilute version of these more energetic and communicable states. Nonfacial as well as facial communication of exuberant moods is more obvious than that displayed in the less effusive "happy" ones, and body language and olfactory communication surely play a far more important role than we now appreciate in the transmission of information, including the nature and intensity of our emotions. We know this is true for other species. Mice, for example, form complex images in their brains about the sex and genetic makeup of other mice on the basis of subtle chemical signals called pheromones. Young male Asian elephants in musth secrete a honeylike odor from their temporal glands, which scientists believe signals youthful and erratic behavior rather than competitive intentions to older males. The temporal-lobe secretions of mature elephants are instead fetid and distinctly different from those excreted by the young males. Being able to correctly distinguish the odors excreted in youth from those of maturity appears to increase harmony between two otherwise potentially competitive groups of elephants. Researchers cite ancient Hindu poetry which speaks of the arrival of bees to "gather sweetness from the temples" of young musth elephants. We do not know that there isn't a like sweet smell of joy and playfulness in humans.

The psychologists Elaine Hatfield at the University of Hawaii and John Cacioppo at Ohio State University have been particularly interested in the specific mechanisms responsible for emotional contagion, as well as the psychological characteristics of people who "catch anxiety" or "catch joy." Evidence is strong that people quickly mimic the physical movements, voices, and facial expressions of those with whom they are in contact and that such "catching" of others' emotions is a universal human phenomenon. Hatfield developed the Emotional Contagion Scale to measure individual differences in vulnerability to emotional contagion; for example, the susceptibility to "catching" anxiety ("When someone paces back and forth, I feel nervous and anxious"), or joy ("When someone laughs hard, I laugh too").

Those who are most susceptible to "catching" others' emotions tend themselves to be more emotionally reactive. They pay close attention to the emotions of others, are adept at reading them, and are more likely to mimic the facial, vocal, and physical expressions. Gender is also a factor. Hatfield and her students interviewed nearly nine hundred men and women from a variety of ethnic groups (Hawaiian, Chinese, African, Filipino, and Korean, among others), as well as from different professional backgrounds (students, military personnel, and physicians), and found that in all groups women were far more susceptible than men to "catching" both negative and positive emotions.

Other researchers have shown that people who are themselves happy are in general more attentive to verbal and nonverbal cues. In particular, they are more attentive to emotional expressions. Depressed individuals, although they are more sensitive to expressions of anger, are far less sensitive and responsive to expressions of happiness.

The effects of expressed joy or sadness show themselves early. Nine-month-old infants, observed as they watch and listen to their mothers express either joy or sadness, look longer at their mothers and express greater joy when their mothers are themselves expressing joy. They also engage in more playful behavior. One-year-olds after watching a videotape of an adult actress portraying either positive or negative emotions, will not only mimic the actress's tone and expression but will also alter their level of joyfulness accordingly. Adults, too, when interacting with someone who is happy tend to imitate that person's smiles, laughter, and positive gestures; at least under experimental conditions, this generates positive emotions in the person exposed to another's happiness.

Laughter, because it is both obvious and pleasurable, is one of the more irrepressible communications our social species has. Although it is by no means exclusively bound to exuberance, laughter is nonetheless its clarion messenger. It is attached to exuberance like a clapper to a bell, and their evolutionary histories progress together in like manner. Laughter is universal in humans and it is innate. Children who are deaf, blind, and mute, for example, and who are therefore unable to imitate the behavior of others, nonetheless laugh when tickled. Darwin believed that laughter was the ultimate expression of joy and an important way for infants and young children to influence their parents' behavior. It signals, "I enjoy this, do it again." The pleasure of the child triggers a respondent pleasure in the parent, who then repeats the behavior, which the child enjoys and continues to encourage. The reverberating delight perpetuates. Laughter creates and strengthens the bonds between parent and child, and among other members of a social group.

Biologists believe that human laughter also signals to other members of the group that it is safe to relax, safe to play. There is evidence for this view in the behavior of nonhuman primates. Chimpanzees and pygmy chimpanzees make panting noises or "play chuckles" and show a "play-face" (formally called a relaxed open-mouth display) when they engage in social play or tickle one another. Tickling, according to Roger Fouts and other primate researchers, is an important behavior in the lives of chimpanzees; chimps that have been taught sign language, for instance, frequently "talk," or sign, about tickling. Their play-face, like laughter in humans, seems to be incompatible with fear or aggression. The chimpanzee "grin-face" (a relaxed bared-teeth display), which is more analogous to human smiling, denotes submission and is associated with affinitive behavior. Affinitive displays, such as smiling, precede and often lay down the emotional groundwork for social play in both human and nonhuman primates. Play rarely *precedes* smiling, however.

Full-throated and contagious laughter probably evolved after our early ancestors scrambled down from their trees in a search for new habitats. Having left their arboreal territories, they required more sophisticated and closer social ties with one another, and faster means of communication; they needed, above all, to cooperate. Smiling and laughter, which in all likelihood evolved as more distinct, less mistakable forms of primate play-grins and play-faces, enabled faster communication across longer distances and helped forge stronger, more obligatory social bonds. The increase in physiological arousal and motor activity intrinsic to laughter made the communication of positive emotional states both more accurate and more difficult to resist. Exuberant laughter is nothing if not hard to ignore and difficult to feign. The ethologists Jan van Hooff, of the University of Utrecht, and Signe Preuschoft, of the Yerkes Primate Lab at Emory University, believe that smiling and laughter had very different origins—smiling as a way to signal nonaggression, laughter in the context of joyful play—and argue that the more hierarchical the primate society, the more unambiguous the display had to be in order to signal a playful release from the usual social status.

The powerful social ties initiated or shored up by laughter would have been vital to early hominids in creating the cooperative behaviors necessary to explore new lands, to hunt, and to raise young together; it would have been critical, as well, to generate the trust necessary to take risks as a group. Laughter, then as now, would have helped to disarm tension and to ameliorate stress and loneliness. It would have prompted closeness: bonded mother to infant, mate to mate, and friend to friend. (Indeed, Rupert Brooke observed nearly a hundred years ago that laughter is "learnt of friends"; and it is in the context of friendship that laughter is most warmly shared: "From quiet homes and first beginning, / Out to the undiscovered ends," wrote Hilaire Belloc, "There's nothing worth the wear of winning, / But laughter and the love of friends.") Laughter, millennia ago as now, would have helped veil the terrors of life. It would have bound individuals to their communities, and its infectiousness—fast sent among its members, body to body, brain to brain—would have allowed the group to respond en masse, not just as individuals, one at a time and disjointedly. Laughter gives rise to yet more laughter and sends out, as it does, well-being to the far corners; it spreads its delights far, fast, and furious.

On occasion, laughter spirals out of control. Epidemics of contagious laughter have been documented by the University of Maryland neuroscientist Robert Provine. One, which he describes in his book *Laughter: A Scientific Investigation*, started in 1962 in Tanganyika, in a mission-run boarding school for girls between the ages of twelve and eighteen. Three girls started laughing. Uncontrollable laughter, crying, and agitation quickly spread to 95 of the 159 students (no teacher was affected). The school was forced to close a month and a half later; it reopened, briefly, but then had to close again after 57 girls were stricken. Before finally abating two and a half years later, Provine writes, "this plague of laughter spread through villages 'like a prairie fire,' forcing the temporary closing of more than 14 schools and afflicting about 1,000 people in tribes bordering Lake Victoria in Tanganyika and Uganda. Quarantine of infected villages was the only means of blocking the epidemic's advance. . . . The epidemic grew in a predictable pattern, first affecting adolescent females at the Christian schools, then spreading to mothers and female relatives but not fathers. . . . The laughter spread along the lines of tribal, family and peer affiliation, with females being maximally affected. The greater the relatedness between the victim and witness of a laugh attack, the more likely the witness would be infected."

Research supports the proposition that laughter has gender-specific aspects. Women laugh more often than men, although men are more likely to evoke laughter in others. Both men and women are far more likely to laugh in the presence of others than when they are alone. This is true, as well, for chimpanzees and college students. In one study of seventy-two undergraduates, psychologists found that students were thirty times more likely to laugh when they were with others than when by themselves.

Contagious laughter is put into service as "new wine" in some Pentecostal revival meetings. One minister, who deliberately induces spreading waves of laughter in his congregants, encourages his followers to let the "Niagara of laughter . . . bubble out of your belly like a river of living water." The "new wine" analogy is apt: laughter causes the release of opiatelike chemicals in the brain; this in turn creates a sense of intense, if transient, well-being. The long-term physical benefits of laughter—increased longevity and immunity, lowered blood pressure—are widely cited but not clearly documented. It is likely, as Provine points out, that at least some of the physical benefits of laughter come about from the advantages conferred by enhanced social bonding.

The biology of laughter is only partially understood. As with so much else that bears relation to exuberance, temperament is involved: extraverts laugh more often and more intensely than introverts. Genetics also plays a role in proneness to laughter. Rat pups, for instance, when tickled on their ribs and bellies respond with a chirping "play vocalization." Rats that chirp the most also play the most; they are also the most eager to be tickled. When scientists interbreed the "chirpiest" rats, they find, four generations of rat later, that the selectively bred young rats chirp twice as often as their great-grandparents.

The physical mechanisms of laughter seem to originate in ancient parts of the brain, but the cortex is involved as well. The stimulation of "pleasure centers" in the brain triggers laughter. Brain scans taken while experimental subjects are listening to jokes show activation in the nucleus accumbens, the same area of the brain that is involved in reward-related behaviors. (Conversely, brain imaging studies show that particularly humorless people are more likely to have damage in that particular part of the prefrontal cortex.) The funnier the joke, researchers find, the more brightly the reward circuitry lights up. Scientists believe that the reward for "getting the joke" is one of evolution's ways of getting our species to exercise and expand more diverse and sophisticated neural pathways. Nonobvious connections are at the quick of wit and social humor; predictability dampens both. We laugh to keep our minds agile and to keep our ties to others fresh and resilient. We laugh because laughter returns us with pleasure to our common humanity; it recommits us to the cooperative roots that sprang into being as we moved from tree to savannah. We need these roots now as we needed them then.

Exuberance attracts and then bonds animal to animal; in doing so, it helps create the emotional ties necessary not only for communities to thrive but for potential breeding pairs to commit genes and energy to mate, reproduce, and raise young together. In the animal world, flamboyant mating dances and the vivid colors and energetic, extravagant rituals of sexual displays are an important part of the excitation of sight, smell, and hearing that attracts sexual partners. We are but one of many of nature's manifold species to first lure a mate with exuberant promise and then engage joy to further entwine us. Cole Porter wrote of this, unforgettably: "Cold Cape Cod clams, against their wish, do it. . . . / In shallow shoals, English soles do it, / Moths, in your rugs, do it"; and, Porter continued, with an image for the ages: "Old sloths who hang down from twigs do it." They all do it, he said, they all fall in love. But it is among our own species—not among moths or sloths—that love is so endlessly celebrated.

Exuberance and passion are critical during the early stages of love, for it is then that an intoxicating attraction must occur, a potential mate must be pursued, and a relationship must be intensely, if not recklessly, explored. "Hot blood begets hot thoughts," wrote Shakespeare, "and hot thoughts beget hot deeds, and hot deeds is love." Passion rides roughshod over hesitating judgment; it dissolves inhibitions. Joy incites a catnip rapture; it provokes play and exploration of another's world. Early love, declared Martin Luther, is "fervid and drunken, blinds us and leads us on." Four hundred years later, Thomas Hardy said much the same thing: "Yea, to such rashness, ratheness, rareness, ripeness, richness," he wrote, "Love lures life on."

Exuberant love is addictive; it excites and infects, and it sends those who experience it out on a quest for more of the same. It not only lures and binds, it teaches. It instructs in the rewards and perils of sex and metes out pleasure for engaging in behavior that perpetuates the species. Love awakens and refreshens. "The simple accident of falling in love," said Robert Louis Stevenson, "is as beneficial as it is astonishing. It arrests the petrifying influence of years, disproves cold-blooded and cynical conclusions, and awakens dormant sensibilities." Love also vouchsafes a time of discovery—of play and expansiveness, of incrementally deepening intimacy—before a more permanent commitment to a partner has to be made.

Exuberant love transforms those swept up in it; it "sets the whole world to a new tune," as William James put it. Things not heard before, emotions not felt before, all come into the lover's field of experience. When the smitten sailor in Gilbert and Sullivan's *H.M.S. Pinafore* sings that only Love had been his tutor, anyone who has been in love instinctively understands. We may or may not learn wisely from love, but we certainly learn. And by our experience we are bound more closely to those who likewise are communicant with the lessons and lure of love.

Romantic love usually settles into a less passionate but more stable relationship. This is an attachment sustained in good measure because of our wiring; pleasure and affinity circuitries wend together in our brains. Receptors for oxytocin, a mammalian neuropeptide that facilitates attachment and social learning, mingle with dopamine receptors in the nucleus accumbens, an area of the brain we have seen to be intimately linked to seeking out and experiencing pleasure. This marriage of affinity and joy is one we share with another unusually monogamous and social mammal—an animal overlooked by Cole Porter, but not by neuroscientists—the prairie vole. This tiny mouselike herbivore pairs for life (which, admittedly, is a fleeting one). As with humans, its oxytocin and dopamine receptors overlap in the nucleus accumbens. When a chemical that blocks oxytocin receptors is injected into the brain of a prairie vole, the animal's typical pair-bonding is replaced by indiscriminate mating behavior. In contrast, the behavior of the montane vole, a meadow mouse that does not establish lasting pair bonds, is more consistent with the approximately 95 percent of mammal species that are not monogamous. Examination of the montane vole's brain by Tom Insel, now director of the National Institute of Mental Health, and his colleagues at Emory University revealed that oxytocin and dopamine receptors are not in the same close proximity as they are in prairie voles.

Exuberant behavior and emotions—whether displayed in love, manifest in laughter and play, or kindled by music, dance, and celebration—have in common high mood and energy. They act on the same reward centers in the brain as food, sex, and addictive drugs, and they create states of mental and physical playfulness. Indeed, brain imaging studies show that brain activation patterns of adult humans looking at or listening to their romantic partners are similar to those taken when the brain is experiencing cocaine-induced euphoria. Insel and other neuroscientists suggest that the neural pathways involved in the pleasurable effects of stimulant drugs may have evolved in part as reward systems for social attachment.

Exuberance also unites members of a group by inducing a synchronous emotional state: it rouses a community to act together and to realize its best and common interests. Passionate leadership at the community or national level, for example, binds us together in essential ways. "Passions are the only orators which always persuade," said La Rochefoucauld. "They are like an act of nature, the rules of which are infallible; and the simplest man who has some passion persuades better than the most eloquent who has none." In times of adversity, inspired leadership offers energy and hope where little or none exist, gives a belief in the future to those who have lost it, and provides a unifying spirit to a splintered populace. At no time in recent history has this been more true than for the Americans and British during World War II, when they were led by two remarkably exuberant men, Franklin Delano Roosevelt and Winston Churchill. Both had come from families of influence and privilege but each had, as well, overcome great personal adversity—Roosevelt had struggled with polio and Churchill with repeated bouts of depression. The same resilience and joie de vivre that helped see Roosevelt and Churchill through difficult illnesses were integral to their success as wartime leaders. Both men were able to draw upon an innate capacity for joy and energy, both found delight in difficult work, and both knew from experience that hardship could be overcome. Both had an infectious wit and optimism, and, observed the British philosopher and historian Sir Isaiah Berlin, they "conspicuously shared . . . [an] uncommon love of life."

To meet Roosevelt, said Churchill, "with all his buoyant sparkle, his iridescence," was like "opening a bottle of champagne." Churchill, who knew both Champagne and human nature, recognized ebullient leadership when he saw it. Roosevelt had utter faith in himself and in the course of his country. "At a time of weakness and mounting despair in the democratic world," said Isaiah Berlin, Roosevelt stood out "by his astonishing appetite for life and by his apparently complete freedom from fear of the future; as a man who welcomed the future eagerly as such, and conveyed the feeling that whatever the times might bring, all would be grist to his mill, nothing would be too formidable or crushing to be subdued." He had "unheard-of energy and gusto . . . [and was] a spontaneous, optimistic, pleasure-loving ruler" with an "unparallelled capacity for creating confidence."

Roosevelt's infectious enthusiasm was commented upon by

almost everyone with whom he came in contact. In his biography of FDR, *A First-Class Temperament*, Geoffrey Ward relates the observations made by a young naval officer when Roosevelt visited his ship just prior to becoming assistant secretary of the Navy. "I can see to this day the new assistant secretary–to–be as he strode down the gangplank to the club float with the ease and assurance of an athlete. Tall . . . smiling, Mr. Roosevelt radiated energy and friendliness." Once aboard the barge, Roosevelt

> showed immediately that he was at home on the water. Instead of sitting sedately in the stern sheets as might have been expected, he swarmed over the barge from stem to stern during the passage to the Navy Yard. With exclamations of delight and informed appreciation he went over every inch of the boat from coxswain's box to engine room. When she hit the wake of a passing craft and he was doused with spray, he just ducked and laughed and pointed out to his companions how well she rode a wave. Within a few minutes he'd won the hearts of every man of us on board, just as in the years to come he won the hearts of the crew of every ship he set foot on. . . . He demonstrated . . . the invaluable quality of contagious enthusiasm.

There was an irrepressible vitality to the British leader as well. Indeed, Churchill's energy wore out even Roosevelt: "I'm nearly dead," Roosevelt said to a member of his cabinet after a few days in the company of Churchill. "I have to talk to the P.M. all night, and he gets bright ideas in the middle of the night and comes pattering down the hall to my bedroom in his bare feet." A friend of Churchill's said of him that "he seemed to have been endowed by fortune with a double charge of life and a double dose of human nature." (Churchill himself remarked, "We are all worms. But I do believe that I am a glow-worm.") He had, according to his friend Brendan Bracken, "tearing spirits—that is, when Winston wasn't in the dumps—a kind of daring, a dislike of a drab existence, a tremendous zest in life."

C. P. Snow, like many others who observed Churchill, believed that his sheer will and optimism led the nation through its grimmest times. Somehow he managed to transfuse hope from his own vast storehouse of hopefulness and energy. His voice, said Snow, "was our hope. It was the voice of will and strength incarnate. It was saying what we wanted to hear said ('We shall never surrender') and what we tried to believe, sometimes against the protests of realism and common sense, would come true." Churchill convinced himself, and then others, through the force of his emotions as well as ideas: "Churchill had a very powerful mind, but a romantic and unquantitative one," observed Snow. "If he thought about a course of action long enough, if he conceived it alone in his own inner consciousness and desired it passionately, he convinced himself it must be possible. Then, with incomparable invention, eloquence and high spirits, he set out to convince everyone else that it was not only possible, but the only course of action open to man." As Churchill himself said of T. E. Lawrence, "The multitudes were swept forward till their pace was the same as his."

Lord Franks, who after the war became ambassador to the United States, spoke of Churchill's ability to give hope to those around him: "I remember early in the war attending a meeting on the roof of the Ministry of Supply when Winston addressed us. I came away more happy about things. He dispelled our misgivings and set at rest our fears; he spoke of his aim and of his purpose, so that we knew that somehow it would be achieved. He gave us faith. There was in him a demonic element, as in Calvin and Luther. He was a spiritual force."

Lord Moran, Churchill's physician, believed that it was passion rather than reason that accounted for much of Churchill's success in leading the nation against Hitler: "He was indeed made for the hour. In the extraordinary circumstances of 1940, with the hopeless inequality of Germany and Britain—or so it seemed—we needed a very unreasonable man at the top. If Winston had been a reasonable man he would not have taken the line he did; if he had been a man of sound judgment he might have considered it his duty to act differently. A sage would have been out of his element in 1940; instead we got another Joan of Arc."

After the war was over, Churchill spoke about inspiration: "I was very glad that Mr. Attlee described my speeches in the war as expressing the will not only of Parliament but of the whole nation. Their will was resolute and remorseless, and as it proved, unconquerable. It fell to me to express it, and if I found the right words you must remember that I have always earned my living by my pen and by my tongue. It was the nation and the race dwelling all around the globe that had the lion heart. I had the luck to be called upon to give the roar." It was the roar of life.

Exuberance is important not only in leadership, that is, in creating a potent link between leaders and groups, it is also indispensable in creating other kinds of social bonds. These bonds are complex, and ancient, and they are forged in many ways. Exuberant dance and music, for example, are universal in human culture, but not unique to it; other animals take their high spirits and energy into rhythmic gamboling as well. Jane Goodall, in a letter from Africa to her family, describes a "rain festival" among chimpanzees, a dramatic response of the troop to nature:

> Suddenly Bare Bum left his troop and galloped, full speed, at the other line, swinging his body from side to side, arms flailing in a scything motion. There was an outburst of panting calls. He moved on up a hill & as he reached the top,

stood up and swiped at a bush with his hand. Well, I thought that was that. But no. They all began to climb trees. Pale Face then turned & began to charge diagonally down the hill. As he went he snatched at the branch of a tree, tore it off, waved it above him, & charged on, dragging it with him. Then he dropped it & climbed a tree. But another large male was on the move, charging down, breaking off his branch. Then two hurtled down, one after the other, leapt into a tree, seized branches, & leapt to the ground—30 ft. at least, taking their branches with them, charging on, reaching a tree, seizing it, swinging round & leaping up into the branches. I think only the males took part in this "Rain Festival." The others staying in their "seats," watching.

I don't think I have ever watched any performance which gave me such a thrill. Mostly the actors were silent, but every so often their wild calls rang out above the thunder. Primitive hairy men, huge and black on the skyline, flinging themselves across the ground in their primaeval display of strength and power. And as each demonstrated his own majestic superiority, the women and children watched in silence, and the rain poured down while the lightning flashed brilliantly across the grey sky.

The "Festival" lasted for thirty minutes, played out in a hillside curve surrounded by trees; it was, Goodall said, a "natural out of door theatre." One can easily imagine in this chimpanzee display the trace beginnings of human dance and music and see in it the dramatic response of a single animal to nature, with cascading effects throughout its social group.

Dance and music are an ancient part of our hominid culture. Over fifty thousand years ago, Neanderthals made flutes from the bones of bears; long before that, mothers sang to their infants and, through that song, parent and child were brought closer together still. Music, the gods knew, transforms. Orpheus sang and turned the course of the rivers; trees stirred from their rootings. Sisyphus took leave of his labors to sit still on his rock and listen to Orpheus; the vulture, under his spell, desisted from ripping at Prometheus's liver. And Jupiter, in awe, took Orpheus's lyre and placed it among the stars.

Music and dance are deeply embedded in the social character of our species. They ignite, infect, and express our collective, complex emotions: music and dance arouse group energies for the hunt, or planting; celebrate the seasons of the gods; and mark, thereby abetting—through communal festivals of passage into adulthood, marriage, and death—the recognition of change within a group. To music we raise our armies and lower our dead.

Music acts quickly on the brain, and fast, simultaneous chemical changes in individual brains within a group make more likely a quick, cohesive response to circumstance. Music and dance, it has been said, paralyze the ability to think logically; it is exactly in this instinctive, rather than more lengthy and fraught cerebral response that impassioned action is taken. "To fling my arms wide / In the face of the sun," wrote Langston Hughes, "Dance! Whirl! Whirl! / Till the quick day is done." The wide-open whirl of dance is a head-thrown-back joy, one that leaps away from life-weariness into a different world. Dance energizes and unites; it quickens, it exhilarates, it liberates.

Exuberant music does likewise. A rapid tempo in major key stirs and loosens. Fear fades a bit. Whether it is a Sousa march or a tarantella, an African drumbeat, a Gilbert and Sullivan tune, a gospel hymn, or a great chorus from Handel, exuberant music lifts up, invigorates, and, for a while at least, brings together those who are listening. Music has an infectious, and on occasion, transformative effect. The trombonist Jack Teagarden described the first time he heard the great Louis Armstrong play: "The [river] boat was still far off. But in the bow I could see a Negro standing in the wind, holding a trumpet high and sending out the most brilliant notes I had ever heard. It was jazz . . . it was Louis Armstrong descending from the sky like a god. The ship hugged the bank as if it were driven there by the powerful trumpet beats."

The effects of jazz were dazzling and disturbing to its creators as well. "It was a breakdown," observed Hoagy Carmichael, "an insane dancing madness brought on by music—new, disjointed, unorganized music, full of screaming blue notes and a solid beat. We pioneers of it all broke down. . . . Jazz maniacs were being born and I was one of them."

For Louis Armstrong, exuberant music was an extension of his natural being. His biographer Laurence Bergreen is convinced that Armstrong's exuberant temperament was an essential element in his genius. He had a "distinctly American brand of optimism and striving," Bergreen writes, but "there was power and even an edge of anger to the laughter. It was a cosmic shout of defiance, a refusal to accept the status quo, and a determination to remake the world of his childhood and by extension, the world at large, as he believed it ought to be." In the end, he says, "it was Louis's animating spirit of joy, as much as his music, that was responsible for his transforming vision." Exuberance lifts mind and soul into freer space: exuberance creates music and is in turn created by it.

Wynton Marsalis, in an interview with Dick Russell for Russell's book *Black Genius and the American Experience*, states that part of Armstrong's genius was that "his sound is both the most modern and the most ancient, like somebody playing outside the walls of Jericho. . . . He brought back a real joy to music. Armstrong loosens up everybody's phrasing, their concept of where to place the beat." Always, Marsalis says, there is a complexity to the emotional experience underlying Armstrong's joy and music: "The question in jazz is how do you make conflicting things harmonious. And achieving that balance is what Pops is basically all about. He could play with tremendous power and grace. In equal measure. He stayed calm within himself and had this great sense of tranquility in his personality. But there was also this barely contained frenzy in his playing, like rice ready to boil over." Armstrong himself said, "They all know I'm there in the cause of happiness."

Music evolved as a "play-space" for the mind, according to Ian Cross of the University of Cambridge. Ambiguities and complex rhythmic patterns create a mental and emotional climate in which conceptual leaps can be taken with no risk to survival. The brain under the influence of music can, as it does in wit and during times of play, experiment and bound about within the confines of a safe environment. It can galumph. Music, like play and laughter, increases emotional arousal and disinhibits; it rewards the participant for passionate engagement in the here and now. Music gives back pleasure in exchange for emotional discovery and involvement. Louis Armstrong, writes Bergreen, "was game for anything. He'd play the slide whistle or any damn fool instrument just to see what kind of a sound he could get from it, and what effect it would have on the music and the crowd. For Louis, that spirit of endless playfulness was the essence of performing."

Music is sometimes used to induce positive emotions in research subjects who take part in studies of problem-solving and imaginative thought; as we have seen, psychologists find that it generates a mental state in which flexibility, memory, and originality increase. Music also increases the likelihood that an individual will remember things important to living in a group. Chanting, drumbeats, and other rhythmic structures imposed upon speech are critical to passing down oral history and group traditions from one generation to the next. They also facilitate tribal remembering of significant information about seasons, plants, and animals; keep alive the rituals of religion; help to recall important details of battles; and even, perhaps, lock in place the memory of the location of natural resources. (We are not alone in the cultural and pragmatic use of rhythm. The acoustic biologist Roger Payne has shown, for instance, that the songs of humpback whales contain refrains that form rhymes. It is possible that these singing whales, like humans, use rhyme as a mnemonic device for complex material.)

Surely man's awed response to nature provoked the first—and then much of the finest—music of joy. Our exultant hymns and great Masses brim with exuberance and praise for the universe and its Creator. The ancient psalms, meant to be sung, exhorted man to make music as a tribute to God upon "an instrument of ten strings, and upon the lute: upon a loud instrument, and upon the harp." The Lord "hath done wonderful things," therefore "shew yourselves joyful unto the Lord, all ye lands: sing, rejoice, and give thanks. / Praise the Lord upon a harp." Not just man but the entire earth should show its joy: "Let the sea make a noise. . . . / Let the floods clap their hands, and let the hills be joyful together before the Lord." The final psalm continues with the invocation of praise through music and dance: "Praise him in the sound of the trumpet," it proclaims, "praise him upon the lute and harp. / Praise him in the cymbals and dances: praise him upon the strings and pipe." All returns to the glory of God: "When I consider thy heavens," sings the psalmist—"The work of thy fingers, the moon and the starres"—man's stance must be one of awe, of celebration. It is easy to imagine, as the Bible records, that David not only sang his tributes to the Lord but danced with abandon before the Ark.

The great joy-filled Christian hymns continue the songs of praise from the Old Testament. It would be difficult to find more exuberant anthems than those filling the Christian hymnal, although despair and anguish are also present in full measure. It is not just the triumphant hymns of Easter or the joyous carols of Christmas that ring with joy. Exuberance, it would seem, is the inherent response of those who are moved deeply by nature and who delight in assigning its glories to a Creator. It is, as well, the response of those who have no such belief but nonetheless exult in the beauty of the world they see around them.

Music activates the same reward systems in the brain that are activated by play, laughter, sex, and drugs of abuse. Brain imaging studies show that pleasurable music creates patterns of change in the dopamine and opioid systems similar to those seen during drug-induced euphoric states. If experimental subjects are asked to listen to music and some are given Naloxone, a drug used to treat addicts by blocking opiate receptors in the brain, and others are given a placebo, those who receive Naloxone report a significant drop in the pleasure they experience while listening to music. Those who are given a placebo do not.

Music not only activates the reward system, it decreases activity in brain structures associated with negative emotions. Music is an expansive pleasure, one that both reflects and generates joy. As the English psychiatrist Anthony Storr has written, "Music exalts life, enhances life, and gives it meaning. . . . Music is a source of reconciliation, exhilaration, and hope which never fails." It is, he argues, "an irreplaceable, undeserved, transcendental blessing."

Exultant states are often a part of religious as well as musical experiences. William James, of course, wrote about this brilliantly. "Man's extremity," he believed, "is God's opportunity," and James brought a sympathetic temperament to the study of those for whom religion exists "not as a dull habit, but an acute fever." There are individuals, he observed, for whom religion in its "highest flights" is an "infinitely passionate" thing. "It adds to life an enchantment . . . [that] is either there or not there for us, and there are persons who can no more become possessed by it than they can fall in love with a given woman by mere word of command." The capacity to experience ecstasy cannot be willed; it is a gift, an ability like a fine wit or a way with shapes and spaces.

The ecstasy associated with religious experiences is transient, more often measured in minutes than in hours, but it shares with exuberance the sense of well-being, expansiveness, joy, upliftedness, and a conviction of significance. Such moments bring with them intense mental and bodily sensations and a feeling that one has entered a new world of meaning. Many things can trigger ecstasy. Marghanita Laski, in her study of secular and religious ecstasies, found that nature, art, sexual love, and religion were by far the most frequent triggers of ecstatic experiences, and she suggested that such ecstasies, in turn, give value to that which triggers them. They serve as points of departure for spiritual journeys, creative quests, or explorations of the mind.

C. S. Lewis, in *Surprised by Joy*, describes his ecstasy in seeing for the first time an illustration in *Siegfried and the Twilight of the Gods*. The long winter, he said, "broke up in a single moment. . . . Spring is the inevitable image, but this was not gradual like Nature's springs. It was as if the Arctic itself, all the deep layers of secular ice, should change not in a week nor in an hour, but instantly, into a landscape of grass and primroses and orchards in bloom, deafened with bird songs and astir with running water." To have this experience once, Lewis says, sets one on a pursuit to recapture what has been and disappeared: "I knew (with fatal knowledge) that to 'have it again' was the supreme and only important object of desire." He felt as a result of knowing "Joy" that he knew nature differently; his knowledge was direct, not apprehended from a distance or learned of from a book. Ecstatic joy was for him an "imaginative Renaissance" that lured him toward his subsequent spiritual journeys. As the Australian banksia plant needs fire to release its seeds, so Lewis needed joy.

The religious impulse, which for some includes the capacity for ecstasy, can be at its best a cohesive force in society. In *Darwin's Cathedral*, the evolutionary biologist David Sloan Wilson argues that religion confers an adaptive advantage on those groups who have it. He believes that the religious impulse is innate and that it makes cooperation, and therefore a common defense and survival, more likely. Faith, he states, "allows you to keep going in the absence of information." Depending upon the reality of the circumstances, this is a good or a not-so-good thing. As the English psychiatrist Henry Maudsley wrote in 1886, the same words in both Hebrew and Greek "denote the ravings of insanity and the often equally unintelligible ravings of the diviner or revealer of divine things."

The exuberant outpourings of Emanuel Swedenborg, the eighteenth-century mystic and scientist, illustrated for Maudsley how religious enthusiasm is differently construed by believers and skeptics: "The visitation [of Swedenborg's hallucinations] was the forerunner of an attack of acute mania, on recovery from which he was what he remained for the rest of his life—either, as his disciples think, a holy seer endowed with the faculty of conversing with spirits and angels in heaven and hell . . . or, as those who are not disciples think, an interesting and harmless monomaniac, who, among many foolish sayings, said many wise and good things, attesting the wreck of a mind of large original endowment, intellectual and moral." (Perhaps weighting one side of the argument would be Swedenborg's fervent belief that he could converse with the inhabitants of all the planets, except for those of Neptune and Uranus, which had not yet been discovered.) Indeed, mania and excited religious states have much in common—euphoria, a sense of intense well-being, and a heightened sensory awareness, among other things—and religion is the most common theme of both manic delusions and hallucinations.

For some, exuberance comes by way of madness or revelation. For others, it is sown into their dispositions as melody is in a songbird. Most, however, experience great enthusiasms only fitfully: when they fall in love, at times of personal triumph or national festivity, at a racetrack, in a bedroom, at war's end, on a playing field, or with a newborn. These occasions are frequent and sustaining enough for most. But not for all. The history of our species shows that we have used every imaginable means to generate even more exuberance.

We are not the only species to seek high moods. Sloths intoxicate themselves by eating fermented flowers and chewing coca leaves, and elephants get high on fermented fruit and vines. Reindeer ingest hallucinogenic mushrooms, water buffaloes graze on opium poppies, and llamas and monkeys, like sloths, ingest the stimulant from coca leaves. Gorillas, wild boars, porcupines, and spider monkeys eat intoxicating or hallucinogenic insects, fungi, berries, and grains. Pleasure-seeking and a desire for novelty must be a part of the reason for this behavior, but the UCLA pharmacologist Ron Siegel suggests that self-medication is also involved. (Elephants, he believes, use alcohol not only as a source of calories and energy but also to relieve the stress produced by having poachers and tourists in their territory. It is not obvious how one could easily test this hypothesis, but it is an intriguing one.)

We, too, have a diverse hankering for intoxicants and hallucinogens. We, too, have been fond of the coca leaf and fermented grains, eaten seeds of the white-flowered morning glory, and enjoyed the magic of wild mushrooms. Many have smoked hashish or tobacco, and others have made ritual drinks of fermented honey and tree bark. A curious few have ingested hallucinogenic caterpillars. Ritual enemas, not to everyone's taste, brought delight to the Aztecs and Mayans, who discovered that such nether-route intoxication was more rapid than drinking or smoking and the side effects were fewer. The Incas used enemas to experience the psychological effects of hallucinogenic seeds, and sixteenth-century Lowland Indians used them to take in tobacco.

There are many nonchemical routes to the high mental states, as well. The ethnobotanist Peter Furst chronicles a rather remarkable variety, including fasting, thirsting, self-mutilation, sleep deprivation (which can also trigger mania in susceptible individuals), exposure to the elements, exhaustive dance, bleeding, immersion in ice water, flagellation with thorns or animal teeth, hypnosis, meditation, rhythmic drumming and chanting, pungent or aromatic scents, and Indian sweat lodges.

The ingenuity of these nonpharmacologic methods notwithstanding, drugs dominate the history of our search for exuberant and ecstatic states. Even the oracle at Delphi who, it is said, spoke for the gods, appears to have owed her prophecies and trances more to earthly intoxicants than to divine inspiration. Indeed, reports the *New York Times* journalist William Broad, the ancient Greeks were the first to suspect that sweet-smelling gases rising up from the floor of the temple might set off the oracle's frenzies. Before prophesying, the oracle breathed in "sacred fumes." Scientists have recently discovered that the Delphi temple sits directly on top of a fault line through which ethylene, a euphoriant gas, escapes. The future had been seen through a vapor.

Drugs and gases can heighten energy and alertness or dull them; they can intoxicate, induce vivid living dreams, stir warlike rages, kill pain, or unite the disparate in a common cause. Drugs bring on exuberant and ecstatic states, as well. Alcohol, of course, which releases dopamine and serotonin in addition to the brain's own naturally occurring opioids, has been used for thousands of years to exhilirate, to disinhibit, and then to numb. But many other substances have also been used.

In the late eighteenth century a truly remarkable gas was discovered by Joseph Priestley. It exuberated. Nitrous oxide, or "laughing gas," gained wide popularity through the experiments of the great English chemist Sir Humphry Davy. Many of these experiments were conducted upon himself. The "pleasure-producing air," he wrote, "absolutely intoxicated me . . . made me dance about the laboratory as a madman, and has kept my spirits in a glow ever since." His friend Samuel Taylor Coleridge, no stranger to the sweet effect of drugs, spoke of its "great extacy," its "voluptuous sensation." Coleridge's fellow poet Robert Southey was even less restrained: "Such a gas has Davy discovered!" he said. "It made me laugh and tingle in every toe and finger tip. . . . It makes one strong, and so happy! So gloriously happy! . . . Oh, excellent air-bag! . . . I am sure the air in heaven must be this wonder-working gas of delight!" Yet another admirer of nitrous oxide said that the sensations experienced under its influence were like the great choruses of *The Messiah* played on the "united power of 700 instruments." Nitrous oxide was taken at dinner parties, and there were "laughing gas evenings" at London theaters. Predictably, P. T. Barnum put together exhibitions for the public and, just as predictably, the crowds flocked in. In the best tradition of science and pleasure-seeking, medical students at Yale administered it to their classmates.

Seeking new sensation was not the only way the "pleasure-producing air" was put to use. William James said that his own experience taking nitrous oxide brought about an intense metaphysical illumination. "Depth beyond depth of truth," he wrote, "seems revealed to the inhaler. . . . No account of the universe in its totality can be final which leaves these other forms of consciousness quite disregarded." Things seemed more "utterly what they are, more 'utterly utter' than when we are sober." One's soul, he said, will "sweat with conviction," and regions of the universe will open. No map would be provided.

Like the results of most drug-induced states, the philosophical insights gained under the influence of nitrous oxide and similar drugs were often more intense than lastingly profound. Sir James Crichton-Browne, in his 1895 Cavendish Lecture, remarked that the thoughts one had while inhaling nitrous oxide were in "nine cases out of ten connected with some great discovery, some supposed solution of a cosmic secret." But such revelations, he wrote in the *Lancet*, usually prove illusory:

> A medical man upon whom my former colleague, Dr. Mitchell, experimented with nitrous oxide gas imagined before becoming unconscious that he had made a most important discovery explaining the whole action of the gas; and Dr. Mitchell himself had repeatedly the same experience, his mind being seized by expansive ideas which, while they lasted, made all dark things clear. . . . We might as well look for phosphorescence on the sea in the blaze of midday sunshine as hope to reproduce such dreamy mental states in the full light of objective consciousness. Nothing but a vague remembrance that they have flashed across the mind remains when waking life is resumed, and endeavors to recall them or grasp them in passing, when not fully futile, are apt to prove ludicrous in their results. I dare say many of us recollect the story of the professor who, having experienced a magnificent thought in the early stages of chloroform inhalation, resolved that he would by one bold sally lay hold of it and so read the riddle of the world. Having composed himself in his easy chair in his study, with writing materials at hand, he inhaled the chloroform, felt the great thought evolve in his mind, roused himself for an instant, seized the pen, wrote desperately he knew not what, for even as he did so he fell back unconscious. On coming to

himself he turned eagerly to the paper, to find inscribed on it in sprawling but legible characters the secret of the universe in these words, "A strong smell of turpentine pervades the whole."

Cocaine—known also as California cornflakes, happy trails, sleigh ride, and nose candy—is another euphoriant. Eaten, sniffed, injected, or inhaled, cocaine quickly causes euphoria by stimulating a part of the brain (the dopamine-containing projection from the ventral tegmental area to the nucleus accumbens) that regulates pleasure. The brain floods with dopamine. The enjoyable effects of cocaine—greater energy and heightened sensitivity of sight, touch, and sound; high mood; increased talkativeness—are not unlike those of the early, mild stages of mania, but they are less intense and of far shorter duration (lasting minutes or hours, not days or weeks). Like most mind-altering drugs, however, cocaine supplies a pleasure freighted with costs. The same neurons that are activated by dopamine and give delight eventually become desensitized to it. Depression and apathy follow. Prolonged cocaine use, which diminishes dopamine functioning, gives support to the general rule that external sources of exuberance are ultimately overruled by the brain's inclination to seek out equilibrium.

Hashish likewise takes back the joy it first so enticingly gives, but the road to ruin is often a glorious one. "Hashish spreads out over the whole of life a sort of veneer of magic," wrote Charles Baudelaire. "Scalloped landscapes; fleeting horizons . . . the universality of all existence, arrays itself before you in a new and hitherto unguessed-at glory." Paradise and heaven, the life of a god itself, are at the user's beckoning: "A wild and ardent shout breaks from his bosom with such force [that it] would bowl over the angels scattered on the paths of Heaven: *I am a God!*"

Théophile Gautier felt no less rapturous. The physician who

gave him hashish told him, "This will be deducted from your share in Paradise," and in Paradise—for a quick while—he was, filled with the "maddest gaiety." "What bliss! I'm swimming in ecstasy! I'm in Paradise! I'm plunging into the depths of delight!" The human frame, he said prophetically, "could no longer have borne such intensities of happiness." Indeed, neither his frame nor Baudelaire's proved an exception to pharmacological law: what goes up must come down. It is a law ignored anew by each generation of pleasure seekers; some pay more for ignoring it, others less. Cocaine, hashish, opium, Ecstasy: all seduce with the promise of rapture or exuberance—and then they collect.

In group celebrations, we find exuberance differently. A love of festivities is universal, observed William James, and in many respects celebration is yet another form of human play. The same acts are experienced more intensely when performed in a crowd than when done alone. In a large and festive group, our actions build in response to those around us; they reverberate and gather energy. Celebrations are not as circumscribed as more ritualized and formal group gatherings; improvisation, playfulness, and exuberance hold sway. We celebrate the end of precarious times—war and winter, for example—and times of great accomplishment: Lindbergh's landing in Paris, a footstep on the moon, or a political candidate's success.

The intensity of exuberance varies of course, depending upon the size of the group and whether the celebration is of a private or a more public nature. Senator George McGovern, the Democratic nominee for president in 1972, contrasts the kind of exuberance he and his crew felt after surviving combat in World War II with the raucous celebration that he and everyone else in the convention hall shared the night he was nominated. As a bomber pilot, he says, he felt exuberant anytime he and his crew survived heavy antiaircraft fire and returned safely to their base. (This was especially true after he had had to land his four-engine B-24 bomber with only two engines operating and on a runway only half as long as needed for a safe landing, a feat for which he received the Distinguished Flying Cross.) A gentler exuberance accompanied his return home after his combat duty was over: "At the end of my 35 missions, flying high over the Atlantic with my crew asleep and [myself] at the controls, I looked at a full moon, lovely white clouds and the ocean below and the war over for me—a wonderful sense of peace, and satisfaction and exuberance came over me. We had done our job well, [although] our navigator had been killed in combat and one of our gunners had been wounded and stayed behind in an Italian hospital. The rest of us were well and healthy and were reflecting on soon seeing our families. It was a quiet exuberance that I'll never forget."

The exuberance McGovern experienced on his nomination in 1972 was very different: "It is difficult for me to imagine a feeling that could transcend the feelings of exuberance that swept my heart and mind and indeed my entire being upon being nominated as the Democratic candidate for President of the United States. I watched the nomination from my hotel in Miami, in a room with a few staffers. Eleanor [McGovern's wife] was on the convention floor with other members of my family. Television captured the exuberance of the crowd for me and greatly added to my own. I believe the greatest feeling for me came the next night when early in the morning (2 a.m.) I went to the Convention for the first time to give my acceptance address. The tremendous applause, the shouts of sheer joy, the demonstrations, including dancing in the aisles, the hundreds of joyful embraces of the delegates—this was the highlight of exuberance for me."

We celebrate national occasions of moment. John Adams, in a letter to his wife, Abigail, written on the third of July 1776, wrote: "I am apt to believe that [Independence Day] will be celebrated by succeeding Generations, as the great anniversary festival. . . . It ought to be solemnized with Pomp and Parade, with Shews, Games, Sports, Guns, Bells, Bonfires, and Illuminations, from one End of this Continent to the other from this Time forward forever more." Adams's use of "solemnized" in the context of bells and bonfires is a telling one. The attachment of joy to the marking of profound events is signal: it ties pleasure to recollection and makes more likely the perpetuation of the occasion thought significant enough to be remembered. Towns are bound together by joy, the country united in common cause and recollection.

The American observance of independence from Britain continues more than 225 years after the fact, with defining elements taken from the earliest celebrations. Eighteenth-century Boston celebrated the nation's independence with cannon fire and gun salutes from the ships in Boston Harbor. A sermon to the state assembly was followed by thirteen toasts, one for each of the new states, proposed by Governor John Hancock; the militia paraded and, at night, fireworks exploded over the city. A century later, the abolitionist and suffragist Julia Ward Howe, author of "The Battle Hymn of the Republic," described the Fourth of July celebrations held in her native New York: "The endless crackling of torpedoes, the explosion of firecrackers and the booming of cannon," she wrote, "made the day one of joyous confusion. . . . It then seemed to be a day wholly devoted to boyish pleasure and mischief." There was, she reported, "a perpetual popping and fizzing [and] shouts of merriment," and, later that night "Roman candles, blue lights, and rockets." On the Fourth of July in 2000, yet another century later, an estimated 5 million people crowded to watch 120 tall ships and forty warships parade past New York Harbor. There was a clear continuity with the first celebrations of the nation's birth: great sailing ships and pealings of church bells, community parades during the day, and fireworks at night.

Fireworks have ignited festive moods for more than two thousand years. A primitive alliance between man and fire, between darkness and light, fireworks are the perfect display of human rejoicing: we send up rockets of light into the sky and they burst into blazing bits of dazzling beauty. Our moods ride with them. Fireworks create magic and bring together those watching them into an ebullient, alert, and awe-filled state; they inspire a shared sense of wonder, of beauty, of excitement. They splash the night world with sound and color. As Barnum knew, the world needs to celebrate; it needs someone to "throw up sky-rockets." It is in our nature to rejoice in and with those who do.

CHAPTER SEVEN

"Forces of Nature"

The pursuit of knowledge is an intoxicant, a lure that scientists and explorers have known from ancient times; indeed, exhilaration in the pursuit of knowledge is part of what has kept our species so adaptive. Early humans survived and then flourished because some took keen notice of the habits of prey and predator while yet others explored to advantage their terrain and the land beyond. A few watched the night skies, traced the movement of the stars, named the constellations, and reckoned the progression of the moon, the sun, and the seasons.

What drew these observers and explorers to their watch and mullings? What pulled them to imagine and wonder about new worlds or new ways of understanding: to count, describe, make sense of, predict? Why did Hipparchus look upward and name the stars while tens of thousands of others slept? What compelled Archimedes to calculate the mathematical properties of spirals and spheres, or Gauss to approach infinity and presume to grapple with it? They had imaginative and audacious minds, certainly. But they also had passion and energy; they took joy in discovering something new. Nature rewards the enthusiastic and curious with excitement in the chase and the thrill of discovery, rewards the intellectually playful with the exuberant pleasures of play. Exuberance in science drives exploration and sustains the quest; it brings its own Champagne to the discovery.

Discovery is undeniably intoxicating and, in its own way, addictive as well, for exuberance experienced in the wake of discovery creates a fresh appetite to discover anew. The history of science is a history of delight in first-seens, first-postulateds, first-came-upons; it is a history of high pleasure in the hunt and of exultation in the netting. Being first is inebriating; it stokes the fire for the next seeking. The English paleontologist Richard Fortey, in his marvelous book *Life: A Natural History of the First Four Billion Years of Life on Earth,* describes the delight:

> The excitement of discovery cannot be bought, or faked, or learned from books (although learning always helps). It is an emotion which must have developed from mankind's earliest days as a conscious animal, similar to the feeling when prey had successfully been stalked, or a secret honeycomb located high in a tree. It is one of the most uncomplicated and simple joys, although it soon becomes mired in all that other human business of possessiveness and greed. But the discovery of some beautiful new species laid out on its stony

> bed provokes a whoop of enthusiasm that can banish frozen fingers from consideration, and make a long day too short. It is not just the feeling that accompanies curiosity satisfied—it is too sharp for that; it arises not from that rational part of the mind that likes to solve crosswords, but from the deep unconscious. It hardly fades with the years. It must lie hidden and unacknowledged beneath the dispassionate prose of a thousand scientific papers, which are, by convention, filleted of emotion. It is the reason why scientists and archaeologists persist in searches which may even be doomed and unacknowledged by their fellows.

Most scientists and explorers relate comparable feelings of excitement after finding something no one else has seen before or understood. Susan Hendrickson, the discoverer of "Sue," the great *Tyrannosaurus rex* named for her, told a *New York Times* reporter, "When you're the first person to see this creature, this magnificent, splendid, awesome creature that no living being has seen for 67 million years, it's a thrill that defies description. It's chemical, physical, emotional—it's a body experience." Elsewhere she elaborated: "It's the thrill of discovery. . . . It's like the high from some drug. It lasts a few minutes. And it's addictive. Those moments are few and far between, but that's what keeps you going." Hendrickson, who has been compared more than once to Indiana Jones, chases her enthusiasms all over the world. She has helped to raise Spanish galleons, found rare butterflies in amber, and participated in underwater expeditions to discover lost cities. To each adventure she brings her exuberance and a mother wit for discovery.

Richard E. Byrd, in *Skyward,* describes the excitement of discovery during his pioneering flight to the North Pole: "We were now getting into areas never before viewed by mortal eye. The feelings of an explorer superseded the aviator's. I became conscious of that extraordinary exhilaration which comes from looking into virgin territory. At that moment I felt repaid for all our toil. . . . We were opening unexplored regions at the rate of nearly 10,000 square miles an hour, and were experiencing the incomparable satisfaction of searching for new land."

The Nobel laureate Max Perutz compares discovery to falling in love or reaching the top of a mountain after a strenuous climb. It is, he says, "an ecstasy induced not by drugs but by the revelation of a face of nature that no one has seen before." The astronaut Buzz Aldrin said that when he and Neil Armstrong were waiting to leave the moon, he became aware that "there was no runway up there. And there certainly wasn't anyone else waiting in line to take off. I was conscious of that, of being first."

The ecstasy of discovery is by no means limited to scientists, of course, although they are the focus of concentration here. Clearly, most individuals—athletes, lovers, artists, businessmen, journalists, parents—will experience moments of delight in discovery; they are threaded into the experimentation of play, individual triumphs in life, and the intensities of love. For artists and writers there is, in addition to the moments of imaginative discovery, a profound pleasure that comes from beauty. When Keats first read Chapman's translation of Homer, he compared his emotions to one who, observing the heavens, sees for the first time what no one has ever seen before: "Then felt I like some watcher of the skies / When a new planet swims into his ken," he wrote. It is not surprising that Keats, who had studied to be a surgeon, turned to the natural sciences for an image of comparable intensity.

Watchers of planets and stars have always been among the most awestruck and greatly curious. The Scottish scientist Mary Somerville, the first woman to publish experimental results in the *Philosophical Transactions of the Royal Society*, wrote in 1831 that "the heavens afford the most sublime subject of study which can be derived from science: the magnitude and splendour of the objects, the inconceivable rapidity with which they move, and the enormous distances between them, impress the mind with some notion of the energy that maintains them in the motions with a durability to which we can see no limits." But, she went on to note, the magnificence of the natural world is met by the discovering powers of man. "Equally conspicuous," she continued, "is the goodness of the great First Cause in having endowed man with faculties by which he can not only appreciate the magnificence of his works, but trace, with precision, the operation of his laws."

For most scientists the excitement of discovery, whether in sky or field, goes back to the early days of childhood and emerges time and again to lure their minds onward. These moments of intellectual joy are what the Harvard psychologist Howard Gardner and others refer to as crystallizing experiences, moments when a young person first falls in love with a particular idea or way of thinking. Intense and early passions often exert their influence over a lifetime, calling to mind C. S. Lewis's declaration that anyone who has experienced joy will try to re-create the circumstances of it. Just as those who have known the euphoria of love, cocaine, or mania, or felt the rush of victory in sports, will attempt to recapture the high, so, too, will those who have felt the delight of discovery.

This was certainly true for Cecilia Payne-Gaposchkin (née Payne), one of the great astronomers of the twentieth century. Payne-Gaposchkin (1900–1979) is credited with essential contributions to astronomy, including the discovery that most objects in the universe are composed primarily of hydrogen. Her dissertation work on the relative abundance of elements in stars, published later as the book *Stellar Atmospheres,* is a classic in astronomy, "an essential step in the scientific demonstration of a philosophical concept: that natural bodies, the stars, the Sun and the Earth are made up of the same stuff." Her second book, *The Stars of High Luminosity,* was also a pioneering study, exemplifying, as one colleague put it, "the bravery and adventure of a mind exploring the unknown."

Payne-Gaposchkin's fervor for science showed itself early. When she was five years old, she saw a meteor and decided on the spot to become an astronomer; she also resolved to begin quickly, "in case there should be no research left" when she grew up. Brought up in England, with a heritage "dominated by women," she was initiated into science as a young girl and sent flying with enthusiasm into a life of science: "The Bee Orchis was growing in the long grass of the orchard, an insect turned into a blossom nestled in a purple star. Instantly I knew it for what it was. My Mother had told me of the Riviera—trapdoor spiders and mimosa and orchids, and I was dazzled by a flash of recognition. For the first time I knew the leaping of the heart, the sudden enlightenment, that were to become my passion. I think my life as a scientist began at that moment. I must have been about eight years old. More than 70 years have passed since then, and the long garnering and sifting has been spurred by the hope of such another revelation. I have not hoped in vain. These moments are rare, and they come without warning, on 'days to be marked with a white stone.' They are the ineffable reward of him who scans the face of Nature."

Other interests also drew Payne-Gaposchkin—she retained lifelong passions for literature, music, theater, and travel—but even as a young schoolgirl she knew that her first love was science: "When I won a coveted prize at the end of the year," she writes in her autobiography, "I was asked what book I would choose to receive. It was considered proper to select Milton, or Shakespeare, or some writer of similar prestige. I said I wanted a textbook on fungi."

In her boarding school, a place whose primary task, she said, was to prepare young women for English society, she taught herself mathematics and science and spurned religious services in order to pursue her individual studies of chemistry and nature. She created a spiritual and intellectual world for herself: "I had, in a sense, converted the laboratory into a chapel. On the top floor of the school (a town house, high and narrow) was a room set aside for the little science teaching conceded to the upper classes. The chemicals were ranged in bottles round the walls. I used to steal up there by myself (indeed I still do it in dreams) and sit conducting a little worship service of my own, adoring the chemical elements. Here were the warp and woof of the world, a world that was to later expand into a Universe. . . . I had yet to realize that the heavenly bodies were within my reach. But the chemical elements were the stuff of the world. Nature was as great and impressive to me as it had seemed when I [had as a child] vowed myself to its service. It overshadowed everything."

Throughout her professional life, Payne-Gaposchkin's exuberance for science was critical to overcoming the obstacles put in her way because she was a woman in an academic world dominated by men. As an undergraduate at the University of Cambridge, she was forced to sit by herself in the front row of the lecture halls (only a few years earlier, women had been required to be chaperoned during lectures), and women were not allowed to receive university degrees. Enthusiasm kept her going. She entered Cambridge in 1919, a time when the study of physics was "pure delight." The Cavendish Laboratory, she said, "was peopled with legendary figures. The great J. J. Thomson [discoverer of the electron] . . . Aston with his mass spectograph, C.T.R. Wilson with his cloud chamber . . . and, looming over all, was the figure of Ernest Rutherford." After hearing a lecture in Trinity College by Professor Arthur Stanley Eddington about the eclipse expedition he had led to Brazil in 1918, she was so excited that "when I returned to my room I found that I could write down the lecture word for word. . . . For three nights, I think, I did not sleep. My world had been so shaken that I experienced something very like a nervous breakdown." (Many people who knew Payne-Gaposchkin over the years were to remark that she often worked to the point of dangerous exhaustion; her mother once observed that her daughter "lived largely on her enthusiasms.") But, she said, Eddington "had opened the doors of the heavens to me."

Later, after hearing a lecture in London by Harlow Shapley, the newly appointed director of the Harvard College Observatory, she intrepidly asked him—the man, she said, "who walked with the stars"—if she could study with him. He agreed, and in 1923 she went to America, where, "in the heady atmosphere of New England," she recalled later, "nothing seemed impossible." She wrote her dissertation (which has been described as "the most brilliant Ph.D. thesis ever written in astronomy") in six weeks, "in a kind of ecstasy," and received the first doctorate in astronomy to be awarded from the Harvard College Observatory.

Because she was a woman, Payne-Gaposchkin remained ineligible for academic appointments at virtually all universities. The discrimination at Harvard was particularly blatant. President Abbott Lawrence Lowell did not believe that women should have academic appointments and therefore, rather than being appointed as an instructor or a professor, she was paid as a technical assistant to Shapley. "Being a woman has been a great disadvantage," she wrote in her autobiography, *The dyer's hand*. "It is a tale of low salary, lack of status, slow advancement. . . . I simply went on plodding, rewarded by the beauty of the scenery, towards an unexpected goal." ("Plodding," it should be noted, would be the last word anyone would use to describe her or her work.)

Her passion for science kept her going, despite the obstacles and overt discrimination. "Astronomers are incorrigible optimists," she wrote in the introduction to her book *Stars in the Making*. "They peer up through a turbulent ocean of atmosphere at the stars and galaxies, forever inaccessible. They speak of million-degree temperatures, of densities smaller than our lowest vacuum; they study light that left its source two hundred million years ago. From a fleeting glimpse they reconstruct a whole history. . . . The drama of cosmic evolution is played out upon a stage that stretches beyond the limits of our vision, at a pace so slow that the span of human history has witnessed no action."

Change in the academic world, while glacial, was perhaps not impossibly slow from an astronomer's perspective. And things did eventually change. In 1952, more than thirty years after beginning her undergraduate studies at the University of Cambridge, she was finally awarded her degree. Four years later, she became the first woman at Harvard to be promoted to full professor and the first to be appointed chair of a department. A minor planet was named for her, and in 1977 she became the first woman to give the Russell Prize Lecture to the American Astronomical Society. A lifetime of joy in scientific discovery came through in her remarks. The reward of a scientist, she said, "is the emotional thrill of being the first person in the history of the world to see something or to understand something. Nothing can compare with that experience. It engenders what Thomas Huxley called the Divine Dipsomania."

Discovery, however divine or intoxicating, is just one aspect of scientific exuberance, however. Science is also driven by curiosity and an enthusiastic restlessness, hastened forward by a drive to explore, a desire to put together the pieces of some pattern of nature. The diversity of scientific inquiry is spectacular, and it is often the most exuberant scientists, the ones who possess the greatest capacity to be easily excited, who pursue their enthusiasms and curiosities over a far-flung range of topics.

"Martin had one characteristic without which there can be no science," wrote Sinclair Lewis in *Arrowsmith*. He had a "wide-ranging, sniffing, snuffling, undignified, unself-dramatizing curiosity, and it drove him on." Martin Arrowsmith was far from alone in his driving curiosity. In scientists before and after, curiosity has compelled not only the intensity of the intellectual pursuit but often its multifariousness as well. Aristotle, for example, wrote not only about metaphysics, logic, politics, and ethics but also about botany, astronomy, psychology, and zoology. Newton studied alchemy as fervently as he did the optics of light and gravitation. Alexander von Humboldt's interests included volcanoes, silver mining, turtle eggs, the movements of tides, botany, bats, missionaries, and Indians. Likewise, "Nature's flaming apostle," the Harvard zoologist Louis Agassiz, took on nearly everything: glaciers, embryology, the anatomy of crickets, fossil fish, the mathematical characteristics of leaves, and the origins of life itself.

Edward Jenner studied not only cowpox but the night-flowering primrose, hibernation in hedgehogs, diseases of the eye and heart, and the habits of cuckoos and earthworms. (His discovery of a vaccination against smallpox brought forth a different order of exuberance, however: "The joy I felt at the prospect before me of being the instrument destined to take away from the world one of its greatest calamities," he wrote, "was so excessive that I sometimes found myself in a kind of reverie.") Charles Darwin, by the time he was eight years old, had a "passion for collecting" and a love of natural history. "I had strong and diversified tastes," he said, "much zeal for whatever interested me, and a keen pleasure in understanding any complex subject or thing." During the voyage of the *Beagle* he worked "to the utmost" from, as he put it, "the mere pleasure of investigation." His love of natural science, he wrote in his autobiography, "has been steady and ardent." Charles Lindbergh, although primarily known as an aviator, had a wide-ranging and questioning scientific mind. Among the many fields he studied were biology, cytology, rocket science, surgery, and physics; he also actively pursued research in organ perfusion, artificial hearts, archaeology, conservation, and animal hibernation.

One of the pleasures of working around scientists is the joy that so many of them take in seemingly small yet important bits of the universe. The articles published in *Science* and *Nature* every week often ask the kinds of questions a child might ask: Are toads right-pawed? Where did the first star in the universe form? How do parrots signal one another? Why do fish not grow impossibly large? Recently, during a flight to California to give a talk in a lecture series at the University of California at Davis, I read through the list of titles of previous years' lectures and found, to my delight, topics such as how bacteria think, the molecular analysis of flower development in snapdragons, ant navigation, why birds sing, and the ecophysiology of crayfish breathing. It is remarkable and quite wonderful that our species asks such questions. It is also inevitable.

We are part and parcel of the universe we inhabit: "hitched" to everything else, as John Muir would put it. "We are not looking into the universe from outside," writes the Harvard biologist George Wald. "We are looking at it from inside. Its history is our history; its stuff, our stuff . . . such a history, that begins with elementary particles, leads perhaps inevitably toward a strange and moving end: a creature that knows, a science-making animal, that turns back upon the process that generated him and attempts to understand it."

Great scientists and explorers—the creatures who look at the stuff of the universe and wonder at their species' ways and origins—tend to be enthusiastic, optimistic and energetic by temperament, although there are, of course, many exceptions. Exuberance serves science well: it helps to overcome the tedium and setbacks intrinsic to scientific work, overrides mental and physical weariness, and makes risk-taking both attractive and probable. Positive mood, because it facilitates creativity and problem-solving, is likewise critical. Exuberance makes science fun, a type of adventure; and adventure, in turn, becomes a part of what scientists and explorers seek. Risk and the possibility of failure fade in comparison.

Charles Lindbergh put the danger of risk in the context of the excitement of flight: "I had been attracted to aviation by its adventure, not its safety," he said, "by the love of wind and height and wings." His fellow aviator Antoine de Saint-Exupéry said much the same thing: "I know nothing, nothing in the world, equal to the wonder of nightfall in the air. Those who have been enthralled by the witchery of flying will know what I mean. . . . 'It's worth it, it's worth the final smash-up.' " Indeed, the dangers of flight often become a part of its pleasures. The pioneer aviator and inventor Alberto Santos-Dumont, young and new to aviation and lost during a violent thunderstorm, said after he was safe again, "There was a fierce kind of joy . . . amid the lightning flashes and the thunderclaps, I was a part of the storm. . . . And when the dawn comes, red and gold and purple in its glory, one is almost loath to seek the earth again, although the novelty of landing in who knows what part of Europe affords still another pleasure." There is, he continued, "the true explorer's zest of coming on unknown peoples like a god from a machine." The pilot, like the explorer, is up against nature and must come to know it in new ways. His business, wrote Saint-Exupéry, "is with the wind, with the stars, with night, with sand, with the sea. He strives to outwit the forces of nature. He stares in expectancy for the coming of dawn the way a gardener awaits the coming of spring. He looks forward to a port as to a promised land, and truth for him is what lives in the stars."

The excitement of adventure can stave off its physical hardship. Saint-Exupéry wrote that his exuberance for flying was such that "I was neither hungry nor thirsty. I felt no weariness. It seemed to me I could go on like this at the controls for ten years. I was happy." The excitement comes back anew. Beryl Markham, the first person to fly solo across the Atlantic from east to west, wrote in her memoir *West with the Night*, "I have lifted my plane from the Nairobi airport for perhaps a thousand flights and I have never felt her wheels glide from the earth into the air without knowing the uncertainty and the exhilaration of firstborn adventure." The certainty of the uncertainty is a rush in its own right.

The promise of adventure can fend off the ennui of conventional life. On the eve of his expedition to search for the source of the Nile, Richard Burton wrote in his journal of escaping world-weariness, the apathy of the known: "Of the gladdest moments in human life, methinks, is the departure upon a distant journey into unknown lands. Shaking off with one mighty effort the fetters of Habit, the leaden weight of Routine, the cloak of many Cares, and the slavery of Home, man feels once more happy. The blood flows with the fast circulation of childhood. Excitement lends unwonted vigour to the muscle, and the sudden sense of freedom adds a cubit to the mental stature. Afresh dawns the morn of life. Again the bright world is beautiful to the eye, and the glorious face of nature gladdens the soul. A journey, in fact, appeals to Imagination, to Memory, to Hope—the sister graces of our mortal being."

Most scientists and explorers throw themselves into their work with enthusiasm, if not an actual sense of adventure, but few exhibit excitement at the legendary level of the Greek mathematician Archimedes, who, having discovered the solution to a problem set for him by the king, reputedly ran naked through the streets of Syracuse shouting "Eureka! Eureka!" ("I have found it! I have found it!") and, it is said by Plutarch, was slain by a Roman soldier because he refused to interrupt his work on a geometry problem he had traced out in the dust. Enthusiasm sweeps up those under its influence, away from the direction of doubt, difficulty, and distraction. Nikola Tesla said, "I do not think there is any thrill that can go through the human heart like that felt by the inventor as he sees some creation of the brain unfolding to success. . . . Such emotions make a man forget food, sleep, friends, love, everything . . . an inventor has so intense a nature with so much of it of wild, passionate quality, that in giving himself to a woman he might love, he would give everything, and so take everything from his chosen field."

Exuberance not only drives people onward, it sustains them in times of drought. David Levy, the discoverer of more than twenty comets, had to wait nineteen years before he found his first one. Why, after a thousand hours of watching the skies to no avail, did he keep on looking? "The point of the search for comets," he says, "is that I love searching for comets." Exploration is its own reward. Good leaders know this intuitively. One hundred and fifty years before David Levy searched the skies, Thomas Jefferson had observed that his close friend and private secretary, Captain Meriwether Lewis, possessed a "passion for dazzling pursuits." Lewis's ardor for exploration, fueled by Jefferson's own scientific enthusiasms and an expansive vision for America, spurred a bold expedition to map the rivers and peoples, describe the wildlife, and give an account of the "soil and face" of the American wilderness. Jefferson knew Lewis's temperament well and understood, as he put it, that "no season or circumstance could obstruct his purpose." Jefferson bet on the power of passion, and he won.

There is in some people a passion to know the ways of nature—whether to discover a comet or to map a continent. Alan Lightman, reflecting on his life as a theoretical physicist, spoke of the all-consuming passion to know, an unrelenting determination to run down the truth: "I miss the intensity," he wrote. "I miss being grabbed by a science problem so that I could think of nothing else, consumed by it during the day and then through the night, hunched over the kitchen table with my pencil and pad of white paper while the dark world slept, tireless, electrified, working on until daylight and beyond." He worked without stop, he remembers, "because I wanted to know the answer. I wanted to know the telltale behavior of material spiraling into a black hole, or the maximum temperature of a gas of electrons and positrons, or what was left after a cluster of stars had slowly lost mass and drawn in on itself and collapsed. . . . I knew that the equations inexorably led to an answer, an answer that had never been known before, an answer waiting for me."

The combination of curiosity and joy so characteristic of scientific work calls to mind the galumphing quality of exuberant play: watching, chasing an idea first up one path and then down another, tussling with competitors, and flat-out exhilaration in the chase. Creative science and play are fun; they promise the unexpected. The nuclear chemist Glenn Seaborg, who, with his coworkers, discovered plutonium and nine other previously unknown elements, said, "I couldn't believe that I was being paid to do what I would have chosen as a hobby. . . . It was exciting just to walk into the lab, full of anticipation that that day I might be the first human being ever to see some unimaginable new creation." Most scientific work is routine; indeed, it is often frustrating, boring, and difficult. Scientists live for the unpredictable turns: the mind's galumphing, like love, is a many-splendored and much-desired thing.

Lewis Thomas, in *The Fragile Species,* describes the playlike quality to the work of the early molecular biologists: "Beginning with the discovery by Avery, MacLeod, and McCarty of DNA as the structure, and the elucidation of its fine architecture by James Watson and Francis Crick, investigators all around the world settled down to play with this new thing . . . it was a long line of extremely hard work, hotly competitive, frustrating to a lot of investigators. Nonetheless, it was the greatest fun for the imaginative winners in the games, one grand game after another. The energy which drove the hard work along was uncomplicated and irresistible: it was the urge to find out how a singularly strange and engrossing part of nature works."

Exuberant play, as we have seen in other mammals as well as in ourselves, creates a more energized and enriched environment in which to imagine, discover, and make connections. Scientists who are enthusiastic and energetic also enrich others with their enthusiasm. They attract people into their intellectual orbits and infect them with their exuberance; they transmit *This is important*, not only to those with whom they work but to scientists in other laboratories as well. Long before the significance of a discovery is fully realized, scientists enthuse; through their enthusings, they heighten awareness of the problem at hand and generate competition, both of which are likely to lead to even faster discovery. Intellectual play, like the play of childhood, is a serious business.

Robert Louis Stevenson, a writer's writer on the subject of adventure and exploration, argued vehemently for the role of play in all types of creative work, especially that of the artist. "The book, the statue, the sonata," he wrote, "must be gone upon with the unreasoning good faith and the unflagging spirit of children at their play. *Is it worth doing?*—when it shall have occurred to any artist to ask himself that question, it is implicitly answered in the negative. It does not occur to the child as he plays at being a pirate on the dining-room sofa, nor to the hunter as he pursues his quarry; and the candour of the one and the ardour of the other should be united in the bosom of the artist." Stevenson's observations are equally true for the scientist.

The delight of finding beauty in the natural world is a further draw of science. This was clear in the lives of John Muir and Snowflake Bentley, and brilliantly apparent in the lectures and writings of Humphry Davy, Michael Faraday, and Richard Feynman. "Science is not everything," said the physicist Robert Oppenheimer, "but science is very beautiful." G. H. Hardy, the father of modern analytic number theory, insisted that beauty was both a lure and a defining quality of mathematics: "The mathematician's patterns, like the painter's or the poet's, must be beautiful. The ideas, like the colours or the words, must fit together in a harmonious way. Beauty is the first test: there is no permanent place in the world for ugly mathematics."

An early jolt of beauty is, for many young scientists, as alluring as the fix of first discovery. The neurologist and writer Oliver Sacks writes of his reaction to learning about the structure of the atom: "Bohr's atom seemed to me ineffably, transcendently beautiful—electrons spinning, trillions of times a second, spinning forever in predestined orbits, a true perpetual-motion machine made possible by the irreducibility of the quantum, and the fact that the spinning electron expended no energy, did no work. And more complex atoms were more beautiful still, for they had dozens of electrons weaving separate paths, but organized, like tiny onions, in shells and subshells. They seemed to me not merely beautiful, these gossamer but indestructible things, but perfect . . . in their balancing of numbers and forces and shieldings and energies."

The exuberance of discovery is often accompanied by a joy in the sheer beauty of its shape or function. James Watson and Francis Crick give separate but like accounts of their delight in determining a fundamental pattern in nature. Shortly after discovering DNA's structure, Watson writes in *The Double Helix*, "we had lunch, telling each other that a structure this pretty just had to exist." Crick, describing an after-dinner talk Watson gave at Cambridge a year or two later, elaborated: "I have seen more than one speaker struggling to find his way into his topic through a haze of alcohol. Jim was no exception. In spite of it all he managed to give a fairly adequate description of the main points of the structure and the evidence supporting it, but when he came to sum up he was quite overcome and at a loss for words. He gazed at the model, slightly bleary-eyed. All he could manage to say was 'It's so beautiful, you see, so beautiful!' But then, of course, it was."

I was fortunate to grow up around exuberance. My father, a scientist and a great enthusiast, was surrounded by ebullient friends and colleagues who were scientists or mathematicians. They were lively and endlessly curious. They were also, for a child, great fun to be around because they were utterly captivated by the same things that enthrall children—stars, fireflies, wind, why a frog is marked the way it is, the reasons snow seems sometimes to crunch and at others to creak—and they laughed a lot and gesticulated wildly when talking about their ideas. They found the physical world fascinating and wondrous, as children do. (Indeed, one of my father's books from graduate school, *Physics of the Air*, which I now own and still read with pleasure, is an amazement of topics—fog, wind gusts and eddies, beaded lightning, snow crystals, whisperings of trees and murmurings of forests, clouds, the twinkling of stars and the luminescence of rainbows—and, even though these natural phenomena are described in terms of often quite incomprehensible diagrams and equations, the very fact that scientists studied such things seemed to me magical when I was young. It still does.) My father and his friends found enormous joy in asking questions, and then more questions, or in just batting ideas about. Best of all, unlike most other adults, they did not curtail their enthusiasms or respond condescendingly to those expressed by children. Exuberance was seen as a natural response to the world, not one to be kept in check.

Science and the enthusiasm of scientists retain their wonder for me. In many ways the passions of the great artists and scientists are not so different, however much has been made of the supposed temperamental divide. Psychologists and historians of science study the discoveries of scientists extensively, but examine their emotions and motivations far less closely. (Of course, scientific discoveries, like works of art, are themselves the important thing. But they are not the only thing.) Yet scientific thought removed from the emotions that drive and nourish it is as desiccated as the heart taken from its vessels of blood.

In 1968 a book was published that detonated the myth of science as a preserve for the calm and detached. James Watson's *The Double Helix* was, for the tens of thousands of young scientists in the making who read it, a fresh and exuberant look into science and intellectual pursuit. It brought to life a world that, in Watson's words, "seldom proceeds in the straight-forward logical manner imagined by outsiders. Instead, its steps forward (and sometimes backward) are often very human events in which personalities and cultural traditions play major roles." He wrote his book, in part, he said, because "there remains general ignorance about how science is 'done.' That is not to say that all science is done in the manner described here. This is far from the case, for styles of scientific research vary almost as much as human personalities. On the other hand, I do not believe that the way DNA came out constitutes an odd exception to a scientific world complicated by the contradictory pulls of ambition and the sense of fair play."

The discovery of the structure of DNA is the defining event of modern biology. It has been described by Sir Peter Medawar as the greatest achievement of science in the twentieth century, and by the renowned Harvard biologist E. O. Wilson as a scientific accomplishment that "towered over all that the rest of us had achieved and could ever hope to achieve. It came like a lightning flash, like knowledge from the gods." Medawar, an immunologist and Nobel laureate, emphasized the elegance of the Watson-Crick solution: "The great thing about their discovery was its completeness, its air of finality . . . if the solution had come out piecemeal instead of in a blaze of understanding: then it would still have been a great episode in biological history but something more in the common run of things; something splendidly well done, but not in the grand romantic manner." Watson himself described his and Crick's discovery as an "adventure characterized both by youthful arrogance and by the belief that the truth, once found, would be simple as well as pretty." The adventure was a quest, the goal was the Holy Grail of biology, and the knights errant turned out to be ferociously competitive, monomaniacal, and right.

The publication of *The Double Helix* rocked the world of academic biology. Some loved it. The mathematician, scientist, and anthropologist Jacob Bronowski said that Watson's account "communicates the spirit of science as no formal account has ever done . . . [it] expresses the open adventure of science; the sense of the future, the high spirits and the rivalry and the guesses right and wrong, the surge of imagination and the test of fact." The sociologist Robert Merton, in a *New York Times* review, wrote that he knew of no other book like it in describing scientists at work, that it was a "wonderfully candid self-portrait of the scientist as a young man in a hurry." Watson, he said, had portrayed honestly what had always been true in history: scientists are intent not only on discovery, but on being first. Richard Feynman wrote to Watson, "You are describing how science *is* done. I know, for I have had the same beautiful and frightening experience."

Alex Comfort, both a scientist and novelist, suggested that Watson might well deserve a second Nobel Prize, for literature. He began his review in the *Manchester Guardian* by stating that there "has never been anything quite like this tactless and truly remarkable book." Generally, he pointed out, "one doesn't write a low-down on the Church while staying in Holy Orders." Comfort captures the book's romance and high spirits perfectly: "The style is elated, and so it should be: there is no experience of human intoxication to equal the solving of a fundamental problem in Cambridge in early spring, when one is in one's twenties. This excitement is transmitted to any reader, even if he thinks DNA is a kind of aircraft glue."

Other reviewers were less kind, not to say vitriolic. Harvard University Press refused to honor its initial commitment to publish the book. Lawsuits were threatened. The editor of *Nature,* the journal that had published the original scientific paper by Watson and Crick, stated that "no fewer than a dozen distinguished molecular biologists had declined an invitation to review the book." Of those who did review it, several were outraged by Watson's depiction of raw scientific competitiveness and his personal arrogance, his caustic appraisal of colleagues (especially Rosalind Franklin), and a seeming abandonment of the grace and ideals of Science. He was accused of character assassination and of portraying a world of intense ambition; his book, it was said, was a "bleak recitation of bickering and personal ambition." The science editor of *The Saturday Review* worried about its damaging effects on immature minds.

Whatever truth there was in some of the criticism, most of it was disingenuous. Science has always been intensely competitive, always marked by heated battles over priority, about who got there first. Among many others, the bitter rivalries of Newton and Leibniz, Lavoisier and Priestley, and Edison and Tesla come quickly to mind. Sir Howard Florey, who received the Nobel Prize for his work on penicillin, observed that priority is an essential concept to scientists. "Like geographical explorers of old," he said, "the scientist likes to be the first to make a discovery, the first to do something." Such competitiveness is scarcely unique to science, of course; it is also integral to love, to the arts, sports, law, politics, and business. Fierce competition, in short, is part and parcel of any field of stake or moment.

Competition is as ancient as the hunt; the same fire that rouses the thrill of pursuit is kept kindled by the joy of victory. There is pleasure in the run, of course, but the high glory is in being first across the line. The biologist Richard Lewontin put it succinctly: "What every scientist knows, but few will admit, is that the requirement for great success is great ambition. Moreover, the ambition is for personal triumph over other men, not merely over nature. Science is a form of competitive and aggressive activity, a contest of man against man that provides knowledge as a side product." The exhilaration of winning is lashed to the rush of discovery.

Watson's portrayal of his fellow scientists is on occasion harsh, but he is also brutally honest about his own behavior. As Medawar writes, "He betrays in himself faults graver than those he professes to discern in others." A lack of tact may betray a friendship, but too much tact may betray the truth. In fact, tact has never been Watson's long suit, but then he has never claimed otherwise. (The editor of *Nature* put reviewers' offended sensibilities and claims of injudicious writing in perspective: "If, of course, his picture is seriously awry, then other people are free to protest and even have a duty to do so. It is not enough simply to resolve never again to invite Professor Watson to tea and biscuits.") Matt Ridley has put it well: "What a much duller—and safer—history DNA would have had without Watson stirring things up."

For those of us who read *The Double Helix* when we were young, however, whatever offense Watson may have caused his fellow scientists was of little consequence; what mattered was the idea that the most famous biologist in the world was saying that science was fun; that science was about asking important questions and taking seriously one's own intellectual life. The book was a classic adventure tale brilliantly told: a zigzag quest, wrong turns, setbacks, new leads, and hot pursuit—part Robert Louis Stevenson, part James Barrie. There was treasure to be found, enemies to fight, exciting terrain to cover, mishaps aplenty, and a shot at glory. It was irresistible.

Impatience in the pursuit of something great, Watson made abundantly clear, was more virtue than vice. Instead of society's usual, more disparaging take on this particular aspect of temperament, Watson was actually saying that speed and passion were essential to the chase. Instead of entreaties to slow down, be patient, be circumspect, someone was acknowledging that impatience is the obverse of exuberance and that exuberance was a good, even necessary thing. "Damn the men of measured merriment!" Martin Arrowsmith had exclaimed. "DAMN their careful smiles!" The ghost of Sinclair Lewis's driven protagonist had found its niche in *The Double Helix*. Intemperance, when coupled with the discipline of scientific thought, was given a far more kindly reading than it usually gets. Francis Crick wrote later that Watson "just wanted the answer, and whether he got it by sound methods or flashy ones did not bother him a bit. All he wanted was to get it *as quickly as possible*. . . . In some ways I can see that we acted insufferably . . . but it was not all due to competitiveness. It was because we passionately wanted to know the details of the structure."

Watson and Crick knew that what they were looking for was as important as it gets, and they did not, or could not, rein themselves in. Maurice Wilkins, their colleague and competitor at King's College, London, was, on the other hand, decidedly lower-key than his Cambridge compatriots. "Maurice continually frustrated Francis by never seeming enthusiastic enough about DNA," wrote Watson. "Francis felt he could never get the message over to Maurice that you did not move cautiously when you were holding dynamite like DNA."

Science, as depicted by Watson, was fast and exciting. It was not a calling for the indifferent, the slow, or the faint of heart. Science needed reason and discipline, of course, but it also required passion; it wanted commitment; it was cutthroat, it was human. "Our characters were imperfect," Watson said not long ago, "but that's life."

When I first thought about writing about exuberance, thirty years after having read *The Double Helix* as an undergraduate, I hoped to capture some of its importance by interviewing several scientists, most of whom I knew personally to be highly exuberant; a few others I knew only through their work. My interest was not in demonstrating that exuberance is essential to good science—clearly it is not; many outstanding scientists are introverted and not demonstrably enthusiastic, and for many others patience and dispassion are essential to the excellence of their work—rather, I hoped to show that for many scientists exuberance plays a critical role in how they think about and actually do their work.

Most of the scientists I interviewed were biologists. In addition to James Watson, I interviewed Carleton Gajdusek, who was awarded the Nobel Prize for his work on unconventional "slow viruses"; Robert Gallo, discoverer of the first human retrovirus, codiscoverer of the AIDS virus, and the only scientist to win the Lasker Award, which is often described as the American Nobel Prize, for both basic and clinical research; Samuel Barondes, a pioneer in the study of a class of proteins called galectins and a leading researcher into the genetic causes of mental illness; Joyce Poole, the scientific director of the Amboseli Elephant Research Project in Kenya; Katy Payne, an acoustic biologist who discovered new ways of understanding how elephants communicate; and Hope Ryden, who studies and writes about wildlife, including beavers, mustangs, and bobcats. I also interviewed two astrophysicists, Robert Farquhar and Andrew Cheng, from the Johns Hopkins Applied Physics Laboratory.

James Watson was an obvious scientist to interview about exuberance. When I asked him to rate both himself and Francis Crick on a hypothetical ten-point scale of exuberance, he said emphatically, "Ten!" then quickly added, "And then some!" He described exuberance, in his staccato, stream-of-consciousness way, as "an obsessive fascination, like religious fanaticism. You have to talk about it. Exuberance flows, it is never slimy. It is close to delirium. There is no feedback, no restraint, no bringing you back."

The greatest thing, Watson said, is the exuberance of sharing beauty or discovery with someone else: "It is necessary to share it. You run around and tell everyone. Shy people are seldom exuberant. It is a state of mind which can only be relieved by communicating the idea. If you are delirious, you have to share it. You have to demonstrate it to other people." When asked if he thought there could be such a thing as solitary exuberance, he said, "Perhaps. I don't know. You have to demonstrate it to other people." The major disadvantage of exuberance, from his perspective, was that "bad people can be exuberant," which makes them more dangerous than they would otherwise be, because they are more persuasive and energetic. In its most extreme form, he says, exuberance is "associated with madness." It can also, he added, "prove too much for your friends to put up with."

When he and Crick discovered the structure of DNA, Watson recalls, both were "bubbling over with exuberance. We *had* to share our ideas, we *had* to talk about it. It was a happy state, virtually delirious." (A scientist who was at the Cavendish in the weeks following Watson and Crick's discovery used similar language to describe their mental state: "Both young men are somewhat mad hatters who bubble over about their new structure," Gerard Pomerat wrote in his diary at the time. "The two chaps," he added, were "certainly not lacking . . . in either enthusiasm or ability.") Watson relates in *The Double Helix* that Crick constantly "would pop up from his chair, worriedly look at the cardboard models, fiddle with other combinations, and then, the period of momentary uncertainty over, look satisfied and tell me how important our work was. I enjoyed Francis's words, even though they lacked the casual sense of understatement known to be the correct way to behave in Cambridge." The next morning, he said, "I felt marvelously alive when I awoke." Crick, meanwhile, had "winged into the Eagle [a Cambridge pub] to tell everyone within hearing distance that we had found the secret of life."

Watson places positive traits such as curiosity and exuberance (which he, like nearly everyone else I interviewed, believes to be more innate than learned) within an evolutionary context. When asked by the novelist Melvyn Bragg why scientists do science, he responded: "I just like to know why things happen and I think that is probably something we have inherited. Curiosity about things, why things happen, can prepare you for how you live in the world. It has great survival value, this sort of curiosity and it is a question of how our curiosity is directed. Many people are very curious about things, are obsessed about things, which you could say have no consequence." (A mind focused on the most critical scientific issues— DNA, cancer, sequencing the human genome, and neuroscience—has marked Watson's scientific career. He has lived James Merrill's injunction that "it's not the precious but the semiprecious one has to resist.")

In his Liberty Medal address in Philadelphia on the first Independence Day of the new millennium, Watson spoke about the survival value of the *pursuit* of happiness, and the importance of having constraints upon that happiness: "Our various brains have been programmed by our genes to initiate actions that keep us alive. Most individuals are only fleetingly happy, say, after we have solved a problem, either intellectual or personal, that then lets our brain rest for a bit. Equally important, happy moods also reward higher animals after they make behavioral decisions that increase their survivability." But, he went on to say, "these moments of pleasure best be short-lived. Too much contentment necessarily leads to indolence . . . it is discontent with the present that leads clever minds to extend the frontiers of human imagination." Happiness, joy, and exuberance exist because they lure us onward and give us respite from our pursuits, but too much pleasure slackens the desire to explore, compete, and make a difference. "Every successful society," Watson emphasized, "must possess citizens gnawing at its innards, and threatening conventional wisdom." He concluded his remarks, "As long as we see happiness ahead, the worries and faults of today are bearable. So in the perfect world we want some day to exist, humans will be born free and die almost happy."

The biologist Robert Pollack has observed that Watson is "always a student, always ready to hear a new idea, always ready to get the picture, always ready to be excited. . . . [He has] an absolute enthusiasm for ideas." His enthusiasm is infectious, albeit yoked to occasional irascibility and brusqueness. The British scientist Lionel Crawford wrote of the intimidating yet inebriating effect of working in Watson's lab, fending off barbs and impatience, keeping up a relentless pace: "It was not only in the lab that patience was in short supply, it was also true of the seminar room. In a seminar by Julian Davies, I remember Jim suddenly standing up about halfway through and saying 'That's enough of that crap Julian, you've got another ten minutes and just give us the facts.' " It was "very different from the deferential attitude of the seminar audiences [Crawford] had been used to in England." But "together with the impatience came a great deal of encouragement and enthusiasm. This is what made our summers in Jim's lab exciting, sometimes exhilarating, and finally exhausting."

When he turned seventy, Watson told a gathering at Harvard

that he thought it would be "very depressing, but it's not really, because there's so much exciting still to hear." At seventy-five he remains an unalloyed enthusiast: optimistic, exuberant, and passionately involved in both science and life. Conversations with Watson gallop through an unbounded range of topics and iconoclastic opinions: the idiocy of one scientist or another, tennis, things Irish, things Scottish, beautiful women, the value of green tea, science (always), the idiocy of most psychiatrists, why fat people are happy, Cold Spring Harbor Laboratory (always), his plan to tell the Pope that if saints get credit for their miracles so should scientists, a fiendish delight in Francis Crick's notion that everyone should be declared legally dead after the age of eighty-five, the satisfactions of revenge, and whether or not, somewhere, we have the new Copernicus in our midst and are unaware of it.

Watson, in full pursuit of an idea, is an unnerving mix of exuberant intuition and deadly logic: one side of his brain lopes ebulliently from thought to thought and the other side applies a quick, remorseless logic to ill-conceived ideas. He is a man of enormous passion. His idealistic but practical mind and heart are drawn far more to the future than to the past. At a scientific meeting in Washington to celebrate the fiftieth anniversary of the discovery of the double helix, Watson talked about the role of passion and reason in the Scottish Enlightenment and touched upon his own lowland Scots ancestry. Passion in the service of reason, he said, was what his scientific life had been all about: it had been at the heart of how a problem in chemistry, the structure of DNA, could be solved by a birdwatcher (himself) and a physicist (Crick). His life, he said, had been all about curiosity and passion.

The next evening, at a book signing in Virginia, Watson's curiosity and restlessness were at full throttle: yes, the double helix was important; yes, the Human Genome Project was important; but *How does the brain work? How do we cure mental illness? How does the golden plover navigate?* His enthusiasm for understanding how the world works is palpable and he remains a case study in his own Fourth Rule for How to Succeed in Science: "Have Fun and Stay Connected. Never do anything that bores you." His experience in science, he says, is that "someone is always telling you to do things that leave you flat. Bad idea."

In his most recent book, *DNA: The Secret of Life,* Watson ends on an optimistic, indeed a near-quixotic note. "I may not be religious," he writes, "but I still see much in scripture that is profoundly true." He quotes from Paul's first letter to the Corinthians—"if I understand all mysteries but have not love, I am nothing"—and says, "Paul has in my judgment proclaimed rightly the essence of our humanity. Love, that impulse which promotes our caring for one another, is what has permitted our survival and success on the planet. . . . So fundamental is it to human nature that I am sure that the capacity to love is inscribed in our DNA—a secular Paul would say that love is the greatest gift of our genes to humanity." Love, certainly, but perhaps exuberance and a passionate curiosity as well.

No one would agree more on the importance of passion in science than Robert Gallo, who for many years was the chief of the Laboratory of Tumor Cell Biology at the National Cancer Institute and is now director of the Institute of Human Virology at the University of Maryland. When asked to define "exuberance," he says that it is, for him, "an over-average bite of life . . . a zest for fun, an overoptimistic view of one's prospects, an overreaction or hypersensitivity to things one does or is involved in." Being exuberant, he believes, "is linked to openness, confidence, overconfidence, and a delight in one's work. It may favor (all things being roughly equal) discovery and, perhaps to the same extent, making mistakes. Its role in research may be inclining some spirits to try to open a field even when they do not always realize it."

Gallo draws upon Jacques Barzun's discussion of Romanticism in *From Dawn to Decadence,* comparing Barzun's concept of passionate love with the intensity some scientists bring to their work. "His concept of this love," Gallo says, "is not only Eros but also much more, crystallizing on the loved one much imagery and the almost perfect joy in its presence. He said it favors youthful feelings, naiveness. Is this a form of exuberance? Aren't some this way with their work in science? Would this lead to more imagination? A tendency to love hypothesis as much as the answer, if a scientist was bent this way." The role of exuberance in other aspects of scientific research, he adds—with considerable personal experience to back it up—may be to supply inspirational leadership, "as long as the leader is not decapitated."

At a more personal level, Gallo—who describes himself as "very highly exuberant" and once told an interviewer that he did not wear shoes with laces because he couldn't wait to get to work and tying his shoes would slow him down too much—talks about exuberance in his own life: "Hyperenthusiasm, the joy of competitiveness, not only with colleagues but with scientific riddles, the great fun in telling your colleagues about your work, the feeling on occasion of 'knowing' one will be right, the energy that fills you, is central to what propels me but I do not know why. Honestly, it is not just winning a race. It is much more internal. The same feelings tell me [even] if I am wrong I will eventually get there."

Resilience is something Gallo understands well and, more than almost anyone I know, personifies. For years he was the subject of unrelenting journalistic and federal investigation into a dispute about contaminated laboratory samples and claims of primacy in the discovery of the AIDS virus, a dispute that cost him hundreds of thousands of dollars in legal fees, near-continuous assaults on his character, and a scandalous number of hours away from his scientific work. (All charges against him were ultimately dropped and a closely related case against a collaborator was dismissed with the pointed comment from the federal appeals board, "One might anticipate from all this evidence, after all the sound and fury, there would be at least a residue of palpable wrongdoing. That is not the case.") Recently the codiscoverer of the AIDS virus, Luc Montagnier, acknowledged that cell cultures in both his and Gallo's laboratories had been contaminated. This critical issue of dispute is discussed further in the chapter notes in a description of events agreed upon by the two scientists.

During the time of the investigation—a nightmare for him, his family, and his laboratory—Gallo kept doing science and somehow managed to regenerate energy and enthusiasm with which to pursue new ideas. Indeed, during the 1980s his was the most referenced scientific laboratory in the world, and it remained, even in the times of greatest controversy, incomparably productive. "He's incredibly resilient," says Anthony Fauci, the director of the National Institute of Allergy and Infectious Diseases. "If you look at the stress and the strain Bob went through over these several years when he was literally under a microscope virtually every day, every week, I think it would have worn down and perhaps broken a lesser spirit."

Gallo's temperament—fiery, exuberant, competitive—fuels his accomplishments even as it inflames criticism. Luc Montagnier, the French codiscoverer of the AIDS virus, stands in stark contrast to Gallo: less controversial, less intense, and less imaginative. To see the two scientists together in person is to see the range of temperaments in science and to witness the opposite ends of exuberance: Gallo gesticulates and laughs a lot, is quick, warm, almost frighteningly intuitive, fast-talking, irritable, mercurial, and fiercely caught up in the topic at hand, whether it is molecular biology or the hills of Rome. Montagnier, on the other hand, is cool, cerebral, distant, and difficult to engage. "We are not alike in our styles, as people or as scientists," Gallo has written in his memoir, *Virus Hunting:* "He is quiet, almost formal, holding his own counsel when competing ideas are being presented. . . . I love the rough-and-tumble of intellectual debate." Gallo is, in fact, far more outspoken than many other scientists about the intense rivalries that exist in science. "It is highly competitive in science," he says. "I cannot tell you that it is more than in business, or more than in politics, but it is more than in some fields, without doubt. . . . Why do we go into it? You think you are good enough to solve problems of nature."

Some combination of competitiveness and natural ebullience kept Gallo and his laboratory going during the years of government investigation, testimony, and innuendo. His exuberance and vivacity are legendary. One scientist, who trained as a postdoctoral fellow in Gallo's lab, said that the two years he spent there were the most exciting of his life and that Gallo's vitality was the uniting force: "He does an excellent job of organizing such a large group of people with diverse personalities, backgrounds, and scientific interests into a cohesive team, and he's able to maintain the energy of the lab via his own personal energy." Gallo's energy and enthusiasm may have ebbed and flowed during the worst years, but they seem always to have come back to revitalize not only him but those in his laboratory. He was asked once by a *New York Times* reporter whether he had been "humbled" by the bruising investigation, an experience that would have brought anyone else to his knees. He responded, "No. Empathetic is a better word: I'm more understanding of other people's hurts and problems. Humbled? No. You get too humble, you lose all your confidence, and then you can't think about science."

Exuberance, according to Gallo, is in part the "capacity to reemerge, not seeing all the reasons why one should not"; it helps to buffer one against hurt and setback. "If I'm down," he says, "exuberance makes it more likely that I'll surely get up and it will be better. There are many ways one can select to avoid hurt, and [I] seem prone to use exuberance." Its other advantages, he believes, are "to make life and work ever so much more fun and to overcome the fear of failure; i.e., to take risks." The disadvantages of exuberance, on the other hand, are that "if one is unaware of it, the possibilities for error increase"; and, he adds, "the appearance of too much joy might solicit some jealousy and even hatred."

In a 1990 *Science* interview, Gallo gave a pell-mell sense of the excitement and competitive fervor in his lab during the first flush of AIDS research in the 1980s: "You know, the juices flow, right? Your mind is thinking and moving and you're saying: 'What's the next question?' We discover HIV goes to the brain—we publish it. What's the target in the brain? Microglial cells—we publish it first. How much does this correlate with the dementia in the brain? We tried to establish that. Is the virus present in plasma? We found it was. . . . When is the virus expressed? . . . What about the genes of the virus: what are their functions? Could we make the blood test better? Can we make an antigen test? We tried, we failed. . . . Those were the things that were on my mind: get rid of the goddamned virus after you figure out how it works."

Like James Watson's, Gallo's are not the words of a dispassionate scientist: curiosity and exuberance drive the chase; competition speeds it onward. Not surprisingly, perhaps, given his temperament, Gallo once remarked, "I have to say that I was never overwhelmed with the need to pipette. I think there are people who need it for the serenity of it all. I am not one who feels this is necessarily a great pleasure." No one has ever accused Gallo of taking life or science at a leisurely pace, and, significantly, the citations for both of his Lasker Awards allude to the pace he set for other scientists. The commendation for the 1982 Lasker Award, given for basic medical research, ends: "To Dr. Gallo for his tenacious and thorough investigations leading to the discovery of the human T-cell leukemia virus and carrying resounding implications that will reshape approaches to cancer much sooner than scientists had expected or humanity had hoped." The 1986 Lasker Award for clinical research acknowledged not only his fundamental scientific contributions, but his essential and energetic leadership: "To a desperate moment of public alarm when physicians lacked any means of treating AIDS patients, Dr. Gallo brought clarity of vision and an invigorating spirit of inquiry that has set a pace for research unprecedented in medical history."

In introducing Gallo at a winter science meeting in Colorado, the Nobel laureate Howard Temin gave the most compelling description of Gallo's temperament, a temperament that, he suggests, is an exuberant life force not far removed from nature itself. Temin said, "I am here only to introduce the keynote speaker. I would like to do it with a very short and simple story. One fine day, when I was a graduate student with Max Delbrück [who received the Nobel Prize for his pioneering research on bacteriophages], a few of us were walking with Max among the pine trees. Suddenly one young but famous (I won't say his name) scientist stopped and asked Max this question: 'Max, why is it that Josh Lederberg [also a Nobel laureate, for his work in bacterial genetics] makes so many discoveries when, in fact, we are so much smarter?' Delbrück paused, looked up at the trees, and then responded: 'Because Josh was born much closer to nature.' With that I introduce Bob Gallo."

Like James Watson, Bob Gallo is simultaneously intuitive and analytical, fiercely competitive, often controversial, and at heart a romantic. Both were influenced when young by the idealistic but driven scientists portrayed in *Arrowsmith* and Paul de Kruif's *Microbe Hunters*. (Gallo says that when he was young, "I saw science as another kind of religion, certainly one that would yield more predictable results if one served it faithfully." *Arrowsmith* and *Microbe Hunters* encouraged this view. Watson says, "I read the books at the same time. I got very excited. They made you want to be a scientist.") Both men love gossip, science, and competition. Both are mercurial and can be intemperate and abrasive; both are exceptionally kind and generous to their friends, impatient, and slow to muzzle their impulses. They are, above all, passionately engaged in science; exuberance is elemental to this and to their interactions with other scientists. They are keen to engage in vehement debate and criticism, and feel strongly that the best way to learn is by arguing with someone who will point out the flaws in an experiment, or the Achilles' heel in an idea.

Watson believes that he and Francis Crick discovered the structure of DNA before their competitors at least in part because no one at Caltech would confront Linus Pauling about his basic error in chemistry and because Rosalind Franklin was isolated from colleagues who might otherwise have given her a different perspective on her crystallography studies. A good scientist, says Crick, "values criticism almost higher than friendship." This is a belief that requires a thick hide and a resilient nature, both of which are aided by an exuberant temperament. Exuberance makes the reaching out to other scientists more likely and makes the rebound from criticism into enthusiasm a near certainty.

The only person I know, scientist or otherwise, who is usually referred to by his peers as a true genius is Carleton Gajdusek, a virologist who received the Nobel Prize for his research on the "slow viruses" (or, as most now believe, prions) responsible for degenerative diseases of the brain. Gajdusek seems always to have loved science and to have been unrestrainedly exuberant. His father, an immigrant butcher in Yonkers, New York, exerted a powerful emotional influence—he had, writes Gajdusek, a "temperament for laughter and ribald fun, lust for life in work and play, music, song, dance and food, and above all, conversation"—and his mother's love of folklore and literature fostered a lifelong passion for studying other cultures and the arts. Before he and his brother had learned to read, they were listening to Homer, Hesiod, Sophocles, Plutarch, and Virgil. As a child he read Scandinavian literature as well as the biographies of scientists such as Marie Curie and Louis Pasteur. Like Watson and Gallo, Gajdusek was influenced by Paul de Kruif's *Microbe Hunters,* although his enthusiasm for the book took him further: he stenciled the names of de Kruif's microbiologists—Leeuwenhoek, Spallanzani, Pasteur, Koch, Reed, Ehrlich—on the steps leading up to his attic chemistry lab, where they have remained for more than fifty years.

Like many scientists, Gajdusek knew from the time he was young that he wanted to study nature. "As a boy of five," he said, "I wandered through gardens, fields and woods with my mother's entomologist-sister, Tante Irene, as we overturned rocks and sought to find how many different plant and animal species of previously hidden life lay before us. We cut open galls to find insects responsible for the tumors. . . . In petri dishes we watched some leaf-eating insects succumb to insecticide poison while others survived." It was his aunt who introduced him to the pleasures of studying the natural world. She was, he says, the dominant influence on his "intellectual curiosity and playful enthusiasm. . . . She made my child's curiosity into a game of investigation and enquiry, and had me doing experiments just after I was a toddler, not just collecting and classifying insects. With passionate zeal she showed me the beauty in complex life cycles of insects in decaying flesh and vegetation, and even in the murders accomplished by ichneumon flies by implanting their eggs into their victims, and in the cannibalism of the praying mantis after mating . . . it was she herself who instilled into me the quest for beauty in nature and taught me to live creatively."

Gajdusek received his medical degree from Harvard in 1946 and set forth on a lifetime of diverse interests and remarkable work. He studied the cultures of the Hopi, Navajo, and Mexican Indians; researched scurvy, bubonic plague, and rabies in Afghanistan, Iran, and Turkey; organized expeditions to the valleys of the Himalaya; and treated diseases in the jungles of South America and the swamps and highlands of Papua New Guinea and Malaysia. Along the way he learned German, French, Spanish, Russian, Slovak, several of the more than seven hundred languages of Melanesia and Papua New Guinea, and Persian, Bahasa Indonesian, and Dutch. In 1954 he went to Australia to do basic and clinical research in immunology and virology in Australian Aboriginal and New Guinean populations.

Gajdusek was impatient and restless: "Everything he possessed spoke of his being peripatetic," said one colleague. "Even standing still, he seemed to be on the move, with top tilted forward, in the breathless posture of someone who never had time enough to get where he had to be." It was while living among the Fore, a Stone Age people of New Guinea, that he and another scientist, Vincent Zigas, discovered kuru, a progressive degenerative brain disease characterized by shivering, pathological laughter, loss of coordination, and, invariably, death.

Kuru had killed thirty thousand people, mostly women and children, in the cannibal population living in the highlands of the central ranges of eastern New Guinea. The pattern of deaths was familial but not genetic, and it took Gajdusek and his colleagues more than five years of tracking down possible causes and studying autopsied brains (which they showed to be "spongiform," or full of holes) to establish that the infectious agent was a previously unknown "slow virus" (later thought to be a prion). Kuru, they demonstrated, was transmitted, often to infants and children, through the ritual eating of the brains of dead relatives. Gajdusek concluded that its transmission took place not just through the eating of the infected brains, but also through the ritual butchery of the dead body. This was carried out by women, who used sharp bamboo instruments while holding their infants on their laps or feeding them. The steaming of body organs was not sufficient to kill the infectious agent and, because the dead were eaten only by close family members, the transmission of the disease followed a familial pattern. Women and children, who ate the brains of the deceased, were far more likely than the men to die of kuru.

Gajdusek's research, which demonstrated beyond doubt the link between the Fore mortuary rituals and kuru, effectively eliminated both the cannibalism practices and kuru. It also led to entirely new ways of understanding the origins and dissemination of infectious diseases. He and his colleagues found that other deadly brain diseases, such as Creutzfeldt-Jakob and bovine spongiform encephalopathy ("mad cow disease"), were caused by similar agents. His research, according to the Nobel Committee, represented an "extraordinarily fundamental advance in human neurology and in mammalian biology and microbiology."

Gajdusek has been described by one of his colleagues as unique in medicine, a man "who combines the intelligence of a near genius with the adventurous spirit of a privateer." This description is as apt for him at the age of eighty as it was when he was a young scientist. Gajdusek retains a level of enthusiasm and adventurousness exhausting to anyone around him. He is legendary for his rapid, nonstop, and expansive monologues. My husband and I once jotted down as many topics as we could remember from a dinner conversation we had had with him. A very partial list included immunology, love, the French, the Americans, the Dutch, suicide, Puritanism and sex, schizophrenia, rat poison, molecular biology, tuberculosis, the FBI, the idiocy of American politics, Melville, Russian explorers, Plato, anthropology, mad cow disease, New

Guinea boys, courage, moods, navigation, linguistics, and meadow mice. Listening to Carleton and remaining indifferent is like staying calm while taking cocaine and listening to Fats Waller. It can't be done. George Klein, a friend and colleague of Gajdusek from the Karolinska Institute in Stockholm, recounts that when Carleton received the Nobel Prize he "exceeded the sacred forty-five-minute limit of the Nobel lecture without the slightest embarrassment, speaking as he did for two hours and ruining the whole afternoon's schedule. Strangely enough, no worried organizer tried to intervene. . . . Whatever had enticed you to attend the lecture, you had no reason to regret coming." People sat in the aisles, mesmerized.

When I asked Gajdusek for his thoughts about exuberance, he replied that, in his view, it was a concept "closely related to manic-depressive psychosis," and that the word had been attached to him often, once defined as "outrageous ardor." Exuberance, he said, is "seductive and closely linked to maturation or the stimulation of motivation. It may engender devotion and love. Like everything associated with seduction, it may be viewed as manipulative and coercive by those fearful of being the subject of desire." Society, he feels, is ambivalent about exuberance and often inhibits its expression. "Enthusiasm, ardor, exuberance are ways of expressing desire, of using it. We should be proud of our ability to do so—our society would make us ashamed."

To be exuberant, maintains Gajdusek, "is to be considered *pas sérieux*—light-headed and flighty!" As a result, exuberance "demands a reverse to be respected and admired—namely withdrawal and contemplation, silent thought. Thus, it is akin to cyclothymia in its necessary swings and reversal. . . . To have force, affect, meaning, requires depth, and that requires subdued colors and sounds and cyclothymia. Manic depression would be the model. . . . Not to show [exuberance] suggests ennui but to have too much is madness!" Surely, he said, exuberance is "linked to life and youth! Those who lack it inhibit its expression."

Exuberance, Gajdusek emphasizes, is infectious: "I often radiate it and from radiators, such as me, it is contagious, communicable. . . . I have recognized that when in my scientific career I have managed to ignite an intense interest and enthusiastic striving for a goal in myself and my coworkers, our team was living 'high,' almost hypomanically, obsessed with our problem and our activity. The feeling was almost like that of piloerection of the scalp with awe or wonder, the thrill of quest." (This is confirmed by the Australian virologist Sir Macfarlane Burnet, who has said that "Gajdusek is quite manically energetic when his enthusiasm is roused.")

Exuberance, for Gajdusek, is associated with fun, play, and creativity. "Creative persons," he told me, "are usually exuberant and never bored. Ennui and noncreativity and plodding exactitude go hand in hand with nonexuberance and fear of originality and change and the untried." Elsewhere, he remarked to a group of scientists who were studying creativity, "I play with ideas. . . . I'm rather confused by having a subject of 'creativity,' and then launching off into a discussion of what, in my opinion, is a very late stage of creative research, after a great deal of creativity is finished—namely, the testing, be it confirmatory or disconfirmatory, of a specific hypothesis. There was little mention of fun and play; these are the maxims on the wall for my forty years in the laboratory; that's all we've ever done, to have fun and play, and all we ever will do." A few years ago, when he was approaching eighty, Gajdusek wrote to my husband in the same spirit: "I look on my current joyful years packed with play and fun and joy and adventure as bonus years little expected." To me he added, "I can never 'finish'! It is fun. I amaze myself."

Samuel Barondes, a molecular biologist and psychiatrist at the University of California–San Francisco, speculates, like Gajdusek and Watson, that some types of exuberant states overlap with mild forms of mania. "To me," he says, "there are two kinds of happiness: a euphoric type of serenity that I became well acquainted with through formal meditation, and which I can now call up at will; and a euphoric type of excitement and anticipation which is what I think of as exuberance. I'm not sure where to draw the line between exuberance and hypomania, but I think the most important difference is in the area of judgment and self-observation. As I think of it, exuberance is still monitored by judgment. In a state of exuberance, judgment is put on hold—but is not turned off completely. In hypomania judgment is napping, but still wakes up periodically to check things out. In mania, judgment is out like a light."

Barondes, who studies the genetics of behavior and psychopathology, believes that exuberance is determined by the environment as well as heredity. "My father was quite exuberant," he says, "and I think I learned some of this from him (and inherited some via his DNA). My father explicitly recommended exuberance as well as serenity. We talked a lot about these matters. Having been raised as an Orthodox Jew of the scholarly and self-questioning type (rather than the more exuberant Hassidic type) has tempered my exuberance very little. I think you can teach people to be a little bit more or less exuberant. But it remains difficult to make a silk purse from a sow's ear." (Most of those I interviewed expressed similar views, namely, that it may be possible to influence the level of exuberance expressed, but not the underlying capacity for it. The child psychologist Ellen Winner, for example, believes that exuberance is inborn. "I think that the environment can kill it," she says, "but I don't think the environment can create it.")

Barondes, like many, believes that exuberance entices scientists toward discovery and helps overcome obstacles. It is, he says, the engine that drives discoveries. "When you feel exuberant"—he characterizes exuberance as "enthusiasm, energy, motivation, optimism, anticipation of wonderful things, and a great deal of pleasure"—"you believe you have a good chance of doing something new and important and exciting. You downplay obstacles and personal limitations, anticipate the joy of doing something special and worthwhile. Exuberance increases risk taking, innovative thinking, and the anticipation of success."

Judgment is critical to separating genuine discovery from deluded enthusiasm. Pipe dreams and obsession must be sorted out from meaningful scientific contributions. Barondes makes this clear: "There is, of course, in all creative activities, a need for a period of sober reflection and corrective skepticism to evaluate the fruits of a period of exuberance. Good scientists get their juice from exuberance, but the judgment that keeps operating in this state is more permissive than scientists can afford. The seemingly great discoveries that are facilitated by exuberance need to be checked by the cold, hard standards that science demands."

Two astrophysicists whom I interviewed at the Johns Hopkins Applied Physics Laboratory, both well known among their colleagues for their exuberance, generally concurred with the biologists' descriptions of exuberance and agreed with them about its importance in their scientific work. They also underscored its potential problems. Robert Farquhar, an expert in orbital mechanics, is internationally recognized for his brilliance in plotting complex routes through space by "gravity assist" techniques and calculating trajectories for missions to comets and asteroids. Farquhar, says one colleague succinctly, "is a genius with celestial pinball." He was mission director for the Near Earth Asteroid Rendevous (NEAR) spacecraft, which initially orbited and then, in early 2001, actually landed on the asteroid Eros. It was the first time a spacecraft had landed on an asteroid or any other small body (Eros, 200 million miles from Earth, is only twenty-one miles long and eight miles wide). Asteroid #5256 is named Farquhar in his honor.

Farquhar says of himself that he is "a classic case study in exuberance. I'm always this way. I can't understand why they pay me to do what I love." He is convinced that exuberance is innate—"It's in you; you either have it or you don't; it is something you carry from cradle to grave"—and that there is a strong link between creativity and exuberance. "If I didn't get very excited by things," he says, "I don't think I would be very creative. My intuitive notions usually have a pretty good feel. Exuberance keeps you going through other people's mistakes. If a computer calculation indicates I'm wrong, my enthusiasm keeps me believing in my own calculations. And it is not unusual for the early computer runs to be wrong." When NASA officials, wary of risking the outcome of the highly successful Eros mission, cautioned against landing the NEAR spacecraft on the asteroid because the landing had not been part of the original flight plan, Farquhar eventually won out. "The bureaucrats didn't want me to risk landing on the asteroid," he says with disbelief. "Why risk failure in an otherwise incredibly successful mission? But I thought that was crazy! How could anyone not want to see what it would be like to land on an asteroid? Bureaucrats drive me nuts."

Like most exuberant people, he sees exuberance as a source of resilience. "You don't sit around and mope, though it may take you a day to regroup. Then you're right back, already working on the next project." Competitiveness—the race to be first and best—is another important component of the link between high energy and enthusiasm. "I'm a very competitive person," he acknowledges, "I like to win. I was really exuberant when I beat the competition [for a NASA proposal to explore comets]. I had sketched out my ideas, then others ran the computer models. I was pretty damned happy when I found out I had the correct solution. I felt great, I never forgot that one. If you get excited you are more motivated and more competitive. You are motivated to work extra hours. Competition makes everything happen."

Farquhar's exuberance, like that of other scientists, is contagious. He loves comets. "It's incredible, exploring the unknown, the nuclei of comets. I love comets, I love the history of comets.

Each comet has its own personality, they're not dull like asteroids. Comets are dynamic, mysterious, beautiful." For Farquhar, there are few disadvantages to being so exuberant except for those times when "you get all excited and then find out your ideas weren't as good as you thought. There is always a low period after a period of exuberance." And, he adds, "people who aren't as exuberant as you are get *really* irritated with you."

Farquhar's colleague at the Hopkins Applied Physics Lab, Andrew Cheng, is even more exuberant than Farquhar. Cheng, a planetary scientist and physicist, inhabits an office as chaotic as Farquhar's is well-organized. Everywhere you look are piles of papers, scientific journals, videotapes, plastic shopping bags bearing the NASA logo, and books about Uranus, Saturn, Pluto and Charon, and Jupiter. His file cabinets are covered with yellow Post-it notes, and a stuffed Sylvester the Cat sits on top of his computer. When he comes into his office he is moving fast, talking fast, and still bubbling over from a meeting about flying airplanes to Mars.

Exuberance, Cheng makes clear, is an indispensable part of his scientific life. "It keeps me alive. I like to have fun, I don't like boredom. Exuberance is necessary, you have to have enthusiasm. Any kind of work involves a lot of tedium, menial tasks, boring tasks. Exuberance allows you to see beyond, to see the goal. You need that kind of emotional makeup to push through the work, to pursue really difficult things. Exuberance stops you from getting discouraged, or not starting in the first place. Science is working out ideas. The majority of ideas don't work out, there are a lot of dead ends. You need to have an exuberant makeup to prevent getting discouraged. Exuberance reduces stress levels." Exuberance also allows you to handle rejection, Cheng points out. "For example, if you put in a big proposal to NASA and it gets rejected, you need resiliency to pick yourself up after that. I have a natural tendency toward exuberance, I am naturally inclined to plunge into things. But rough times always come." Work, he emphasizes, is inherently stressful. "Stepping back, relaxing, enjoying, not getting all wound up or spinning your wheels, this keeps you from wearing yourself out.

"It is hard for me to empty my mind," Cheng explains. "Exuberance helps in focusing on the task, otherwise I am scattered everywhere else. If I am intensely involved or absorbed I can go for hours. I don't want to stop. You have to make work feel like play, you have to find out what's pleasurable. Exuberant people get excited to do things that have not been done before, to measure things that have not been measured before." Exuberance, he believes, is innate: "It can't really be learned, but it can be influenced." When asked specifically what he feels like at his most exuberant, Cheng describes the experience in terms of the intensity of war: "It's like soldiers in war. When people have been in combat they describe the experience as the most intense they have ever had. The defining moment. The most they have ever felt alive. Exuberance during scientific work is the most intensely experienced time. It is an emotional high, like euphoria; the outside world doesn't exist. It is moment to moment, one hundred percent absorption, like a religious experience."

Cheng, like Robert Gallo and Carleton Gajdusek, emphasizes the contagious nature of enthusiasm in interacting with other scientists. Exuberance, he states, "gives you the ability to get other people excited. This is especially important in [space] missions or other projects, to inspire a team of people who are on a tight budget, to get them committed in the same way you are. Also, because science is supported by the American public, you need to be able to communicate to them why you are excited."

Resilience, Cheng believes, is a vital benefit of exuberance. He talks about the ebbs and flows of scientific creativity, and how exuberance restores life when intensely productive times are followed by fallow ones: "I just had an article [about the asteroid Eros] in *Sci-*

ence and now I'm experiencing a kind of postpartum depression. There's a very uncomfortable time: *What's next?* I haven't made a commitment to a new idea or a new project; it's very similar to clinical depression, it's hard to focus. But inevitably I will get wrapped up in something new." Having said this, he went over to his computer and showed me close-up shots of the asteroid's surface. With each one he became more and more animated. "Look at those flat-bottomed craters!" he yelped in delight. "They look like they have been lubricated by fluids, but you can't have fluids on an asteroid. So I'm beginning to get intrigued. I'm moving on to this." He goes back to his computer, waving his arms about with great gusto. "It looks like mud flows but it can't be mud flows. Great!" He pauses briefly to shift gears. "There's *another* project I'm beginning to get interested in, it's this airplane-to-Mars thing. *Anyone* would find that great! So you start getting interested again. And you get excited."

More than any scientist I interviewed, Andrew Cheng also articulated the perils of great enthusiasms. As someone who rates himself as a "ten" on a ten-point scale, he made it clear that exuberance has not always served him well. "I have suffered from exuberance, from being scattered, a lack of focus," he says. Conflicting enthusiasms caused him to switch scientific fields several times, from high-energy astrophysics to space physics, to particles and fields, and finally to planetary science. (His various professional fields have honored him in many ways, however; among other things, Asteroid #8257 is named Andycheng.) Much like Gajdusek, who observed that exuberant people run the risk of being regarded as *pas sérieux,* Cheng reflects that "very exuberant scientists, such as Carl Sagan, make themselves vulnerable to ridicule. Too much exuberance turns other scientists off. Scientists are natural skeptics, they want to see behind the flash and drama. Show me the data, show me the calculations." An excitable person, he acknowledges, "may not get it right. Many make lots of mistakes."

Exuberance also presents the danger of overcommitment: "You don't focus sufficiently on a task, don't get things done, there are too many projects. You get excited. And you start forgetting meetings and ignoring your other responsibilities. And you start getting other people mad at you. If you're working on a mission, you can't do this. You need to work toward a launch date and those deadlines aren't flexible. Planets have to be aligned just right. You have only one correct date for a launch."

Cheng, despite his acute sensitivity to the drawbacks of exuberance, is a thoughtful advocate on its behalf. He draws parallels between emotional and cosmic exploration: exuberance takes you on journeys you would not otherwise take. "People need to be absorbed in something outside of themselves, something important. Exploration, like exuberance, lets people go out of and then back into their everyday lives. It makes life more interesting, life more worth living. The immediate benefit of going to Mars or to an asteroid is that it gets people excited. It is part of what defines the character of a country, that you are exploring, leading the way. It keeps a country, like a person, from stagnating."

I was interested in looking at the role exuberance plays not only in the lives of basic scientists, such as astrophysicists and molecular biologists, but also in those of scientists who do their work in the field. For this I turned to three women whose scientific work and writings I have greatly admired over the years: the elephant biologists Katy Payne and Joyce Poole, and the naturalist Hope Ryden.

Katy Payne, a research associate in the Bioacoustics Research Program of Cornell University's Laboratory of Ornithology, is best known for her studies of the complex and changing songs of humpback whales, as well as for discovering that elephants communicate by infrasound. She was brought up on a farm in Ithaca, New

York, and recollects a childhood of books and music, and of parsnips, pigs, apple pies, pumpkins, and White Leghorn hens. Nature was her teacher, and before long she was questioning back. As a young girl, for example, she wondered whether she could tell maple trees apart by the taste of their syrup. Dipping her fingers into the syrup buckets hanging from the trees, she tested the idea, and she has kept on questioning and testing things ever since.

Like Cecilia Payne-Gaposchkin, she describes an ecstatic response to nature in her childhood: "I remember my first encounter with myself, on a high day in late summer," she writes in her hauntingly beautiful memoir, *Silent Thunder*. "Standing alone in a field where wildness crowded up yellow and green against our garden and house, I said out loud, 'This is the happiest day of my life and I'm eleven.' I raised my skinny arms to the blue sky and noticed them, and my ragged cuffs, and a mass of golden flowers that was hanging over me. Their color against the sky made my heart leap. Since then I have seen the same yellows, green and blue in van Gogh's harvest paintings and heard the same hurrahing in Hopkins's harvest poem, but my hurrahing, that made me inside out with exuberance, was for wildness."

Payne, who is a Quaker, quotes a Quaker saying that "the water tastes of the pipes" to illustrate her belief that "there is no such thing as an indifferent observer." Certainly, she is someone who has been profoundly influenced by sounds and pipes. She recalls hearing the chords from a pipe organ when she was thirteen years old: "The organ was alive. In a powerful combination of voices it was introducing the great chorus that opens the second half of Bach's *Passion According to St. Matthew*. . . .The organist pulled out the great stop and the air around me began to shudder and throb." Years later, sitting beside the elephant enclosure at an Oregon zoo, she heard a "faint rumble . . . like the feeling of thunder but there had been no thunder," and suddenly she remembered the church organ and wondered, Were the elephants calling to each other in infrasound? Payne and her colleagues carried out acoustical studies, which established that the elephants were, in fact, communicating with sounds too low for humans to hear. Their discovery revolutionized the field of elephant communication. Payne has carried out other important studies of African elephants since, including research on how elephants organize their societies over long distances; together with Joyce Poole, she is compiling an elephant dictionary. She and another colleague are using acoustic techniques to monitor the health and behavior patterns of forest elephants in the Central African Republic and Ghana.

Payne is a great advocate for science: "Happy fishermen," she says of herself and her fellow biologists, "we stand together on the pier casting and reeling, mesmerized by a certain shimmer visible from this or that angle, enchanted by the concentric rings here and there, imagining things just below the dimpled surface. If something leaps into the air, we all lift eyebrows together as if we were one great being with one eyebrow—*that?* Is it alone, or part of a school? What else lies below?"

When I asked Payne about exuberance, she replied: "Moods in general, and exuberance as a stage in a mood, play a huge role in my life as a biologist as well as in the rest of my life." She defines exuberance as "part of a continuum (enthusiasm—exuberance—ecstasy) and I [see] my life as a biologist as part of a continuum that includes the artistic side of my life as well: for an observer, the two are inseparable. The moment of intense observation of any kind of truth or beauty is a peak experience: I sense myself as a part of what I am seeing, hearing, feeling, smelling and therefore have no doubts about it; all boundaries evaporate and all desire for control, which is after all irrelevant when one is living the truth. Part of why I love reality is that I know that nobody can ever understand it fully . . . the full answer is never at hand: everything is always evolving . . . my favorite feelings about what is unknown stem in part from this sense of beauty, and of the presence of steady, life-giving things around me, and these in turn stem partly from the river of happiness in which I am fortunate enough to swim at times. I really understand why people with manic-depressive illness don't want to curb the flow . . . the definition of illness must be arbitrary in any case, because the value of moods also has its continuum."

Payne describes the psychological experience of exuberance as a progression from active intoxication to a more transcendent peacefulness. Initially, she says, she feels "[h]igh, energetic. Excited. Passionate. Uninhibited. Garrulous if with someone else, ecstatic if alone." The exuberance progresses into "a state of serenity, during which I especially like to be alone—to lie in a hammock looking up into the leaves or stars, thinking and glorying and giving thanks for the wholeness and beauty of life."

Joyce Poole, the scientific director of the Amboseli Elephant Research Project in Kenya, is a colleague of Katy Payne's and, like her, an excellent writer. She has been a field biologist for nearly thirty years and has made key contributions to our understanding of elephants, including pioneering research on musth (behavior which is marked, in males, by high levels of testosterone and secretions from the temporal gland). She has also studied elephant feeding, reproductive, and social behavior; their vocal and olfactory communication patterns; the effects of poaching on the social structure of elephant families; and elephant genetics. An American who fell in love with Africa when her father was director of the Africa Peace Corps program, she was, she writes, an emotional child "quick to laughter, but also quick to anger and easily wounded and saddened. With my father's spirit of adventure and sense of principle and my mother's emotions, I frequently became deeply upset by the injustices of life."

While Poole was still a graduate student at the University of

Cambridge, her research on musth was acknowledged by a major award from the American Society of Mammologists and published in *Nature*. She finished her dissertation in record time because, as she put it, "I went into an intensely manic state, working until 2:00 a.m. most mornings and returning to the office again at 6:00 a.m." Poole took her formidable energy and scientific abilities back to Africa, where she became a consultant to the World Bank and the African Wildlife Foundation. In the early 1990s she was the coordinator of the Elephants Programme, working directly under Dr. Richard Leakey, for the Kenya Wildlife Service. Her ongoing research with Cynthia Moss in the Amboseli National Park, the longest study of wild elephants ever undertaken, follows the lives of more than a thousand individual animals.

Like many scientists, Poole describes her work as all-absorbing: "Cynthia and I spent the long days watching elephants and the evenings recounting elephant gossip by candlelight," she writes. "It was like reading a long novel that you don't want to end. I became so engrossed in the elephants' lives that on Sundays, when it was my duty to guard the camp, I used to will them to take the day off, too, so that I wouldn't miss any important events in their lives."

She remains transfixed by elephants: "What do elephants think about?" she asks. "What kind of emotions do they experience? Can they anticipate the future? Do they contemplate the past? Do they have a sense of self? A sense of humor? An understanding of death?" Other elephant people, she says, "suspect I believe I am, in fact, an elephant. In some ways, perhaps, they are right. Like Africa, the elephants take hold of your spirit." In her wonderful memoir, *Coming of Age with Elephants,* Poole's final acknowledgment is to the elephants she has studied over the years, "for giving me a life of meaning and days of joy." A delight in their company and an anguish in their destruction is palpable in her writing.

When I asked Poole which experiences from her work had

given her the greatest sense of exuberance, she replied: "Well, certainly the major discoveries that I have made give me, still, a great sense of buoyant and exuberant feelings. But also the day-to-day new ideas, or the recurrence of old ideas, about why elephants do the things they do, and their development into a plausible theory, give me a sense of exuberance. Some days I have none of these feelings; other days I may feel carried along by a fast-flowing stream of exciting ideas."

Sometimes something as simple as being able to identify a particular elephant at a distance of two kilometers will trigger a momentary sense of exuberance, she says, or being able to calm an attacking elephant by simply using her voice. The exuberance of being in the presence of elephants themselves is infectious. Poole describes the physical and psychological state of exuberance in vivid terms, with the observational eye of a good scientist. "I feel a rush of what I think of as adrenaline," she says. "I think rapidly, thoughts falling one over the other so fast that I can hardly keep up. My heart feels as though it is beating faster. Sometimes my body almost trembles with excitement. For some reason I am very aware of my hands and fingers—a tingling sensation in my body that seems concentrated there—perhaps this is because often I am writing or driving and so my hands are the 'doing' part of my body in these situations and therefore I am more aware of them. I am totally 'in the moment,' completely focused, effervescent, buzzing, bubbling, high, on top of the world, I can do anything. I CAN."

Like the scientists who study creativity and wonder whether creativity leads to exultation or exultation propels creative work, Poole is uncertain about the relationship between moods and imagination. "It is hard to say what triggers what," she observes of the overlap between her exuberance and times of discovery. "Do I have a glimmer of a good idea, and that stimulates the exuberance that is so associated with the initial 'spark' and the development of the idea? Or am I already in a creative phase when the glimmer of an idea strikes and the 'exuberance' flows? I don't know."

Hope Ryden, who studies and writes about many animals—among them coyotes, mustangs, bobcats, and beavers—distinguishes between the more energetic moments of joy she experiences at times of creativity and discovery and the quiet pleasure she feels when observing animals in the wild. "Of course," she says, "my response to what I see must be quiet, so as not to disturb my subjects. . . . Much of what I do is wait for something to happen, and exuberance is not compatible with that inactivity. When the action begins, however, I do feel a surge of excitement that is akin to exuberance. Nevertheless, when the animals learn to tolerate me and go about their business, my emotions settle and I watch quietly. What I feel, as I note odd bits of animal behavior, might best be described as enchantment and pleasurable empathy for my subjects who are so busily engaged in their struggle for survival."

The quiet mood can change abruptly, however: "I experience moments of exuberance when I see something never before recorded, or when I see something that contradicts a prevalent notion that has long been accepted as fact. At such times my competitive nature asserts itself and I feel jubilant." She feels that exuberance, as she experiences it—"a pleasurable upwelling of excitement, a speeding up of thought and action"—is not generally useful to her in the field. "While in this state," she explains, "the photographs I take are likely to be over- or underexposed or I may fail to notice that the film isn't properly threaded in my camera and so is not taking up. A little exuberance, while a highly pleasurable state, goes a long way for me. Mostly I try to tamp it down to something more like the pleasurable absorption I felt as a child while engaged in certain activities, such as building a sand castle. I have to conclude that watching wildlife is like play. The child in me loves to do it."

. . .

Science is beholden to the exuberance and curiosity typically associated with the adventurous spirit of childhood, and owes its furtherance to the temperamental quirks of individual scientists. James Watson, speaking at the Nobel Prize ceremonies in 1962, ended his remarks by talking about the human side of science:

> Good science as a way of life is sometimes difficult. It often is hard to have confidence that you really know where the future lies. We must thus believe strongly in our ideas, often to the point where they may seem tiresome and bothersome and even arrogant to our colleagues. I knew many people, at least when I was young, who thought I was quite unbearable. Some also thought Maurice was very strange, and others, including myself, thought that Francis was at times difficult. Fortunately we were working among wise and tolerant people who understood the spirit of scientific discovery and the conditions necessary for its generation. I feel that it is very important, especially for us so singularly honored, to remember that science does not stand by itself, but is the creation of very human people. We must continue to work in the humane spirit in which we were fortunate to grow up. If so, we shall help insure that our science continues and that our civilization will prevail.

We are beholden to the human side and enthusiasm of scientists; their passion in the pursuit of reason is heady and requisite stuff indeed.

CHAPTER EIGHT

"Nothing Is Too Wonderful to Be True"

Richard Feynman depressed, observed a colleague, was "just a little more cheerful than any other person when exuberant." He was, as well, a preternaturally original thinker, irrepressibly curious, and one of the great teachers in the history of science. For me, Feynman's legendary brilliance as a teacher is his most remarkable legacy, but then I am prejudiced. My grandmother, mother, aunt, and sister were teachers, and almost everyone else in the family—my grandfather, my father, an aunt, a great-uncle, a nephew, three cousins, my brother, and I—are, or were, professors. To teach well, I heard early and often, is to make a difference. To teach unusually well is to create magic.

It is a magic often rooted in exuberance. Great teachers infect others with their delight in ideas, and such joy, as we have seen—whether it is sparked by teaching or through play, by music, or during the course of an experiment—alerts and intensifies the brain, making it a more teeming and generative place. Intense emotion also makes it more likely that experience will be etched into memory.

Horror certainly does. We remember with too much clarity where we were and what we were doing when we first saw the hijacked planes fly into the World Trade Center and then, minutes later, as yet another crashed into the outer wall of the Pentagon. No one who saw this will forget. But joy, differently, also registers. We recollect moments of great pleasure and discovery—watching on a remarkable July night as the spider-legged lunar module dropped onto the moon's surface, for instance, or listening for the first time to Beethoven's *Missa Solemnis;* falling in love; playing through the long evenings of childhood summers—for nature has supplied us with the means to absorb the essential and to recall the critical. Teachers are among the earliest and most powerful of these means of knowing.

To teach is to show, and to show persuasively demands an active and enthusiastic guide. This must always have been so. Emotional disengagement in the ancient world would have been catastrophic if the young were to learn how to hunt, how to plant, and how to track the stars. The stakes were high: to motivate their wards would have been requisite for the first teachers—the parents and shamans, the priests and the tribal elders. Passing on knowledge about the behavior of the natural world from the emerging sciences of astronomy, agriculture, and medicine would have been among a society's first priorities. The great natural scientists, such as Hippocrates and Aristotle, Lucretius and Hipparchus, not only studied the ways of man and of the heavens but also taught what they knew to others. In turn, a few of those whom they taught observed the world in their own and slightly different ways, added to the stock of what was known, created a new understanding, and passed it on. "All vigor is contagious," said Emerson, "and when we see creation we also begin to create." When we see a glimpse of creation through the eyes of an enthusiast, our imaginative response is intensified.

Joy is essential to this. The pleasures of play, which attend and abet learning in young children, must perpetuate if they are to continue to reward inquisitiveness as children grow older. Thoreau wrote of his boyhood: "My life was extacy. In youth, before I lost any of my senses—I can remember that I was all alive. . . . This earth was the most glorious musical instrument, and I was audience to its strains. To have such sweet impressions made on us, such extacies begotten of the breezes. I can remember how I was astonished. I said to myself, —I said to others, —'There comes into my mind or soul such an indescribable, infinite all absorbing, divine heavenly pleasure, a sense of elevation & expansion—and [I] have had nought to do with it. I perceive that I am dealt with by superior powers. This is a pleasure, a joy, an existence which I have not procured myself—I speak as a witness on the stand and tell what I have perceived.' " Responsiveness to the natural world gives rise to knowledge; it begets curiosity and exploration.

Nature is the first tutor. No one remains untouched or unschooled by the earth, seasons, and heavens. In *The Education of Henry Adams,* Adams writes that to a New England boy "Summer was drunken." Each of his senses, he says, was assaulted and charged. Smell was the strongest, the unforgettable scents of "hot pine-woods and sweet-fern in the scorching summer noon; of new-mown hay; of ploughed earth; of box hedges; of peaches, lilacs, syringas; of stables, barns, cow-yards; of salt water and low tide on the marshes." And then came taste. The children, said Adams, "knew the taste of everything they saw or touched, from pennyroyal and flagroot to the shell of a pignut and the letters of a spelling-book." Smell and taste, touch and sight, are, of course, the ancient mammalian paths of learning: a whack of paws, a pungent scent, a shriek: all teach fast and enduringly. Adams writes that even sixty years out of childhood he could revive on his tongue the memory of his spelling book, "the taste of A-B, AB," and with ease and a kind of sweet ferocity could recall the colors taught him by the summer days. A peony and a sense of blue, he recollected late in life, would always be for him the sea near Quincy and "the cumuli in a June afternoon sky." The happiest hours of his education were, he said, "passed in summer lying on a musty heap of Congressional Documents in the old farmhouse at Quincy, reading . . . and raiding the garden at intervals for peaches and pears. On the whole [I] learned most then." Like most, he learned best under the spell of nature.

Science, like the arts, is rooted in the desire to understand and then create; society requires that this desire be transferred to succeeding generations. It is teachers who convey it, especially exuberant ones. Exuberance is a dispositional thread in the lives of great teachers and scientists, and most of them, in turn, kindle enthusiasm and curiosity in their students. Great scientists who are also extraordinary teachers—Humphry Davy, Michael Faraday, and Richard Feynman, for example—give us a glimpse into how a passionate temperament and a love of discovery fire up enthusiasm in those they teach. (It goes both ways, of course. "The justification for a university," said Alfred North Whitehead, "is that it preserves the connection between knowledge and the zest for life, by uniting the young and the old in the imaginative consideration of learning." The enthusiastic young act upon those by whom they are taught.)

Not all brilliant scientists are spellbinding teachers, of course. Newton famously was not. When he lectured, a Cambridge contemporary declared, "so few went to hear Him, & fewer y^t understood him, y^t oftimes he did in a manner, for want of Hearers, read to ye Walls." Newton was temperamentally unsuited to be a great teacher. He was by nature solitary and wary of others, and whatever passion he had for his work he never directly transmitted to the students who might have been inspired by it.

But Newton's ideas, of course, were a different matter. One hundred years later, even poets were caught up in the intellectual excitement generated by his discoveries. Lord Byron, whose personality was the opposite of Newton's in nearly every particular, was fascinated by the observations of Newton and other scientists: "When Newton saw an apple fall," Byron wrote marvelously in *Don Juan*,

he found
In that slight startle from his contemplation
. . . .
A mode of proving that the earth turned round
In a most natural whirl called 'Gravitation;'
And this is the sole mortal who could grapple,
Since Adam, with a fall, or with an apple.

(It is a pity Byron never taught, for the same vivacity and wit that Newton lacked Byron had in glut.)

Byron and his fellow poets were not alone in being swept up in the nineteenth century's excitement at scientific discovery. In Europe and America the public paid rapt attention to the remarkable progress of scientists as they began to unravel the workings of the natural world. Scientific knowledge was widely disseminated by popular writers and journalists, and the impact of science on daily life was readily apparent from advances in agriculture, technology, and medicine. Interest in science was further galvanized by public talks on scientific topics.

Indisputably, the dynamic lectures given by Sir Humphry Davy at the Royal Institution, which had been established in 1799 to encourage the dissemination of scientific knowledge to the general public, were among the most influential and widely attended in history. Davy's enthusiastic teaching had a lasting and profound impact on scientists and nonscientists alike. He was, says one biographer, "a star"; when he lectured "buckles flew, stays popped." Those who knew Davy personally were struck by his vivacity and unstoppable enthusiasm. He "talks rapidly, though with great precision," said one observer, "and is so much interested in conversation that his excitement amounts to nervous impatience and keeps him in constant motion." Robert Southey concurred, noting, "I have never witnessed such indefatigable activity in any other man." He was, remarked more than one acquaintance, a flurry of ideas. Michael Faraday, who began his scientific career as Davy's laboratory assistant, said that Davy's temperament was passionate and restless; so, too, was his mind, which, according to the president of the Royal Institution, was "naturally ardent and speculative."

Davy's restless liveliness, a catalyst to his scientific imagination and integral to his gifts as a teacher, had its drawbacks: he was often irritable and impatient and difficult to satisfy. An acquaintance, later his biographer, remarked that fly fishing, perhaps not surprisingly, was ill-suited to Davy's disposition: "The temperament of Davy was far too mercurial; the fish never seized the fly with sufficient avidity to fulfill his expectations, or to support that degree of excitement which was essential to his happiness."

Davy's scientific achievements were as fundamental as they were diverse. He showed that by using electricity, chemical compounds could be decomposed into their constituent parts, and he isolated, for the first time, the elements sodium, calcium, barium, potassium, and magnesium. He also demonstrated, despite entrenched scientific opinion to the contrary, that iodine and chlorine were elements; he described in detail the physiological effects of many gases (including, as we have seen, nitrous oxide), wrote influential scientific texts, and invented a widely used safety lamp for miners. (Davy refused to take out a patent on the lamp because, he said, the sole object of his invention was to "serve the cause of humanity." His public-spiritedness is this regard is not unlike that of Benjamin Franklin, who chose not to patent the lightning rod or the Franklin stove, and of the Stevensons in Scotland, who declined to patent their inventions in lighthouse technology.)

Davy's response to his scientific discoveries was unconstrained exuberance. When he first isolated potassium and saw the "minute globules of potassium burst through the crust of potash and take fire as they entered the atmosphere," Davy's cousin reports, he "could not contain his joy—he actually bounded about the room in ecstatic delight . . . some little time was required for him to compose himself sufficiently to continue the experiment." Discoveries, declared Davy, "are like blessings of heaven, permanent and universal . . . by learning man ascendeth to the heavens and their motions, where in body he cannot come."

Such excitability, together with Davy's literary sensibilities and scientific accomplishments, were enough to attract the fascination not only of Byron, who engaged in animated dinner conversations with him about volcanoes and gases, but also of Percy Bysshe Shelley and his wife, Mary, who consulted Davy's *Elements of Chemical Philosophy* as she wrote *Frankenstein*. Samuel Taylor Coleridge was also captivated by Davy and declared that he had "never met so extraordinary a young man." Had he turned his mind to it, said Coleridge, Davy would have been the "greatest poet of his age." Almost everyone recognized Davy as one of the preeminent teachers of his time; the Church of England, impressed by his passion and sense of theater in the lecture hall, implored him to become a preacher.

Davy was, in fact, all the things he was said to be—preacher, teacher, poet, and scientist—but nowhere more impressive than when he lectured to the public about chemistry. His talks at the Royal Institution were a sensation. The lecture theater, which held a thousand people, was nearly always packed by the time he arrived. Carriages blocked the streets nearby, and the crowds were so great in Albemarle Street that, in order to accommodate traffic, it had to be converted into a one-way thoroughfare, the first ever in London.

A London tanner who attended the talks gives a sense of the excitement they generated: "The sensation created by his first course of Lectures at the Institution, and the enthusiastic admiration which they obtained, is . . . scarcely to be imagined. Men of the first rank and talent,—the literary and the scientific, the practical and the theoretical, blue-stockings, and women of fashion, the old and the young, all crowded—eagerly crowded the lecture room." Davy infected the audience with his passion for science. He was a natural storyteller, as he had been even when very young: "I was seized with the desire to narrate," Davy wrote about his childhood. "I gradually began to invent, and form stories of my own. Perhaps this passion has produced all of my originality." Certainly passion fueled his curiosity and his enthusiastic talks on science. His lectures were punctuated by his own surges of enthusiasm as well as by the dramatic explosions of gases that he created in order to demonstrate principles of chemistry and electricity.

Humphry Davy, like his apprentice Michael Faraday and, a century and a half later, Richard Feynman, stoked enthusiasm in his listeners and was in turn stoked by theirs. And, also like Faraday and Feynman, Davy was an impassioned teacher because he was indefatigably curious and excitable. His zest for learning ignited those who heard him speak. He transferred to his audience his love of the natural world and his profound appreciation for its enchantments. He gave his first lectures at the Royal Institution when he was twenty-three, and he spoke then, as later, about the beauty of natural law: "The appearances of the greater number of natural objects are originally delightful to us, and they become more so when the laws by which they are governed are known," he said. "The study of nature, therefore, in her various operations must be always more or less connected with the love of the beautiful and sublime." Davy was by birth an enthusiast, who loved science, knew its beauties, and conveyed without check his enthusiasm to those he taught.

One of those ignited by Davy's exuberant lecturing was Michael Faraday, the son of a blacksmith, who had taught himself science by reading books on chemistry and physics. He attended Davy's lecture series at the Royal Institution when he was scarcely twenty and took such lucid, fastidious notes, nearly four hundred pages of them, that when he sent them to Davy and asked for a position as his assistant, Davy took him on. During the decades to follow Faraday did the work that established him as one of the greatest scientists in history (Einstein said that there were four physical scientists who towered above all others: Galileo, Newton, Faraday, and Maxwell). He was the first to conceptualize magnetic fields and the first to discover the laws of electromagnetic conduction and rotation, as well as the laws of electrolysis. Faraday, in short, gave the world its fundamental understanding of electricity.

But Faraday was also a great teacher. Like Davy, he believed passionately that the ways and findings of science should be communicated to the public and that preparation for public lectures should be rigorous. He wrote a "Manual for the Lecturer," which presented in painstaking detail the desired physical layout of lecture halls—their ventilation systems, entryways, and exits, and the arrangement of the experimental apparatus to be used in teaching demonstrations. Lecturers, he believed, owed it to their audiences to be not only prepared but also stimulating. "A lecturer should exert his utmost effort to gain completely the mind and attention of his audience," he wrote. "A flame should be lighted at the commencement and kept alive with unrelenting splendour to the end."

Faraday's first talk at the Royal Institution was widely acclaimed as brilliant. Later he would come to be known as a supreme showman, the "Prince of Lecturers" in Victorian England. Charles Darwin and Charles Dickens were among the thousands who flocked to hear him, and the painter J.M.W. Turner consulted him about the chemistry of pigments. He founded the Christmas Lectures for Juveniles, a prestigious series which continues to this day, broadcast by the BBC. He gave his first Christmas lecture in 1827, and his last in 1860, with such titles as "The Chemistry of Coal," "Water and Its Elements," and "Atmospheric Air and Its Gases." The talks were wildly popular, and none more so than his most famous lecture, "On the Chemical History of a Candle," which was first delivered to an audience of young people in 1849. There were "bangs, flashes, soap bubbles filled with hydrogen floating upwards, and other spectacular effects" and, in a demonstration not unlike the drama to be provided by Richard Feynman during the twentieth-century NASA *Challenger* hearings, Faraday "placed two vessels made of half-inch-thick iron, and filled with water, in a freezing solution, then went on lecturing until the vessels exploded."

Explosions, flashes, and bubbles were only part of Faraday's genius at lecturing, however. Like that of so many other enthusiasts—Snowflake Bentley, John Muir, Richard Feynman, Humphry Davy—Faraday's work and teaching were fueled by an exuberance for science. He, like they, maintained the joy in discovery that children have in play; he taught best, as children learn best, because he had the ebullient inquisitiveness so central to youthful exploration. "The Christmas lectures brought out the boy in Faraday," writes his biographer James Hamilton, "his 'wonderful juvenility of spirit' being never far from the surface: 'Hilariously boyish upon occasion he could be, and those who knew him best knew he was never more at home, that he never seemed so pleased, as when making an old boy of himself, as he was wont to say, lecturing before a juvenile audience at Christmas.' The fun came to a climax in many ways, most noisily perhaps in Faraday's flinging a coal scuttle full of coal, a poker and tongs at an electro-magnet to demonstrate the powers of magnetism." Faraday said simply, "I claim the privilege of speaking to juveniles as a juvenile myself."

Faraday punctuated his lectures not only with explosions and dramatic hurlings but with bursts of his own enthusiasm. "Wonderful" and "beautiful" emerge time and again to describe his delight in nature. In Faraday's attempts to link gravity, magnetism, and electricity he wrote memorably, "Nothing is too wonderful to be true" (a statement he qualified by the less often quoted, "if it be consistent with the laws of nature, and in such things as these, experiment is the best test of such consistency"). Indeed, an irrepressible sense of wonder finds its way into his talks. During the sixth and final lecture in his Christmas series "On the Chemical History of a Candle," for instance, Faraday's language is charged with that sense of wonder, of beauty and joy: nature is wonderful, beautiful, and curious.

"*Wonderful* it is," he said to his young audience when explaining the conversion of carbon into carbonic acid, and "*wonderful,*" he said again, is the change provided by respiration, the life and support of plants and vegetables. He continued, during the candle's burning: "What a *wonderful* change of carbon must take place . . . what a *curious, wonderful* change." Powdered lead burning in air, he said, is a "*beautiful* instance" of chemical affinity; Japanese candles do not start into action at once, they wait for years or ages, and this waiting is a "*curious and wonderful* thing." "See how *beautifully* [the candles] are colored," he went on. They are "most *beautifully* shaped" and, if watched carefully, there is a "*beautifully* regular ascending current of air." Rubies and diamonds cannot rival the remarkable beauty of a candle's flame, he enthused; the light is "*glorious.*" The use of "beautiful" and "wonderful" continues: a process of nature is not just a process, said Faraday, it is a "*beautiful* and simple process." The analogy between respiration and combustion is "*beautiful and striking,*" he tells his audience; "*what an extraordinary notion,*" what a "*most curious and beautiful* [thing] it is to see." Faraday's language, if stripped of "wonderful," "curious," "extraordinary," and "beautiful," which I have italicized for emphasis, would be plain speech indeed; it would also have been speech unlikely to infect an audience with the enthusiasm for science that he so deeply felt. Nature, he knew as well as anyone, was never too wonderful to be true.

A century after Faraday's last Christmas lecture at the Royal Institution, Richard Feynman gave three lectures at the University of Washington in Seattle. The imagination of nature, he told his audience, is always far greater than the imagination of man. He contrasted the scientific view of the natural world to the one contrived by the ancients: their belief that the earth was the back of an elephant that stood on a tortoise that swam in a bottomless sea, he said, was the result of imagination. But truth is even more marvelous: "Look at the way we see it today. Is that a dull idea? The world is a spinning ball, and people are held on it on all sides, some of them upside down. And we turn like a spit in front of a great fire. We whirl around the sun. That is more romantic, more exciting."

Feynman's exuberance about science, about life, and about discovery, together with his rapier, intuitive intellect made him a celebrated teacher. Robert Oppenheimer, who worked with him at Los Alamos, described him in 1943 as "the most brilliant young physicist here, and everyone knows this . . . and [he is] an excellent teacher." Later, when trying to recruit him to Berkeley, Oppenheimer specifically cited Feynman's lucidity as a teacher and recommended him as "a rare talent and a rare enthusiasm." Those who heard him teach would resoundingly echo Oppenheimer's early observations. The physicist David Goodstein, vice provost of the California Institute of Technology, where Feynman was on the faculty, said, "I think Dick was a truly great teacher, perhaps the greatest of his era and ours." For Feynman, he said, "the lecture hall was a theater, and the lecturer a performer, responsible for providing drama and fireworks as well as facts and figures." His graduate students agreed. Laurie Brown, who went on to become a professor of physics and astronomy, said that Feynman stressed creativity: "He urged each of us to create his or her universe of ideas. . . . It was excitingly different from what most of us had been taught earlier."

Feynman was an exuberant teacher in every way. His enthusiasm and curiosity spilled over onto those whom he held captive in his orbit. He thought the quest to know the laws of the universe was the most exciting adventure a person could undertake. "I'm an explorer," he once said, "I like to find out." The natural world was to him wonderful, beautiful, and an object of endless play.

Feynman was the ultimate scientific galumpher. Science was fun for him, and he made science fun for those he taught. Freeman Dyson observed, "I never heard him give a lecture that did not make the audience laugh." For Feynman, laughter, excitement, and scientific imagination were inseparable. His sister said of their childhood that there was "this excitement in the house, this great love of physics. . . . The feeling of excitement was in the house all the time, in my brother and my father. So I just grew up with it."

His mother, Feynman wrote, taught him that one of the highest forms of understanding was laughter; in like vein, his thesis adviser at Princeton observed that for Feynman, "discussions turned into laughter, laughter into jokes and jokes into more to-and-fro and more ideas." All his life, his sister said, Feynman did physics for fun: "When people asked him how long he worked each week, he really couldn't say, because he never knew when he was working and when he was playing." Not surprisingly, perhaps, Feynman was the "favorite adult playmate" of the children of his colleagues.

Even at his last public appearance, just a few months before he died of cancer, a fellow physicist was amazed by Feynman's joy in talking about teaching high school physics: "As I watched, I realized I was witnessing something extraordinary. Feynman's energies grew as he responded to question after question. The outside corners of his eyes were creased by the smiles that played over his face as he talked about physics. His hands and arms cut through the air with increasing vigor. . . . It was the enjoyment he exuded as he stood there talking physics with an eager, receptive group of physics teachers that moved me. It was an enjoyment I could feel. . . . I had the feeling that I was standing on holy ground."

Feynman utterly enjoyed the idea of things. Science, he said, is "done for the excitement of what is found out. . . . It is almost impossible for me to convey in a lecture this important aspect, this exciting part, the real reason for science." You cannot understand science, he emphasized, unless you "understand and appreciate the great adventure of our time . . . [this] tremendous adventure . . . [is] a wild and exciting thing." Life, for Feynman, was the fabulous pursuit of truth and beauty and joy:

> It is a great adventure to contemplate the universe, beyond man, to contemplate what it would be like without man, as it was in a great part of its long history and as it is in

> a great majority of places. When this objective view is finally attained, and the mystery and the majesty of matter are fully appreciated, to then turn the objective eye back on man viewed as matter, to view life as part of this universal mystery of greatest depth, is to sense an experience which is very rare, and very exciting. It usually ends in laughter and a delight in the futility of trying to understand what this atom in the universe is, this thing—atoms with curiosity—that looks at itself and wonders why it wonders. Well, these scientific views end in awe and mystery, lost at the edge in uncertainty, but they appear to be so deep and so impressive that the theory that it is all arranged as a stage for God to watch man's struggle for good and evil seems inadequate.
>
> Some will tell me that I have just described a religious experience. Very well, you may call it what you will.

Like John Muir, who could not keep "glorious" from popping out of his inkwell, and Snowflake Bentley, who didn't even try, Feynman's exuberant affair with nature bubbled over into his writings and into the lecture hall. "The world is so wonderful," he exclaimed, using the word that dotted Faraday's talks at the Royal Institution. To trace out the origins of man and the universe, Feynman believed, is to be held spellbound; it is to stand in awe, to fill with delight: "Where did the stuff of life and of the earth come from? It looks as if it was belched from some exploding star, much as some of the stars are exploding now. So this piece of dirt waits four and a half billion years and evolves and changes, and now a strange creature stands here with instruments and talks to the strange creatures in the audience. What a wonderful world!"

When the chairman of a conference attended by Feynman proclaimed that scientists should teach what is known about science and not the "wonders of science," Feynman passionately disagreed. "I think we should teach them wonders," he insisted, "that the purpose of knowledge is to appreciate wonders even more. And that the knowledge is just to put into correct framework the wonder that nature is." The thrill and mystery of nature, he said, "come again and again," and with more knowledge comes "deeper, more wonderful mystery." (Einstein spoke of this love for the mysterious as well. "The most beautiful and deepest experience a man can have is the sense of the mysterious," he said in Berlin, in 1932. "It is the underlying principle of religion as well as all serious endeavour in art and science. He who never had this experience seems to me, if not dead then at least blind." Einstein, like Feynman, felt that the mystery of nature carried with it a trace of religion: "To sense that behind anything that can be experienced there is something that our mind cannot grasp and whose beauty and sublimity reaches us only indirectly and as a feeble reflection, this is religiousness. In this sense I am religious.")

Feynman knew—his life was a testament to it—that exploration leads to pleasure and that such pleasure, in turn, leads to more discovery; that wonderful mysteries lure one "to penetrate deeper still . . . we turn over each new stone to find unimagined strangeness leading on to more wonderful questions and mysteries—certainly a grand adventure!" He addressed Thoreau's pessimism about education—that it "makes a straight-cut ditch of a free, meandering brook"—with the expansive love and optimism of a wondering mind. (Carl Sagan, another great teacher of science, agreed that to teach science is to teach awe. "Not explaining science seems to me perverse," he once said. "When you're in love you want to tell the world." When we understand science, he continued, "we're moved—because in its encounter with Nature, science invariably conveys reverence and awe.")

Feynman dazzled his audiences with intellectual cape-work and held them riveted by wit and energy. Mostly, he infected audiences with his own joy in discovering the "beautiful things" of nature. Exuberance is beauty, declared Blake, and so Feynman showed it to be. "I'm delighted with the width of the world!" he said, and he found great delight in sharing his joy with others: "I love to teach. I like to think of new ways of looking at things as I explain them, to make them clearer—but maybe I'm not making them clearer. Probably what I'm doing is entertaining myself."

The world outside the lecture halls of Caltech needs enthusiastic teachers more than did the math and science students kept enthralled by Richard Feynman. In *The Water Is Wide*—a devastating, often hilarious, but finally heartbreaking account of a year fighting dim-witted education bureaucrats while trying to teach impoverished African-American schoolchildren living on an island off the South Carolina coast—the novelist Pat Conroy makes this point painfully clear. When Conroy first met with the children assigned to him, he found that they had been taught next to nothing. They were essentially illiterate; their expectations of life were bleak and their notions of themselves worse. The school system had given up on them, if indeed it had ever tried at all. Many of his students could not recite the alphabet, he discovered by the end of his first day of teaching, and several could not spell their own names. The litany of neglect went on: most of the children thought John F. Kennedy was the first president of the United States and they "concurred with the pre-Copernican Theory that the earth was the center of the universe"; two children did not know how old they were.

Conroy went through an early and understandable period of complete discouragement: "What could I teach them or give them that would substantially alter the course of their lives? Nothing. Not a goddam thing. Each had come into the world imprisoned by the water binding them to their island and by a system which insured his destruction the moment he uttered his first cry by his mother's side." Learning not only was not fun, it was at its best drudgery; at worst it came with whippings and humiliation. It seemed impossible to instill joy in such forgotten children. Conroy was saved, in the end, by his own enthusiasm for life and an emerging belief that a "pep-rally" method of education could work. He would transfuse his energies into his students: "All right, young cats," he told his young students, "we are about to embark on a journey of knowledge."

Their journey was led by a wildly enthusiastic young teacher who believed that learning could be fun, that "life was good, but it was hard; we would prepare to meet it head on, but we would enjoy the preparation." Conroy used everything he could find to cast a spell, to pique curiosity, to encourage exploration: he and his students listened to music, went to the theater, and watched films; they took field trips and talked about witches. They danced; they sang. They were, Conroy exorted them, "going to have more twenty-four-carat fun than any group in the long history of mankind." His expansiveness pushed them not just onward but upward. He railed against the philosophy of his predecessor, which had been in effect: Keep them busy. We are not here to have fun. We are here to educate. (Henry Adams would have recognized this philosophy. "The chief wonder of education," he had said, "is that it does not ruin.")

What was perhaps of most lasting value, Conroy tried to give his students a sense of the glory of their own island; how to observe and learn from the natural beauties of its plants and sea animals; how to know and to love the beach and gardens. He tried, he writes, to teach them to "embrace life openly, to reflect upon its mysteries, rejoice in its surprises, and to reject its cruelties." He failed to do much of what he set out to do, of course; perhaps this is inevitable in shooting for the moon. But a teacher need not do all that may at first seem essential: if he is exuberant, some of the joy will stick; if optimistic, some of the hope will linger; if delighting in the adventure of life, a bit of that excitement will obtain.

To teach unusually well is indeed to create magic. Teaching is a first and noble calling. "Lord, I am a teacher and a coach," proclaims Pat Conroy's protagonist at the end of *The Prince of Tides:* "That is all and it is enough."

Parents are our first and most enduring teachers. My father was studying for his Ph.D. at the University of California–Los Angeles when he looked up and noticed a quotation from Michael Faraday carved into an archway over the physics building. He was completely enchanted and insisted on taking me to see it as well. I loved the quotation as much as he did and often sought it out when I was a student and then a professor at UCLA. Years later, however, with a jadedness I didn't entirely feel, I wrote that life hadn't always been so wonderful for Faraday—mentioning, among other things, his nervous breakdowns—and that his observation, in any event, was demonstrably untrue. My father read my book, said he felt otherwise, and sent me a note. He would like to share with me, he said, Faraday's "affirmation." Enclosed with his note was a gold locket from Tiffany; it was inscribed, *"Nothing is too wonderful to be true—Michael Faraday."*

CHAPTER NINE

"We Should Grow Too Fond of It"

There are dead ideas and cold beliefs, wrote William James, and then there are hot and live ones. When an idea "grows hot and lives within us," he believed, everything must recrystallize around it. The exuberant life, bursting as it does with feverish beliefs, is one of constant recrystallization; in this lies much of its value, complexity, and potential danger.

Passionate enthusiasms, I have argued throughout this book, are as essential to survival as they are indispensable to imagination and social change. Passions bring to our attention the overlooked; they compel commitment of time and heart. They persuade by sheer dominance of the emotional and mental field, for they are, as the Harvard scholar Philip Fisher observes, "monarchical" in their power: they drive out other states of being, even those that are themselves extreme. Man's exhilaration in war, for example, serves in part to impede its terror; curiosity countervails fatigue and setback; and the thrill of the chase acts to overwhelm the hunter's fear of what he hunts.

That which is most deeply felt is also most powerfully expressed to others. "We cannot write well or truly but what we write with gusto," said Thoreau. " The body the senses must conspire with the spirit—Expression is the act of the whole man. that our speech may be vascular." But our beholdenness to passion assures a darker side. Exuberance can veer sharply into disturbing territory. Champagne enchants, but it also intoxicates more quickly than stiller wines: heed glides into heedlessness as effortlessly as the silk chemise drops to the floor. The things that excite contain the capacity for excess and the potential to shame or devastate. Enthusiasm shares a border with fanaticism, and joy with hysteria; exuberance lives in uncomfortable proximity to mania. Exuberance, as Shakespeare wrote of music, "hath such a charm / To make bad good, and good provoke to harm." Thwarted or deviant enthusiasms, once provoked, are powers to reckon with.

The fever of passion itself is not the difficulty, argued William James; rather, trouble lies in the nature of the passion and how well it holds up to the light of day. "Surely the fever process as such is not the ground for our disesteem," he wrote. "For ought we know to the contrary, 103° or 104° Fahrenheit might be a much more favorable temperature for truths to germinate and sprout in, than the more ordinary blood-heat of 97 or 98 degrees. It is the disagreeableness itself of the fancies, or their inability to bear the criticisms of the convalescent hour."

Disagreeable fancies are irksome at best and calamitous at worst. Too ardent or misdirected exuberance creates mayhem for the individual and exposes others to the possibility of mishap, if not actual danger. Unchecked, enthusiasm runs roughshod over reason and intrudes into the private emotional territory of others, imposing, as it goes, its own energy and tempo. Exuberance whips its way in, dominant, and forces itself upon those trapped in its eddy. At its best, it is infectious and enlivening; at its worst, it stifles the ideas and feelings of the less exuberant. Not everyone delights in delight, especially if it is not their own, and few wish to have their moods hijacked by those of others. Sustained or nuanced social interactions are difficult in the presence of great exuberance, and indiscriminate enthusiasm hinders the discernment necessary to sort out true friend from possible foe. The lack of fixity creates discomfort and mistrust: the mobility of mind and attachment that is artistically helpful may not prove an asset in other circumstances. Like Browning's Last Duchess, who had "A Heart—how shall I say?—too soon made glad, / Too easily impressed; she liked whate'er / She looked on, and her looks went everywhere," the exuberant are easily engaged. And exuberance is, in its very effusiveness, liable to misconstruction and suspicion, often misinterpreted as sexual interest when none is intended, or as implying a more sustained emotional commitment than is warranted by the high spirits that, however persuasive, may prove to be transient or directed in any number of places.

Those who are most exuberant are often acutely aware of the toll their enthusiasms can take. The late J. Carter Brown, who was for many years the director of the National Gallery of Art in Washington, D.C., was a great enthusiast who attributed much of his passion for art to his exuberant temperament. Brown's family motto is "Gaudeo" ("I rejoice"), and those who knew him appreciated how true to its bearer that was. After Brown's death in 2002, Mark Leithauser, the chief of design at the National Gallery, said of him that "there was this restless creativity, this striving. . . . He loved what he was doing and that was infectious to everybody around him." One day, Leithauser recalled, as he was walking by the Mall and the Washington Monument, he saw Brown "pull up in his BMW and stop right in traffic, and jump out of the car and run among the cherry blossom trees, just taking them in, enjoying them." Paul Richard, an art critic for the *Washington Post,* noted "something buoyant, almost boyish about [him]." Brown, he said, "thought sprightly thoughts." He was a man of both grace and zest.

Carter Brown was mindful, however, that not everyone found his energy to their liking (although most who knew him certainly did). His tendency, as he put it, to "lope into others' pastures" was, he acknowledged, not infrequently experienced as "grating." Brown, who could no more keep his enthusiasm in check than an otter can keep to the riverbank, believed that his exuberance was an integral part of his leadership of the National Gallery, but he was also aware that it caused envy in some and made others feel overwhelmed. Brown said he tried to slow down his speech and to keep his long arms and hands from waving into the "emotional space" of other people, but that it was an uphill fight. He was a WASP who did not always keep to the constrained ways of the WASP. But then, as the Harvard theologian Peter Gomes has observed, Carter Brown was also a preacher. "His text was beauty," said Gomes, "and his parish was the nation."

Richard Feynman spoke of a similar daunting-by-dint-of-enthusiasm phenomenon in his role as a teacher. Although a brilliant lecturer, he found that the eclipsing power of his enthusiasm made it difficult at times to allow his research students the room to develop their own ideas. "I've put a lot of energy into my students," he said, "but I think I wreck them somehow." He was constitutionally incapable of standing back from an intriguing scientific problem; on more than one occasion, he took up the dissertation topic or other intellectual problem assigned to a student, wrapped his formidable brain around it, and solved it more quickly or elegantly than the student could. Feynman was a magician, according to the mathematician Marc Kac, and magicians "seldom, if ever, have students because they cannot be emulated and it must be terribly frustrating for a brilliant young mind to cope with the mysterious ways in which the magician's mind works." The very enthusiasm and curiosity that made Feynman's ideas and lectures magical were, for at least some of his graduate students, also intimidating. Feynman was, as Rabbit had said of Tigger, the sort "who was always in front when you were showing him anywhere."

Exuberance disturbs in other ways as well. Without a counterweight or discipline, it can be dazzling scattershot: excitement without substance, all fizz and no gin. When enthusiasm lacks a fuller emotional or intellectual context, it lends credibility to those, more circumspect, who contend that high spirits and unrelenting optimism are intrinsically shallow, *pas sérieux,* lacking the gravitas of the tragic or heroic, wanting in grandeur, not struggling with profound issues of humanity, not contending with the shadows cast by death. Exuberance is not an inward-looking state; it looks upward and forward, rarely to the past. Disquieting emotions are overpowered by the excitement of the idea or the moment; the past cedes its territory to the present and future.

Leon Wieseltier, in his remarkable book *Kaddish,* derides what he sees as the American preoccupation with moving on, "closure," tidying up painful experiences and memories. He is not speaking of exuberance, but his concern, the danger of disregarding the essential lessons of the past, is germane: "Americans really believe that the past is past," he writes. "They do not care to know that the past soaks the present like the light of a distant star. Things that are over do not end. They come inside us, and seek sanctuary in subjectivity. And there they live on, in the consciousness of individuals and communities." The forward thrust of exuberance, like closure, risks leaving behind an essential past.

"Rainbows flowered for my father in every sky," wrote Wallace Stegner of his exuberant father. "Led by pillars of fire and cloud," the older Stegner was a "boomer, a gambler, a rainbow-chaser, as footloose as a tumbleweed in a windstorm." Because his father was so restless, Stegner knew constant motion as a child. It was only as he grew older that he fully realized the drawbacks of unrest and appreciated the importance of his mother's desire for a sense of place, for settling long enough to know the land and other people. The swirl of movement was exciting, but it was in putting down roots that he came to know himself and to understand the attraction and failures of the American West.

Western expansion, Stegner believed, was a case study in unfettered exuberance, one that had led to reckless exploitation of the land. The dangers from the "come-all-ye enthusiasm" and restless wanderings of men like Stegner's father were real. Moving away from the past may have been essential, but it was action that carried a cost. "Complete independence, absolute freedom of movement, are exhilarating for a time," cautions Stegner, "but they may not wear well." It is, he continues, "probably time we settled down. It is probably time we looked around us instead of looking ahead. . . . History was part of the baggage we threw overboard when we launched ourselves into the New World. We threw it away because it recalled old tyrannies, old limitations, galling obligations, bloody memories. Plunging into the future through a landscape that had no history, we did both the country and ourselves some harm along with some good. Neither the country nor the society we built out of it can be healthy until we stop raiding and running, and learn to be quiet part of the time, and acquire the sense not of ownership but belonging. . . . Only in the act of submission is the sense of place realized and a sustainable relationship between people and earth established."

Many who are not exuberant are wary of those who exhibit unbridled enthusiasm, not only because of their insensitivity to personal and social history, but because their interests are strewn so widely as to be diffuse, themeless, and shallow. Exuberants, they believe, are easily excited—magpies who swoop down on glittering things—but lack the discipline or capacity to discriminate the meaningful from the superficial. Christopher Irmscher, in *The Poetics of Natural History*, makes this point about the ever-exuberant P. T. Barnum. Barnum's museum, hugely successful with the public, brimmed over with items that had caught Barnum's enthusiastic eye over the years—more than 100,000 "curiosities": a snuffbox made out of bits of the pulpit used by John Knox; a dried mermaid; the head of a fossil elk; anacondas; a Chinese summer tea house; a mammoth ox; and a "curious mortuary memorial" to Washington and Lincoln, which was ten feet high and made from more than two million seashells. But, as Irmscher notes, "[t]he principle of Barnum's museum is the profusion of sights, not the achievement of insight into a predetermined order of things." Even recurrent themes, he observes, "point less to an order intrinsic to the collection than to a meaning superimposed on it by the avuncular collector." An exuberant temperament by no means leads inevitably, or even usually, to shallowness, but the potential is there. Subtleties and sustained thought can be lost in a swirl of vivacious moods and energies, just as monomaniacal enthusiasm can limit awareness of and sensitivity to the perspectives, needs, and contributions of other people.

The exuberant brain is a hopping, electric place, a breeding ground for both invention and rashness. It is by nature impatient, certain, and high on itself; inclined to action rather than reflection; overpromising; and susceptible to dangerous rushes of adrenaline. The exuberant mind is also disinclined to detail, error prone, and vulnerable to seduction. All people, said Walter Bagehot, are most credulous when they are most happy; for someone who is exuberant, self-deception is just the next mountain over from credulousness. All seems possible, much seems essential, and unwarranted optimism feels fully warranted. Self-deception can then move, by conscious intent or not, into the deception of others. ("It is unfortunate, considering that enthusiasm moves the world," said the Earl of Balfour, "that so few enthusiasts can be trusted to speak the truth.")

However scamming or unsustainable the promise, we are easily swept up by contagious enthusiasm. People will gladly suspend disbelief for sizzle, dash, and a chance of escape. P. T. Barnum knew this to his marrow, as did Meredith Willson when he created the character of Professor Harold Hill, the razzler-dazzler of *The Music Man*. A traveling salesman who lives off the unrealized dreams of others, Hill sprinkles bright bursts of color over the drab of life and promises the undeliverable. He sweeps into River City and convinces the town, thanks to his verve and their credulousness, that without a boys' band, which he assures them he can provide (though he has no intention of doing so), its young people are headed straight to hell ("Ya got trouble, folks, right here in River City. Trouble, with a capital 'T' and that rhymes with 'P' and that stands for pool!"). Hill, like Barnum, is exuberant and charismatic, "a bang beat, bell-ringin', big haul, great go, neck-or-nothin, rip roarin', ever'time-a-bull's eye salesman." (The "fella sells bands. Boys Bands," explains one of his competitors, "I don't know how he does it but he lives like a king, and he dallies and he gathers, and he plucks and he shines and when the man dances, certainly, boys, what else: the piper pays him.") *The Music Man* captures the dupability of a town in need of pizzazz, as well as the persuasive powers of exuberance—"I spark, I fizz"—but it also makes clear the desire of the audience to suspend its own disbelief and to be persuaded, for an evening, that love can transfuse character into a con man, to believe that the exuberance to which they have responded is more than empty promise and undelivered dreams.

Reason takes a particular holiday when it comes to financial schemes and crazes. "Not only fools but quite a lot of other people are recurrently separated from their money in the moment of speculative euphoria," writes John Kenneth Galbraith in *A Short History of Financial Euphoria*. Such "recurrent speculative insanity," he believes, follows a predictable course: "Some artifact or some development, seemingly new and desirable—tulips in Holland, gold in Louisiana, real estate in Florida, the superb economic designs of Ronald Reagan—captures the financial mind or perhaps, more accurately, what so passes. The price of the object of speculation goes up. . . . This increase and the prospect attract new buyers; the new buyers assure a further increase. Yet more are attracted; yet more buy; the increase continues. . . . Something, it matters little what—although it will always be much debated—triggers the ultimate reversal. Those who had been riding the upward wave decide now is the time to get out." The bubble bursts; disaster follows. Built into such speculative episodes, Galbraith asserts, is euphoria, "the mass escape from reality, that excludes any serious contemplation of the true nature of what is taking place." Such "irrational exuberance," a term made famous by Chairman of the Federal Reserve Board Alan Greenspan in a 1996 talk about the stock market, is defined by the Yale economist Robert Shiller as "wishful thinking on the part of investors that blinds us to the truth of our situation."

In the high-flying world of boom, bubble, and bust, no speculative mania has been more dazzling than that for the tulip, a seventeenth-century mania that truly blinded investors to truth. Neither the first nor the last flower to incite preposterous financial trading—hyacinths and ranunculus and gladioli have had their day; a single dahlia was exchanged for a diamond in nineteenth-century France; red spider lilies set off a speculative craze as recently as 1985 in China, where, according to the historian Mike Dash, the most desirable lilies sold for prices "equivalent to no less than three hundred times the annual earnings of the typical Chinese university graduate"—but nothing has claimed the public's imagination like the Dutch tulip hysteria.

Charles MacKay, in *Extraordinary Popular Delusions and the Madness of Crowds,* wrote about tulip mania, among other all-consuming enthusiasms: "In reading the history of nations, we find that, like individuals, they have their whims and their peculiarities; their seasons of excitement and recklessness, when they care not what they do. We find that whole communities suddenly fix their minds upon one object, and go mad in its pursuit; that millions of people become simultaneously impressed with one delusion, and run after it, till their attention is caught by some new folly more captivating than the first." And, he remarks: "Men, it has been well said, think in herds; it will be seen that they go mad in herds, while they only recover their senses slowly, and one by one."

The passion for tulips was remarkable. People sold their houses and other possessions in order to invest in tulip bulbs; eventually, tulips were themselves used as cash. The "aberrant enthusiasm," as Galbraith would have it, spread like a fever, and to a degree unimaginable. But why the tulip? It has neither the perfume nor the beauty of the rose, MacKay noted, and lacks the beauty of the sweet pea. But it does have variegation—"the world can't show a dye but here has place," he quotes one poet—although any particular variation's beauty, growers soon found, was singularly difficult to perpetuate. Such trouble in cultivation together with the tulip's beauty, MacKay believed, were at the heart of the rising wave of irrational purchase and investment: "Many persons grow insensibly attached to that which gives them a great deal of trouble," he observed dryly. "Upon that same principle we must account for the unmerited encomia lavished upon these fragile blossoms." Anna Pavord, writing 150 years later in her book *The Tulip*, makes a similar point. The fact that the flower could "break," that is, change color unpredictably and often magnificently, added to the tulip's allure. Only rarely would these broken, multicolored flowers, which were also the most extravagantly bought and traded, promulgate a like variation. The breaks were caused by an aphid-spread virus, which, because it weakened the flower, sowed further uncertainty.

How irrational did the tulip mania become? Abraham Munting, a Dutch horticulturist who wrote a contemporaneous account of the tulip craze, records that at one point a single bulb was exchanged for twenty-four tons of wheat, forty-eight tons of rye, four fat oxen, eight fat swine, twelve fat sheep, two hogsheads of wine, four tons of beer, two tons of butter, a thousand pounds of cheese, one bed, a suit of clothes, and a silver drinking cup. A lone bulb of "Semper Augustus," the most glorious and renowned tulip, traded for a sum that, according to Mike Dash, would feed, clothe, and house an entire Dutch family for half a lifetime, or be sufficient to purchase one of the "grandest homes on the most fashionable canal in Amsterdam for cash, complete with a coach house and an eighty-foot garden." Beauty was indeed beyond rubies.

Such insanity could not and did not last. Gradually people cottoned on to the reality that their speculation was built on sand, or at best the soil of the Dutch tulip fields, and, as the comprehension spread, prices inevitably dropped; then they plummeted. The tulips returned to being objects of beauty, no longer ones of folly.

The dangers of unchecked enthusiasms go beyond irrational financial behavior. The haste and vehemence of exuberance can trample reason and lay siege to the truth. Sinclair Lewis's young doctor, Martin Arrowsmith, irritated by the vapid exuberance of his colleague, Dr. Pickerbaugh, exclaims, "I must say I'm not very fond of oratory that's so full of energy it hasn't any room for facts," and rails against Pickerbaugh's superficial science and relentlessly excited but hollow public-health crusades, noting how justifiably bored the public would be at "being galvanized into a new saving of the world once a fortnight." Not enough, perhaps, separates crusade from tirade and intense enthusiasm from fanaticism. Recklessness springs naturally from overoptimism. Left to its own highly persuasive devices, exultant mood will nearly always trump rational thought. It is in the amalgam of fever and reason that genius lies. Passions are like fire and water, observed the journalist Sir Roger L'Estrange more than three hundred years ago: they are good servants but poor masters. Passion kept on a loose bit serves its master far better than if it is left unbridled. Brakes are necessary; the exuberant mind must preserve the capacity to take a dispassionate measure of itself and the object of its zeal. When it does not, the consequences can be devastating.

Exuberance becomes dangerous when the goal is reprehensible, the means suspect, or the delight indiscriminate. Enthusiasts may be more interested in the problem to be solved than in the ethical issues which obtain in the wake of a solution. The Australian physicist Sir Mark Oliphant, who worked on the atomic bomb and then expressed severe misgivings about it, commented, "I learned during the war that if you pay people well and the work's exciting they'll work on anything. There's no difficulty getting doctors to work on biological warfare, chemists to work on chemical warfare and physicists to work on nuclear warfare." Curiosity and enthusiasm can be put to terrible ends.

Richard Feynman, who worked on the Manhattan Project, candidly related his delight on hearing about the atomic bomb explosion over Hiroshima: "The only reaction that I remember—perhaps I was blinded by my own reaction—was a very considerable elation and excitement, and there were parties and people got drunk. . . . I was involved with this happy thing and also drinking and drunk and playing drums sitting on the hood of . . . a Jeep and playing drums with excitement running all over Los Alamos at the same time as people were dying and struggling in Hiroshima."

Feynman's response is disturbing, complex, and more understandable than one would like: he and his colleagues, many of them exuberant by temperament, had been working single-mindedly on an intellectually fascinating problem; the atomic bomb, they believed, would end the war quickly, and they knew it was imperative to develop the technology before the Germans did. The scientists had been living and working together for years, walled off from the rest of society, completely focused on science and the success or failure of their project. The achievement of their goal would have been far less likely had they not been as enthusiastic and committed as they were. Still, Feynman's unabashed delight is unnerving; more than anyone in the world, he and his fellow scientists knew the devastation they wrought on Hiroshima.

"All invasive moral states and passionate enthusiasms make one feelingless to evil in some direction," concluded William James. War uniquely provokes such passions. Madame de Staël, a great proponent of enthusiasm, wrote that war "always affords some of the enjoyments of enthusiasm. . . . The martial music, the neighing of the steeds, the roar of the cannon, the multitude of soldiers clothed in the same colors, moved by the same desire, assembled around the same banners, inspire an emotion capable of triumphing over that instinct which would preserve existence; and so strong is this enjoyment, that neither fatigues, nor sufferings, nor dangers, can withdraw the soul from it. Whoever has once led this life loves no other." Robert E. Lee said simply, with the perspective of someone who had been there, "It is well that war is so terrible: we should grow too fond of it."

Paul Fussell, in *The Great War and Modern Memory*, describes war as a baptismal experience holding the possibility of resurrection, "a transformation of the personality." It is for some a romantic quest filled, as Eric Auerbach observed, with miracles and danger rituals and enchantment; but war is also, in Tennyson's words, a "blood-red blossom with a heart of fire." At its most intense, war is a complex mesh of emotions and behaviors sustained at fever pitch: fear and rage, atrocity, madness, exultation, shame, and courage.

War brings out a disturbing side of human passion, including that of violent exuberance. Clearly, for some—no one knows how many—the thrill of pursuit and pleasure in the kill are truly intoxicating. Theodore Roosevelt, like many before and after him, exulted in his brief experience of war: "All men who feel any power of joy in battle," he wrote, "know what it is like when the wolf rises in the heart." Of his charge up San Juan Hill, he said, "we were all in the spirit of the thing and greatly excited"; the men were "cheering and running forward between shots." Roosevelt himself, wrote one friend, "was just revelling in victory and gore." Henry Adams noted in Roosevelt a disquieting combination of energy and restlessness: "Power when wielded by abnormal energy is the most serious of facts," he wrote, "and all Roosevelt's friends know that his restless and combative energy was more than abnormal . . . he was pure act." Roosevelt lived, said Adams, in a "restless agitation that would have worn out most tempers in a month." It may be that such exhilaration in the hunt, such pleasure in the fight are necessary in order for war to occur at all. Certainly it must confer advantage to those whose energy and enthusiasm are needed to lead the battle and sustain the fight.

For some, combat is the most intense experience of their lives:

the stakes are the highest for which they will ever play, and their senses are alert as they will never be again. During the Civil War, one young officer, who in civilian life was an attorney from Indiana, wrote home to his fourteen-year-old sister after fighting at Chickamauga Creek, a battle that resulted in more than 35,000 Union and Confederate casualties. In response to her question of how it felt "when in the hottest of a battle," he wrote that as soon as a soldier begins firing at the enemy he becomes animated. He forgets danger: "The life blood hurries like a race horse through his veins, and every nerve is fully excited. . . . His brain is all alive; thought is quick, and active, and he is ten times more full of life than before." Despite his reason's awareness that he might die, the soldier's feelings seem to "give the lie to it. He seems so full of life that it is hard for him to realize that death is so near." Adrenaline assures the attention and quickness essential to survive; a sense of vitality provides the denial necessary to continue fighting.

In *The Pity of War,* the Oxford historian Niall Ferguson cites the experiences of soldiers for whom the intensity and adventure of war made all else pale. One said that war was the greatest adventure of his life, "the memories of which will remain with me for the remainder of my days, and I would not have missed it for anything." Another compared war to a mistress: "Once you have lain in her arms you can admit no other." He missed, he said, the "living in every nerve and cell of one's body." Chris Hedges, a *New York Times* reporter who covered wars in the Balkans, Central America, and the Middle East, said that for him war became a "potent addiction." He describes an unforgettable intensity: "There is a part of me—maybe it is a part of many of us—that decided at certain moments that I would rather die like this than go back to the routine of life. The chance to exist for an intense and overpowering moment, even if it meant certain oblivion, seemed worth it in the midst of war."

For some, the excitement of war is in the hunt or in the sheer exuberance of killing. The Vietnam War correspondent Laura Palmer interviewed American helicopter pilots a month after the Christmas bombing of Hanoi in 1972. These were, she acknowledged, some of the most disturbing interviews she had ever done. One of the pilots who, like the others, did not want to return home, told her, "I've been having a blast, killing dinks, chasing them. Getting shot at. It's fun. It's exciting. If you ever do it sometime, you'll like it." Another pilot, asked what it felt like to kill someone, said: "First, when they start shooting at you, it feels good. . . . It's fun chasing 'em. You go flying over a fighting position or something else and a *2001* space odyssey comes flying up at you, tracers and everything. It's neat. Looks like the Fourth of July. We let 'em shoot at us first. Then we kill 'em." Palmer remarked that until she had interviewed these pilots, "I never knew there was more than one way to die in Vietnam. You could die fast, or you could die slow. After endless nights, who can be blamed for finally befriending the dark?"

The American helicopter pilots were by no means unique in their delight in killing. The screenwriter William Broyles, Jr., a Marine officer in Vietnam and later the founding editor of *Texas Monthly* and a past editor of *Newsweek*, wrote a harrowing account of his combat experiences. He loved war, he said, "in strange and troubling ways." Some feel an excitement in war which is real and odious. After one battle, when many enemy soldiers had been killed and their naked bodies had been piled together for removal, Broyles saw a look of "beatific contentment" on his colonel's face, a look "I had not seen except in charismatic churches. It was the look of a person transported into ecstasy. And I—what did I do, confronted with this beastly scene? I smiled back, as filled with bliss as he was. That was another of the times I stood on the edge of my humanity, looked into the pit, and loved what I saw there."

Joanna Bourke, a historian at the University of London, writes in *An Intimate History of Killing* that commanding officers are praised if they manage to sustain a "joy of slaughter" in their troops. She quotes soldiers who found, in sticking the enemy with a bayonet, an "exultant satisfaction" and others who described the "sickening yet exhilarating butchery" as a "joy unspeakable." She cites a particularly chilling passage from Henry de Man's book *The Remaking of a Mind*, published shortly after World War I. "I had thought myself more or less immune from this intoxication [of slaughter]," de Man wrote, "until, as trench mortar officer, I was given command over what is probably the most murderous instrument in modern warfare. . . . One day . . . I secured a direct hit on an enemy encampment, saw bodies or parts of bodies go up in the air, and heard the desperate yelling of the wounded or the runaways. I had to confess to myself that it was one of the happiest moments of my life." He added that he could have "wept with joy."

The perversity of war is nowhere more eloquently addressed than by T. E. Lawrence in his masterpiece, *Seven Pillars of Wisdom*. He writes of war's brutalizing influence and its intimate, proximate delights: "The everlasting battle stripped from us care of our own lives or of others'. . . . We lived always in the stretch or sag of nerves, either on the crest or in the trough of waves of feeling. . . . Gusts of cruelty, perversions, lusts ran lightly over the surface without troubling us; for the moral laws which had seemed to hedge about these silly accidents must be yet fainter words. We had learned there were pangs too sharp, griefs too deep, ecstasies too high for our finite selves to register. When emotion reached this pitch the mind choked; and memory went white." The disappearance of moral law encouraged a descent in which Lawrence found comfort. "I liked the things underneath me and took my pleasures and adventures downward. There seemed a certainty in degradation, a final safety. Man could rise to any height, but there was an animal level beneath which he could not fall. It was a satisfaction on which to rest."

The excitement of killing is not limited to our species; indeed, there are many examples of it among our closest kin, the great apes. In *Demonic Males: Apes and the Origins of Human Violence*, Richard Wrangham and Dale Peterson give a particularly graphic example of chimpanzees attacking another of their kind:

> It began as a border patrol. At one point they sat still on a ridge, staring down into Kahama Valley for more than three-quarters of an hour, until they spotted Goliath [an elderly chimp and former member of the group to which the attackers belonged], apparently hiding only twenty-five meters away. The raiders rushed madly down the slope to their target. While Goliath screamed and the patrol hooted and displayed, he was held and beaten and kicked and dropped and bitten and jumped on. At first he tried to protect his head, but soon he gave up and lay stretched out and still. His aggressors showed their excitement in a continuous barrage of hooting and drumming and charging and branch-waving and screaming. They kept up the attack for eighteen minutes, then turned for home, still energized, running and screaming and banging on tree-root buttresses. Bleeding freely from his head, gashed on his back, Goliath tried to sit up but fell back shivering. [He] was never seen again.

Ritualistic behaviors leading up to contagious excitement are common among young mammals and, as we have seen, often preparatory to group activities which entail risk and require social cohesion. Play followed by hunting is widespread in animals, whether African wild dogs, apes, or humans. The rewards for competing or killing have to be intensely pleasurable in order to excite the dangerous behaviors required to survive, and they necessarily involve highly activated if not overtly exuberant states. This is a disturbing but not unexpected reality. War is the ultimate test of survival; in its ultimate stakes is the possibility of ultimate pleasure.

Chuck Yeager, who was the first pilot to break the sound barrier and is believed by many to be the greatest American aviator ever, describes the ancient delight of the hunt: "That day [during World War II] was a fighter pilot's dream. In the midst of a wild sky, I knew that dogfighting was what I was born to do. It's almost impossible to explain the feeling: it's as if you were one with that Mustang, an extension of that damned throttle. You flew that thing on a fine, feathered edge, knowing that the pilot who won had the better feel for his airplane and the skill to get the most out of it. . . . Concentration was total; you remained focused, ignoring fatigue or fear. . . . You fought wide-open, full-throttle. With experience, you knew before a kill when you were going to score. Once you zeroed in, began to outmaneuver your opponent while closing in, you became a cat with a mouse. You set him up, and there was no way out: both of you knew he was finished. You were a confident hunter and your trigger finger never shook. . . . When he blew up, it was a pleasing, beautiful sight. . . . The excitement of those dogfights never diminished. For me, combat remains the ultimate flying experience."

If war is to be used to defend territory or governments, there must be enthusiasm for it. Yeager, whose name means "hunter," was an enthusiastic warrior and immensely valued for it by his country. Conviction is demanded, and a measured fanaticism requisite, if the young are to be asked to die for a cause. But exuberance for war must be held in restraint lest it become atrocity, and passion for a cause must be kept in line lest the enthusiast turn fanatic.

Who is to say, however, when the line has been crossed? In the heat of war, in the face of death, can any line be made inviolable? The passions of war require a complex use of spur and bit in ways that those of ordinary times do not. With war, observed T. E. Lawrence, "a subtle change happened to the soldier. Discipline was modified, supported, even swallowed by an eagerness of the man to fight. This eagerness it was which bought victory in the moral sense, and often in the physical sense, of the combat. . . . Eagerness of the kind was nervous, and, when present in high power, it tore apart flesh and spirit." To incite emotions to fever pitch during war required a counterpoised use of restraint in times of peace, argued Lawrence: "To rouse the excitement of war for the creation of a military spirit in peace-time would be dangerous," he said, "like the too-early doping of an athlete. Consequently discipline, with its concomitant 'smartness' (a suspect word implying superficial restraint and pain) was invented to take its place."

Not all warriors respond to discipline: two of the most renowned military leaders of the twentieth century, General George S. Patton and General William "Billy" Mitchell, famously did not. In their defiance lies the permeable border between fanaticism and visionary leadership; in their vehement enthusiasm lies the realization that a great man is not necessarily a good man.

Patton and Mitchell had much in common. Both were born to privilege; both were passionate and scathingly intolerant of those not sharing their enthusiasms. Thwarted exuberance tipped easily into anger and self-righteousness. They were fiery leaders, impetuous, vain, and brave beyond question; both were incendiary advocates for innovative warfare. Both were regarded by their superior officers as loose cannons, publicly rebuked for insubordination, and at times considered mentally unstable (indeed, both the Patton and Mitchell families had histories of mental instability). Temperamentally they were far better suited to war than to peace. They led with exuberance and they misled with its excesses. Neither was willing to settle for anything other than his own vision of how armies and men ought to perform. They sought glory and they made history.

Patton's enthusiasm for war was early and unqualified. "God but I wish there would be a war," he wrote to his future wife while still a cadet at West Point. War, he said, was life. He expressed his passion for combat, and his disdain for those who disagreed, in verse:

When the cave man sat in his stinking lair,
With his low browed mate hard by;
Gibbering the while he sank his teeth
In a new killed reindeer's thigh.

Thus he learned that to fight was noble;
Thus he learned that to shirk was base;
Thus he conquered the creatures one and all,
And founded a warrior race.

. . . .

They speak but lies these sexless souls,
Lies born of fear of strife
And nurtured in soft indulgence
They see not War is Life.

They dare not admit the truth,
Though writ in letters red,
That man shall triumph now as then
By blood, which man has shed.

It isn't Yeats, but the point is clear. Patton saw himself as a warrior first, last, and always. (The "always" was quite literal: he believed that, in previous lives, he had been, among others, a Roman soldier, a Viking, and a soldier in Napoleon's army.) In a blood-rousing passage written in his West Point notebook, he exclaimed: "You have seen what the enthusiasm of men can mean for things done . . . you must do your damdest and win. Remember that is what you live for. Oh you must! You have got to do some thing! Never stop until you have gained the top or a grave." The passion for winning extended to battles on the playing fields as well. To his son many years later, Patton wrote: "You play games to *win* not lose. And you fight wars to win! That's spelled W-I-N! . . . Polo games and wars aren't won by gentlemen. They're won by men who can be first-class sonsofbitches when they have to be. It's as simple as that. No sonofabitch, no commander." The leader, Patton believed, commanded by an "all-pervading, visible personality. The unleavened bread of knowledge will sustain life, but it is dull unless seasoned by the yeast of personality."

At its best, Patton's personality served him and the United States well. His tactical brilliance and exuberance were legendary. General Dwight Eisenhower said of Patton that he was "one of those men born to be a soldier, an ideal combat leader whose gallantry and dramatic personality inspired all he commanded. . . . His presence gave me the certainty that the boldest plan would be even more daringly executed. . . . George Patton was the most brilliant commander of an army in the open field that our or any other service produced." But the separation between gainful exuberance and bloodthirst eroded over time. His remarks during World War II went from the merely reckless—"This is a damn fine war," he said, "I hope to God I get killed up front somewhere"; once, he jumped from his car shouting, "Where are the damned Germans, I want to get shot at!"—to speeches that were unhinged, indefensible, and arguably at least in part responsible for the slaughtering of prisoners of war under his command. In August 1942 he gave a fiery speech to his troops: "We're going to go right in and kill the dirty bastards," he exhorted them. "We won't just shoot the sonsofbitches. We're going to cut out their living guts—and use them to grease the treads of our tanks. We're going to murder those lousy Hun bastards by the bushel." He was no more circumspect in a speech delivered not long after: "We'll rape their women and pillage their towns," he said, "and run the pusillanimous sonsofbitches into the sea."

At some point, Patton ceased to inspire his men to greatness and instead quite possibly inflamed them to wanton killing. Intemperate behavior disintegrated into the unpardonable. He was forced to apologize publicly for slapping a shell-shocked soldier, and later, after his behavior became increasingly erratic and his speeches even more inflammatory, Eisenhower relieved him of his command of the Third Army. Exuberance had shifted into deranged hatred and zealotry.

General Sir Alan Francis Brooke, the chief of the British Imperial General Staff, believed that Patton was a "dashing, courageous, wild and unbalanced leader, good for operations requiring thrust and push but at a loss in any operation requiring skill and judgment." Dwight Macdonald denounced Patton as "an extreme case of militarist hysteria." "Compared to the dreary run of us," he added, "General Patton was quite mad." S.L.A. Marshall concurred: "I think he was about half mad. Any man who thinks he is the reincarnation of Hannibal or some such isn't quite possessed of all his buttons." Patton's biographer Ladislas Farago said much the same: "I am convinced," he wrote, "that he was, if not actually mad, at least highly neurotic."

Mad or not, Patton was an uncommon leader who used his passions both well and ill. When he died in December 1945, the *New York Times* gave a sense of his complexity: "He will be ranked in the forefront of America's great military leaders"; a legend, he was "a strange combination of fire and ice. Hot in battle . . . a profound and thoughtful military student." He was not, they said, "a man of peace."

Neither was General Billy Mitchell, the military aviation pioneer whose much ridiculed and seemingly grandiose visions of the future of airpower turned out to be correct. His advocacy for a separate branch of the military dedicated to the tactical use of airpower led directly to the formation of the United States Air Force, but his dream was realized at the cost of his career. His exuberance about the possibilities of flight hardened into what critics perceived as fanaticism, and his volatile temperament gave them the cause they needed to prevent his ideas from being put into action. It is unclear, however, given the intransigence of his opposition, that another temperament would have been as effective or persuasive.

Mitchell, the son of a U.S. Senator, was the top combat airman of World War I, having, in a single campaign against the Germans, commanded nearly fifteen hundred aircraft. He believed, contrary to his superior officers in the Army and Navy, that airpower would make Navy battleships obsolete; his enthusiasm was absolute and his public statements were commensurate with his enthusiasm. To prove his point, he ordered his pilots to sink the reputedly unsinkable former German warship *Ostfriesland;* it took them only a few minutes. Mitchell had made his point, but few wanted to hear it and fewer still appreciated his attitude. "Those of us in the air," he wrote, "knew we had changed the methods of war and wanted to prove it to the satisfaction of everybody."

Mitchell vociferously and repeatedly made his arguments for air supremacy and the need to equip the United States with a modern air force. His passions lay in the future, while those of most in the military were bound up in the armies and navies of the past and present. His love for flight was matched by his disdain for the "older services," an unrelenting attitude that antagonized and ruffled many. "Napoleon studied the campaigns of Alexander the Great and Genghis Khan," he commented. "The navies draw their inspiration from the Battle of Actium in the time of the Romans, and the sea fight of Trafalgar." But, he said, in the development of airpower "one has to look ahead and not backward, and figure out what is going to happen, not too much what has happened."

The future of defense was in the air, Mitchell proclaimed over and over. "The competition will be for possession of the unhampered right to traverse and control the vast, the most important, and the farthest reaching element of the earth, the air, the atmosphere that surrounds us all, that we breathe, live by, and which permeates everything." His fierce exuberance about airpower was influenced strongly by his sense that the stakes were high, but many felt that his certainty of belief was arrogant. Mitchell believed that those who took to the air were a special breed, a belief that got no argument from pilots but incited a great deal of resentment in others. "Bold spirits that before wanted to 'go down to the sea in ships,' now want to go up in the air in planes," he wrote. "The pilots of these planes, from vantage points on high, see more of the country, know more about it, and appreciate more what the country means to them than any other class of persons."

The pilot was not only privy to a view of the country that others could only envy, he was also of a select breed that called for a different disciplinary standard. "The old discipline, as conceived and carried out by armies and navies throughout the centuries," maintained Mitchell, "consists in the unhesitating obedience by a subordinate to the orders of his superior. . . . With the aviator, however, the keenest, best educated, most advanced kind of man has to be selected." Unhesitating obedience may suffice for the average man—it was only, after all, "within the last generation that most of the men composing armies could read or write"—but not for the aviator. Indeed, according to Mitchell, "[t]he [Army's] General Staff was trying to run the Air Service with just as much knowledge about it as a hog has about skating."

These sentiments, heartfelt and publicly expressed, were provocative and meant to be. Mitchell was incapable of keeping his concerns and enthusiasms to himself. In 1925, however, he went too far for the military to look the other way. After the air arm had a series of flying accidents, which occurred in the context of a consistent pattern of underfunding, Mitchell gave a prepared statement to six reporters. He said, among other things, "I have been asked from all parts of the country to give my opinion about the reasons for the frightful aeronautical accidents and loss of life, equipment and treasure. . . . These accidents are the direct result of the incompetency, criminal negligence and almost treasonable administration of the national defense by the Navy and War Departments." He was court-martialed, of course, on charges of "conduct to the prejudice of good order and military discipline and in a way to bring discredit upon the military service." He was tried and convicted of insubordination and sentenced to a five-year suspension from active duty without pay or allowances. He resigned from the Army in 1926 and died, essentially in exile, ten years later.

Perhaps a less passionate man, or a less ambitious one, would have been able to bring about change in a less tumultuous manner. Probably not. The strength of Mitchell's convictions gave them a prominence they would not otherwise have had, and it is unlikely that his provocativeness alienated many people not already likely to be alienated. Certainly he was controversial. Certainly he was an enthusiast who crossed over the line into crusaderhood, if not fanaticism. But he was right on vital fronts. He was correct about the centrality of air power to the national defense; he predicted, in 1924, that the Japanese would one day bomb Pearl Harbor and then the Philippines; he rightfully and passionately warned that Alaska and the Pacific would play critical roles in military strategy. He predicted traveling into interstellar space. Time bore him out. In 1947 the U.S. Air Force was created; Mitchell was posthumously awarded a special Medal of Honor and promoted to the rank of major general.

Mitchell, like Patton, was a complicated, enthusiastic, and angry man. Alfred Hurley, in his book *Billy Mitchell: Crusader for Air Power,* speaks of this complexity: "He erred in believing that the realization of his vision would justify his tactics. Those tactics included his denial of the integrity of an often equally dedicated opposition, his substitution of promises for performance, and his failure to sustain the kind of day-to-day self-effacing effort that builds any institution, whether military or otherwise." But, Hurley concludes, "Americans might well regard Mitchell as one of the extraordinary men in their history, one who employed some remarkable gifts and unusual energy in trying to alert his countrymen to the promise of aviation. Indeed, every age has had its crusaders—men like Mitchell whose relentless insistence on the correctness of their beliefs ultimately destroyed them. In the interim, however, their zeal also sustained them in combating the antagonism of the shortsighted." Mitchell's was a passionate life in pursuit of reason, but few saw it that way at the time. A more temperate man would have been less grating, but he would not have been Billy Mitchell.

It is obvious that not all exuberance is tethered to reality, nor is it always put to the common good. Strong passions, like fire, can civilize or kill. Daedalus, believing he was according his son the freedom of the air, made him wings of wax and feathers and taught him how to fly. He warned Icarus of the dangers of the sun's heat, but, writes Ovid, the son

began to feel the joy
Of beating wings in air and steered his course
Beyond his father's lead: all the wide sky
Was there to tempt him as he steered toward heaven.
Meanwhile the heat of sun struck at his back
And where his wings were joined, sweet-smelling fluid

Ran hot that once was wax. His naked arms
Whirled into wind; his lips, still calling out
His father's name, were gulfed in the dark sea.

Icarus brought daring to the air and found joy and death; Daedalus brought caution. Both count. Daedalus lived, but Icarus is the stuff of legend. "Who cares that he fell back to the sea?" asked Anne Sexton: "See him acclaiming the sun and come plunging down / while his sensible daddy goes straight into town."

The raptures of the air and sea can lead to peril. In 1965, the *Gemini 4* astronaut Ed White spent nearly half an hour somersaulting, floating, and space-walking outside his spacecraft. He was so euphoric that his fellow astronaut Gus Grissom, tracking him from Mission Control in Houston, worried about his safety. The astronauts Alan Shepard and Deke Slayton give their account of what happened next: " 'Gemini Four,' Gus called in a stern voice, *'get back in.'* [Crew mate James] McDivitt called to White, still frolicking outside. 'They want you to get back in *now.*' Ed White didn't want to return to the cockpit. 'This is fun!' he said exuberantly. 'I don't want to come back in, but I'm coming.' " The diver and adventurer Hans Hess has written of the seductiveness of deep-ocean exploration. Below a depth of 160 feet, he says, a "deep sea intoxication" occurs. "One loses all misgivings and inhibitions. The abyss below becomes a pleasant walk. Why not? A little bit further—why not? And then suddenly comes the end, without one even being aware of it. Death catches the diver in a butterfly net whose mesh is so soft that it closes in on him unnoticed."

Most strong enthusiasms are of no danger, and they add color not only to the lives of those who hold them but to those in their presence as well. Several years ago I was asked by *Nature* to review *Eccentrics*, a captivating book by David Weeks and Jamie James that discusses at length the enthusiasms of notable eccentrics—a woman who collected 7,500 garden gnomes, for instance, and a nineteenth-century man who, having been expelled from both Westminster and Harrow, spent £500,000 on alcohol in less than twenty years and kept, at one time, two thousand dogs, which he fed Champagne and steak. At one of his dinner parties he dressed in full hunt regalia and rode on the back of a bear; the latter, perhaps not surprisingly, was less amused than the guests and ate part of his rider's leg. In a tale for our materialistic times, the authors also recount the story of a man who moved to Sherwood Forest, dressed up in green, carried a longbow, and called himself Robin Hood. Prior to taking up his new identity he had, appropriately enough, earned his living by installing automatic cash-dispensing machines.

Many of the individuals portrayed by Weeks and James had symptoms of grandiose and delusional thinking, which are associated with mania, yet almost all were happy with their lives. They were exuberant, unusually assertive, and remarkably curious as well. Few were shy or timid. The authors struggled with the distinction between "vivid imaginings" and delusions; it was "no easy matter," they concluded, "to clarify the line of demarcation between eccentricity and mental illness."

Eric Hansen, in *Orchid Fever,* also describes the thin line separating enthusiasm from pathological exuberance. He interviewed one orchid grower who had started off with a single windowsill plant but then, he told Hansen, "[p]retty soon I decided I wanted another orchid. First a red one, then a pink one, then I had to have a white one with spots. . . . I couldn't stop. . . . Now I have a 2,200-square-foot greenhouse with about 200,000 plants." Another enthusiast talked about how his first wife couldn't handle his obsession: "One morning she sat him down at the breakfast table and explained that he would have to choose between his orchid collection and their marriage. 'That's the easiest decision I'll ever make,' he told her. 'You're out of here, baby!' "

Once a person has been "properly seduced" by the sight or

scent of an orchid, Hansen says, "he or she has little choice but to collect or buy the plant, take it home, build a special enclosure for it, feed, water, and groom the thing, and then dote over the plant for years. These people usually park their cars on the street because their garages are filled to capacity with potting tables surrounded with commercial quantities of cork bark, oyster shell, crushed dolomite, sphagnum moss, twist ties, horticultural charcoal, several different types of fertilizers, insecticides, pots, baskets, respirators, protective rubber clothing, Perlite, tree ferns, and gardening stakes. These orchid people . . . cater to the needs of their beloved orchids with a single-minded devotion that blurs the line between love and lunacy."

There is no absolute border between a delight in life and a delight that is a manifestation of a potentially pathological state such as mania. Sir Humphry Davy, in discussing the dangers of unskeptical enthusiasm, put it well: "In a person of irritable temperament, when the organization is perfect, and the body in a state of perfect health, the spirits rise sometimes almost to madness, trains of pleasurable ideas producing raptures introduce themselves into the Mind, all the ideas of self excellence are enlarged. . . . He imagines himself to be the peculiar favourite of the Deity. . . . This enthusiasm tho' rare at first[,] promoted and indulged in becomes at length habitual[,] and increased in a very high degree is absolute Insanity."

Where does exuberance end and mania begin? What is eccentricity, or simply a normal variation in temperament, and when does it tip over into irrational exuberance and psychopathology? We do not know. The edges of mania may be exhilarating, as Clifford Beers relates in *A Mind That Found Itself*—"It seemed as though the refreshing breath of some kind Goddess of Wisdom was being blown gently against the surface of my brain. . . . So delicate, so crisp and exhilarating was it that words fail me in my attempt to describe it"—but mania itself, as Beers knew from his long months in the asylum, is a dangerous place. A slightly fevered and manic brain may be adaptive: James Watson, for example, when asked about why the genes for manic-depressive illness survive in the gene pool, responded, "Survival might often depend on not if we think two and two is four, but on being slightly wild. Because life is just much more complicated than when we try to organize it. And so a brain which is slightly disconnected from reality might be a good thing. I think when we do science we see that a little madness does help, and you propose bizarre things which everyone says can't be true. Conceivably what you need is sometimes just to start up with a different set of facts." But too much fever destroys the brain.

Normal exuberance can escalate into pathological enthusiasm, anger, or even mania. Those who have what Emil Kraepelin called a "manic predisposition" are not only extraverted, cheerful, and overly optimistic, they also possess highly unstable and irritable moods. Indeed, those most inclined to exuberance are often most subject to despair and hopelessness. These dark sides of exuberance both help and hinder: if enthusiasm switches quickly to wrath or is bound too often to impetuous action, many of the dangers we have discussed are made more likely. If melancholy gives a humanizing perspective to exuberance, however, there is less risk of hazardous behavior and shallow thought. As we shall see, a close familiarity with both exuberance and despair may lead to a profound understanding of human nature, as well as an ability to more complexly express it in the arts and sciences.

Moderation in strong emotions is not always easily come by. Lucretius observed two thousand years ago that the destructive motions "can never permanently get the upper hand and entomb vitality for evermore. Neither can the generative and augmentative motions permanently safeguard what they have created. So the war of the elements that has raged throughout eternity continues on equal terms." More often than not, the war is a struggle within as much as without: opposing natures reside within the same person; positive forces of mood and energy alternate or coexist with threatening or nihilistic ones. Exuberance does not stand alone; it exists in a landscape of other emotions and circumstance.

For some, intense and opposing emotions are an integral part of their temperament or mood disorder. Cyclothymia, for example, is a form of manic-depressive illness that consists of short cycles of depression and mild mania interspersed with periods of normal mood and behavior. Usually exuberant individuals may be subject to days or weeks of low energy and dejection. These cyclothymics, wrote the German psychiatrist Ernst Kretschmer, "have a soft temperament which can swing to great extremes. The path over which it swings is a wide one, namely between cheerfulness and unhappiness. . . . Not only is the hypomanic disposition well known to be a particularly labile one, which also has leanings in the depressive direction, but many of these cheerful natures have when we get to know them better a permanent melancholic element in the background of their being." The elements remain in flux: "The hypomanic and melancholic halves of the cycloid temperament relieve one another," continued Kretschmer, and "form layers or patterns in individual cases, arranged in the most varied combinations."

Moods are mutable. They swing into and out of one another, for joy and grief inhabit close and traversable lands. New Orleans funeral marches begin with dirgeful music during the procession to the cemetery but, on the return journey, exuberance bounds back. As the music critic Ben Ratliff describes it, "the drummer takes the muffling handkerchief off the snare drum and lays down some opening rolls, and the songs turn up-tempo and cheerful, with a wildly multileveled polyphonic chatter that we all know is closer to the usual experience of life." Any song, says Jelly Roll Morton, can be played as "blues" or "joys."

It is not always easy to believe that the same individual might, as Byron put it, have under the same skin "two or three within." Those apparently exuberant are less often recognized as having a darker side, and those cast as doomed or depressive may never be seen for the liveliness they actually possess. Virginia Woolf is remembered more for her madness and suicide than for her vitality, despite the testament of her friends to her animation and dazzling laughter. Christopher Isherwood recalled that "listening to [Woolf] we missed appointments, forgot love-affairs, stayed on and on into the small hours," and Nigel Nicolson spoke of her vivacity: "One would hand her a bit of information as dull as a lump of lead," he said. "She would hand it back glittering like diamonds. I always felt on leaving her that I had drunk two glasses of an excellent champagne. She was a life-enhancer." Elizabeth Bowen, while recognizing Woolf's ultimate fate, put her suicidal depression in the context of the rest of her life:

> I was aware, one could not but be aware, of an undertow often of sadness, of melancholy, of great fear. But the main impression was of a creature of laughter and movement. In the Diary she says, "I enjoy everything I do." Do you remember?—it was on a good day. And her power in conveying enjoyment was extraordinary. And her laughter was entrancing, it was outrageous laughter, almost like a child's laughter. Whoops of laughter, if anything amused her. As it happened, the last day I saw her I was staying at Rodmell and I remember her kneeling back on the floor—we were tacking away, mending a torn Spanish curtain in the house—and she sat back on her heels and put her head back in a patch of sun, early spring sun. Then she laughed in this consuming, choking, delightful, hooting way. And *that* is what has remained with me. So I get a curious shock when I see people regarding her entirely as a martyred . . . or defi-

nitely tragic sort of person, claimed by the darkness. She ended, as far as we know, in darkness, but—where is she now? Nobody with that capacity for joy, I think, can be nowhere. And it *was* joy.

Yet it remains difficult to keep in mind the duality of moods. The prevailing mood tends to dominate the emotional landscape, both for the person and for those in its sphere of influence. James Barrie's private secretary, Cynthia Asquith, described the effect of his inconstant moods on those around him: "No pen could convey how widely Barrie varies. One day he looks so weary, sallow, lack-lustre, that had I to 'do' him in our analogy game I'd compare him to a full ash-tray and an empty ink bottle; the next day he may look positively tingling with health—alert, aglow . . . of course, such a strong—such an overwhelmingly strong—personality has the most terrific effect on others. When he's grey ashes, he's devastatingly depressing. It's almost impossible to fight against the influence. On the other hand, on his good days, he's so alive, so full of charm—more than charm—a kind of benign wizardry, that it makes one feel well and happy."

It was Barrie's fellow Scot Robert Louis Stevenson, however, who wrote most brilliantly on the duality of moods and human nature, articulated so clearly the beholdenness of light moods to darker elements, and made explicit that gaiety carries a price. Stevenson was drawn to the darker sides of human character—evil and despair and malignant hypocrisy—but he was as aware of the joy of adventure, the delights of friendship, and the saving, high pleasures of life.

Stevenson was, by all accounts, a vivacious, immensely charming, and mercurial man. Edmund Gosse said that "gaiety" was his cardinal quality. "A childlike mirth leaped and danced in him," Gosse wrote. "He seemed to skip upon the hills of life. He was simply bubbling." Another friend, Sidney Colvin, said that the "most robust and ordinary men seemed to turn dim and null in presence of the vitality that glowed in the steadfast, penetrating fire of the lean man's eyes, the rich, compelling charm of his smile . . . and lively expressiveness of his gestures." Within an hour of first meeting, he said, Stevenson "had captivated the whole household. . . . He sped those summer nights and days for us all as I have scarce known any sped before or since." Stevenson's stepchildren agreed. Belle Osbourne wrote of him that he "brought into our lives a sort of joyousness hard to describe," and Lloyd, her brother, observed that he was "so gay and buoyant that he kept every one in fits of laughter." The poet Andrew Lang remarked that Stevenson so "excited a passionate admiration and affection" in people that they "warmed their hands at that centre of light and heat."

Stevenson, said Henry James, had "substance and spirit"; he was someone whom "life carries swiftly before it and who signal and communicate, not to say gesticulate, as they go. He lived to the topmost pulse." When Stevenson died, James wrote: "He [had] lighted up a whole side of the globe, and was in himself a whole province of one's imagination . . . he had the *best* of it—the thick of the fray, the loudest of the music." James Barrie, stricken by Stevenson's death, felt his absence acutely: "When I came to London," he said, "there was a blank spot in it; Stevenson had gone. It could not be filled till he came back, and he never came back. I saw it again in Edinburgh the other day. It is not necessarily that he was the greatest, I don't think he was the greatest, but of the men we might have seen he is the one we would like best to come back."

But Stevenson was also described by those who knew him as excitable, high-strung, and inclined to restlessness, moodiness, and fits of rage. He was quick to the boil and wept easily. "It was not in Louis to remain long in any mood," said Edmund Gosse; "he was as restless and questing as a Spaniel." Andrew Lang, too, noticed

Stevenson's erratic nature. There was, he said, "a sort of uncommon celerity in changing expression, in thought and speech." His legendary restlessness was summed up most graphically by Henry Adams, who said that Stevenson "seems never to rest, but perches like a parrot on every available projection, jumping from one to another, and talking incessantly." Keeping to his bird analogy, but switching species, Adams wrote to another friend that Stevenson looked like "an insane stork, very warm and very restless." An acquaintance of Stevenson's in Samoa concurred: "He was as active and restless as if his veins had been filled with quicksilver."

W. E. Henley wrote of Stevenson that he was as "mutable as the sea, / The brown eyes radiant with vivacity . . . / A spirit intense and rare, with trace on trace / Of passion, impudence, and energy." Another friend said that "there were two Stevensons . . . this strange dual personality . . . I have seen him in all moods . . . chatting away in the calmest manner possible; and I have seen him become suddenly agitated, jump from that table and stalk to and fro across the floor like some wild forest animal . . . his face would glow and his eyes would flash, darkening, lighting, scintillating, hypnotising you with their brilliance and the burning fires within." Stevenson had, in short, a febrile temperament.

The intensity and variability of Stevenson's moods—his not infrequent black depressions and his contrasting exuberance—certainly contributed to his understanding of the underbelly of delight. His temperament was peculiarly tuned to not only the darker side of human nature and its ready accessibility but to a first-hand knowledge of man's multiplicity of selves. Stevenson's own fluctuating and wildly disparate moods made him especially sensitive to the ambiguities, shadings, and inconsistencies of human enthusiasms and, indeed, of life itself. "It is in vain to seek for consistency or expect clear and stable views," he wrote. "In this flux of things, our identity itself seems in a perpetual variation. . . . All our attributes are modified or changed; and it will be a poor account of us if our views do not modify and change in a proportion."

Stevenson's close knowledge of dark and inconstant moods inevitably influenced his work. It provided him a keen sensitivity to mood states of all kinds, and enhanced his genius for portraying their nuances. It also gave him a hard appreciation for the seductiveness of uninhibited states of mind. Stevenson's intimate acquaintance with contrary and unpredictable moods did not account for all, or even perhaps most, of his perspective on life. But to underestimate it is to underestimate Stevenson himself; it is, as well, to underestimate the raw, knowing, and deeply human power of his greatest writings.

Robert Louis Stevenson (1850–1894) was born in Edinburgh, an only child, and reared in a deeply Calvinist household. "My childhood," he wrote, "was in reality a very mixed experience, full of fever, nightmare, insomnia, painful days and interminable nights." He used to lie and hold his breath, he said, in a state of "miserable exaltation." His ancestors were lighthouse engineers who designed and built scores of lighthouses along the coast of Scotland. (When a young man, Stevenson briefly contributed to the family tradition—to the "towers we founded and the lamps we lit"—by delivering to the Royal Scottish Society of Arts a medal-winning paper about a new form of intermittent light for lighthouses.) Several others in his family were, like Stevenson, subject to violent mood swings and nervous breakdowns. "The family evil [is] despondency," he said. "Our happiness is never in our own hands. We inherit our constitution . . . we may be so built as to feel a sneer or an aspersion with unusual keenness . . . we may have nerves very sensitive to pain, and be afflicted with a disease very painful."

Stevenson's paternal uncle suffered from melancholia and had a psychotic breakdown; his cousin had severe recurrent depressions. Stevenson's father suffered from depression as well, but also experienced periods of great enthusiasm and vitality. He had, his son wrote, "a profound essential melancholy of disposition," as well as a "most humorous geniality." He was "passionately attached, passionately prejudiced; a man of many extremes—liable to passionate ups and downs"—but his "inmost thoughts were ever tinged with the Celtic melancholy." The biographer Frank McLynn describes Stevenson's father as showing the classic symptoms of manic-depression, including extreme changes in his moods and behavior: "At one moment he would be skipping about and telling strange stories to the servants," and "at another he would become almost catatonic and immobilised with the 'black dog.' "

When Stevenson was seventeen he went through a period of depression and was given medication for severe nightmares. He spoke of being at one with the twenty-four-year-old poet Robert Fergusson, who had died in 1774 in the Edinburgh Bedlam: "ah! what bonds we have—born in the same city; both sickly, both vicious, both pestered, one nearly to madness, one to the madhouse. . . . You will never know, nor will any man, how deep this feeling is: I believe Fergusson lives in me." In 1872 and 1873, when Stevenson was in his early twenties, he struggled to come to terms with his lack of religious belief, a struggle that was exceptionally painful for the dissension it caused with his parents. The problem was confounded by his father's explosive moods and Stevenson's growing tendencies in the same direction.

"I'm getting tired of this whole life business," he wrote in January 1873. "Let me get into a corner with a brandy bottle; or down on the hearthrug full of laudanum grog; or as easily as may be, into the nice wormy grave." Stevenson was in no better spirits by the following September. "You will understand the wearying, despairing, sick heart that grows up within one," he wrote, "and how the whole of life seems blighted and hopeless and twilight." He was diagnosed as suffering from "acute nerve exhaustion" and ordered to France to recuperate. He was, he said, "wretchedly nervous" and felt a "fever of restlessness." Although he was still only in his early twenties, he reported that he felt like an old man: "If you knew how old I felt," he wrote. "I am sure this is what age brings with it, this carelessness, this disenchantment, this continual bodily weariness; I am a man of seventy; O Medea, kill me, or make me young again!"

In *Ordered South,* Stevenson gives a signal account of his depression: "The world is disenchanted for him. He seems to himself to touch things with muffled hands, and to see them through a veil. His life becomes a palsied fumbling after notes that are silent when he has found and struck them. He cannot recognise that this phlegmatic and unimpressionable body with which he now goes burthened, is the same that he knew heretofore so quick and delicate and alive." This is one of the most succinct and best descriptions of depression I know.

Stevenson's black moods afflicted him intermittently for most of his adult life. Five years after the attack that precipitated his treatment in France, he said, "Black care was sitting on my knapsack; the thoughts would not flow evenly in my mind; sometimes the stream ceased and left me for a second like a dead man." "The ill humours," he wrote on another occasion, "got uppermost and kept me black and apprehensive." His struggles with ill humors continued. In an 1881 letter to a friend who had complained of insomnia, Stevenson described the dull sluggishness and oversleep which so characterize some forms of deep depression: "Insomnia is the opposite pole from my complaint; which brings with it a nervous lethargy, an unkind, unwholesome, and ungentle somnolence, fruitful in heavy heads and heavy eyes at morning. You cannot sleep; well, I can best explain my state thus: I cannot wake. Sleep . . . lingers all day, lead-heavy, in my knees and ankles. Weight on the shoulders, torpor on the brain."

Stevenson wrote that he at times wished he were dead, that

there was "devilish little left to live for." His mood swings were seasonal, with his depressions usually occurring in the fall and winter; they often followed periods of high energy and mood. Only after he moved to Samoa toward the end of his life did he report a steadying in his moods. "Half of the ills of mankind might be shaken off without doctors or medicine by mere residence in this lovely portion of the world," he wrote. Later he reaffirmed this belief: "Health I enjoy in the tropics. The sea, islands, the islanders, the island life and climate, make and keep me truly happier. These last two years I have been much at sea, and I have *never wearied*."

For most of his life, however, Stevenson felt the brunt of shifting moods and an inconstant temperament. A year and a half before he died, in response to a humorous self-portrait sent to him by James Barrie, Stevenson summed up his own disposition: "Drinks plenty. Curses some. Temper unstable . . . Given to explaining the universe. Scotch, sir, Scotch." The phrase "temper unstable" is critical to understanding not only Stevenson but also one of his most important works, *The Strange Case of Dr. Jekyll and Mr. Hyde,* for it is a book that deals with the multifariousness of moods and mental states in all of us.

Dr. Henry Jekyll, a distinguished physician and heir to a large fortune, has, he says, "every guarantee of an honourable and distinguished future." The worst of his faults is a "certain impatient gaiety of disposition, such as has made the happiness of many, but such as I found it hard to reconcile with my imperious desire to carry my head high, and wear a more than commonly grave countenance before the public. Hence it came about that I concealed my pleasures." Man, he said, is a multiplicity of conflicting selves:

> I thus drew steadily nearer to that truth, by whose partial discovery I have been doomed to such a dreadful shipwreck: that man is not truly one, but truly two. I say two, because

> the state of my own knowledge does not pass beyond that point . . . man will be ultimately known for a mere polity of multifarious, incongruous and independent denizens. . . . I saw that, of the two natures that contended in the field of my consciousness, even if I could rightly be said to be either, it was only because I was radically both.

No sooner had *Jekyll and Hyde* been published than it was subject to a variety of moral, social, and psychological interpretations. It has been construed as, among other things, an indictment of Victorian hypocrisy, an extension of earlier literary concepts of doubles, a portrayal of repressed homosexuality, and an archetypal telling of the primitive conflict between good and evil. No single interpretation ever suffices for a great piece of writing, of course, but one cannot help but be struck by Stevenson's extraordinary portrayal of starkly contrasting moods, the spreading, staining force of evil, and the incapacity of Hyde's vileness to exist outside of the sustaining petri dish of Jekyll's life and values. The description of Jekyll's mood state after he drinks the potion that transforms him into Hyde is telling: he feels "younger, lighter, happier"; there is "something strange in [his] sensations, something . . . incredibly sweet." Images run in his head "like a mill race." He experiences, most intoxicatingly, a "solution of the bonds of obligation." He knows himself to be "tenfold more wicked" and the thought of it, he says, "brace[s] and delight[s] me like wine." Hyde's love of life, Jekyll relates, is "screwed to the topmost peg"; his faculties seem sharpened, and his spirits "more tensely elastic." He feels a pervasive "contempt of danger." The temptation to again take the potion, to "spring headlong into the sea of liberty," is inevitable as well as deadly. Jekyll's at first bidden, then unbidden, mutations into Hyde bring him a joy he has not known, but it comes tied to perfidy and death.

The juxtaposition of the exuberant and the malignant is potentially dangerous, but a balance between the two can provide ballast and gravitas. Excessive lightness can be given a grace note by the dark, as melancholy and mania can give each other depth and height. To make use of despair is an ancient gift of the artist: to learn from pain; to temper the frenzied enthusiasm; to rein in the scatter, the rank confidence, and the expansive ideas generated during times of unchecked exuberance. Melancholy has a way of winding in the high-flying expectations that are the great gift of exuberance but its liability as well; it forces a different kind of looking. "In these flashing revelations of grief's wonderful fire," wrote Melville, "we see all things as they are; and though, when the electric element is gone, the shadows once more descend, and the false outlines of objects again return; yet not with their former power to deceive." Melancholy forces a slower pace, makes denial a less plausible enterprise, and constructs a ceiling of reality over skyborne ideas. It thrusts death into the mental theater and sees to it that the salient past will be preserved.

Exuberant ideas benefit from skepticism and leadshot. Whether the ballast comes from melancholy, from law or social sanction, from an astringent intellect or the incredulity of others, discipline and qualm are conducive to getting the best yield from high mood and energy. "Write with fury, and correct with flegm," said Thoreau. "Keep your early enthusiasm," Louis Pasteur advised his fellow scientists, "but let it ever be regulated by rigorous examinations and tests. . . . Worship the spirit of criticism. . . . Without it, everything is fallible; it always has the last word. . . . When, after so many efforts, you have at last arrived at a certainty, your joy is one of the greatest which can be felt by a human soul."

CHAPTER TEN

"It Is Not Down in Any Map"

The New England Puritans cannot have been as bloodless as we have been led to believe. Certainly the Pilgrims who sailed on the *Mayflower* in 1620 were resilient. Enough of them also carried an improbable optimism—bound, in turn, to ingenuity and vigor—to plant the seeds of commerce and revolution. Whether or not to cross the Atlantic—having fled England, now to leave Holland—had been the subject of intense moral and pragmatic debate. The cautious had indisputable cause to be so; those who were hopeful about settling a new land had to persuade the reluctant to set sail for the edges of a continent about which they knew little and which they were right to fear.

William Bradford, the governor of Plymouth Colony, wrote an account of the dispute leading up to the *Mayflower*'s departure. It was not, he said, a decision made in the "giddy humor by which men are sometimes transported to their great hurt and danger" but, rather, one entered into under the pressure of dire circumstance. Any voyage to the "vast and unpeopled countries of America," argued those who opposed it, would be subject to many dangers: there would be "casualties of the sea . . . miseries of the land, [and they would be liable to] famine and nakedness and the want, in a manner, of all things." The change of air, diet, and drinking water would infect their bodies with "sore sicknesses and grievous diseases," and there would be "savage people" who would "delight to torment men in the most bloody manner that may be; flaying some alive with the shells of fishes, cutting off the members and joints of others by piecemeal and broiling on the coals, eat the collops of their flesh in their sight whilst they live." As Bradford succinctly put it, "The very hearing of these things could not but move the bowels of men to grate."

But those set on going to America answered back. "All great and honourable actions are accompanied with great difficulties and must be enterprised and overcome with answerable courages," they argued. "It was granted the dangers were great, but not desperate. The difficulties were many, but not invincible." The new world would be dangerous and many would die, but that, in itself, was no reason to stay put, stagnant, with little to hope for from the future. Their ends were "good and honourable," they reasoned, "their calling lawful and urgent." It was not beyond fairness to expect the blessing of God in their continued pilgrimage and, though they should lose their lives, "yet might they have comfort in the same and their endeavours would be honourable."

Most did lose their lives. More than half of the passengers who arrived at Cape Cod aboard the *Mayflower* died within the year. They lacked food, safe water, and shelter; yet somehow they persevered. They farmed, fished, cleared forests, and laid down a government. They built schools, established trade and commerce, and created a university whose intellectual influence has been second to none. They and those who came after them prospered. Settlements flourished, new colonies came into being. Optimism, laced with desperation, had paid off.

It is in the nature of a questing species to move on, however, and it was not long before their descendants left New England to explore new lands for themselves. Some were restless and others poor, but they, with many of their countrymen, were caught up in the swell of the nation's westward expansion. The nation was afoot. Those who were the most enthusiastic and enterprising, who possessed a stupendous energy and the will necessary to take on the mountains and cross the prairies, would prove to have an incalculable advantage over the more faint of heart. All of the force an exuberant temperament could bring to bear—vitality, optimism, an expansive mind to imagine what the wilderness might one day be; expansive moods to offset despair, rebounding energies—would be called upon by those who moved west to pit their resources against those of nature. Exuberance came into its own—needed and selected for—as a vital feature in the American character.

"This land was an enigma," writes Willa Cather in *O Pioneers!* "It was like a horse that no one knows how to break to harness, that runs wild and kicks things to pieces." The Nebraska land "wanted to be let alone, to preserve its own fierce strength, its peculiar, savage kind of beauty, its uninterrupted mournfulness." The prairie belonged to the pioneer who could know it, could believe in it, and could trump its mournfulness with heart and resourcefulness. "A pioneer should have imagination," says Cather, "should be able to enjoy the idea of things more than the things themselves." Pioneers should be optimistic; they should have one foot, and much of their heart, in the future.

Cather's heroine, Alexandra, has this imaginative affinity with the future; she has been hard tutored in the limits set by nature but keeps joy in what she can bring to the land through hard work and intelligence, and by putting into the new land traces of life from an earlier world. To the wild larkspur and cotton and wheat, she brings pumpkins and rhubarb, gooseberries, zinnias, and marigolds. She plants apple, mulberry, and apricot trees, as well as orange hedges; and she sets beehives in the orchards. She sees opportunity where others see unbroken lands or nothing at all; her boldness and belief make possible that which seemingly was not. In short, she brings to the possibilities of the land her own exuberance of ideas, beliefs, and hope. She, like Vachel Lindsay's expansive, wandering Johnny Appleseed, carries life to the frontier and into the future:

In a pack on his back,
In a deer-hide sack,
The beautiful orchards of the past,
The ghosts of all the forests and the groves—
In that pack on his back,
In that talisman sack,
To-morrow's peaches, pears and cherries,
To-morrow's grapes and red raspberries,
Seeds and tree-souls, precious things,
Feathered with microscopic wings
. . . .
Love's orchards climbed to the heavens of the West
And snowed the earthly sod with flowers.

. . . .

He saw the fruits unfold,
And all our expectations in one wild-flower written dream.

The most significant thing about the American frontier, proposed the historian Frederick Jackson Turner, was that it lay "at the hither edge of free land." As long as there was land ahead, there was cause for optimism. There was also need for it. The early pioneers, said Turner, were an essentially hopeful people: "As they wrested their clearing from the woods . . . as they expanded that clearing and saw the beginnings of commonwealths, where only little communities had been, and as they saw these commonwealths touch hands with each other along the great course of the Mississippi River, they became enthusiastically optimistic and confident of the continued expansion of this democracy. They had faith in themselves and their destiny. And that optimistic faith was responsible both for their confidence in their ability to rule and for the passion for expression. They looked to the future."

From the harsh and unpredictable conditions of life on the frontier came certain traits that would, according to Turner, mark the American pioneer: a "coarseness and strength combined with acuteness and inquisitiveness," a "practical, inventive turn of mind," a "restless, nervous energy . . . and withal that buoyancy and exuberance which comes with freedom—these are traits of the frontier, or traits called out elsewhere because of the existence of the frontier." Unexplored land required active imagination, energy, and a belief that insurmountable problems were surmountable. America, said Edward Harriman, the owner of the Union Pacific Railroad, had been developed by pioneers "flush with enthusiasm, imagination and speculative bent"; its success, he contended, was owed to individuals who had seen into the future and "adapted their work to the possibilities."

For the pioneer who could work backbreakingly hard, improvise nimbly under pressure, and imagine a sustaining harvest before the land had yet been seeded or even cleared, the West promised open opportunity. If the pioneer would strip himself of assumptions and habits better suited to the drawing rooms of the Atlantic seaboard, the West would deliver him into new space and freedom. It would reward expansive ideas with expanding horizons, value enthusiasm over restraint, and encourage entrepreneurial will over mindless hewing to established ways. Such freedom of spirit and movement was pealed out by Whitman in his "Song of the Open Road":

From this hour I ordain myself loos'd of limits and
imaginary lines,
Going where I list, my own master total and absolute,
. . . .
I inhale great draughts of space;
The east and the west are mine, and the north and the south
are mine.
I am larger, better than I thought.

His was the exuberance held by those bound for the West.

Neither freedom nor the West was easily won, of course. The struggle for the land cost greatly in both lives and sanity, a reality darkly and beautifully told in O. E. Rölvaag's classic saga of the Dakota prairie, *Giants in the Earth*. It is a tragedy, a story of the fullness and failures of human temperament played out against the killing moods of nature. The Norwegian immigrants Per Hansa and his wife, Beret, move west from Minnesota to settle in the Dakota Territory, bringing with them different dreams, energies, and imaginative capacities. Beret is fragile, filled with biblical foreboding, and unseverably tethered to what she has left behind. Her initial reaction to the prairies is bleak and it remains so: "*Here some-*

thing was about to go wrong. . . . How will human beings be able to endure this place? she thought. Why, there isn't even a thing that one can *hide behind*!"

Per Hansa, on the other hand, is physically vigorous and exuberant, and he lives in and for the future. His response to the land is one of passion: "This vast stretch of beautiful land was to be his—yes, *his*. . . . His heart began to expand with a mighty exaltation. An emotion he had never felt before filled him and made him walk erect. . . . 'Good God!,' he panted. 'This kingdom is going to be *mine*!' " Beret sees only a punishing land and a graceless, godless existence; she cannot imagine being a meaningful participant in its development. "This formless prairie had no heart that beat," she despairs, "no waves that sang, no soul that could be touched . . . or cared. . . . How *could* existence go on, she thought, desperately."

Life, however threatening, however impossibly hard, is different for her husband. "Where Per Hansa was, there dwelt high summer," it was said. He imagines his land not as it is but as it will become; his resilience matches the prairie's:

> Now it had taken possession of him again—that indomitable, conquering mood which seemed to give him the right of way wherever he went, whatever he did. Outwardly, at such times, he showed only a buoyant recklessness, as if wrapped in a cloak of gay, wanton levity; but down beneath all this lay a stern determination of purpose, a driving force. . . .
>
> As Per Hansa lay there dreaming of the future it seemed to him that hidden springs of energy, hitherto unsuspected even by himself, were welling up in his heart. He felt as if his strength were inexhaustible. And so he commenced his labours with a fourteen-hour day. . . . [H]e accordingly lengthened the day to sixteen hours, and threw in another

> hour for good measure. . . . [A] pleasant buoyancy seemed to be lifting him up and carrying him along; at dawn, when he opened his eyelids, morning was there to greet him—the morning of a glorious new day.

Per Hansa has "such a zest for everything" that he scarcely sleeps; in the blaze of his first prairie summer, he "plow[s] and harrow[s], delves and [digs]"; he builds a house, weaves fishing nets, and plants saplings and potatoes. "He was never at rest, except when fatigue had overcome him and sleep had taken him away from toil and care. But this was seldom, however; he found his tasks too interesting to be a burden; nothing tired him, out here. . . . Per Hansa could not be still for a moment. A divine restlessness ran in his blood; he strode forward with outstretched arms toward the wonders of the future, already partly realized. He seemed to have the elfin, playful spirit of a boy; at times he was irresistible; he had to caress everything that he came near. . . . But he never could be still. . . . Endless it was, and wonderful!"

To Beret, however, the facts of the prairie are "unchangeable—it was useless to juggle with them, or delude oneself; nothing but an eternal, unbroken wilderness encompassed them round about, extending boundlessly in every direction." The desolation of the land, she feels certain, "called forth all that was evil in human nature." The bleakness of the prairie and the assault of nature—its ferocious summer storms and unrelenting winter blizzards, locusts that ravage their crops, prairie fires, famine—as well as Indian attacks, take their toll on her and the other settlers. Beret goes mad: she "heeded not the light of the day, whether it might be grey or golden. [She] stared at the earthen floor of the hut and saw only night round about her. . . . [S]he faced only darkness. She tried hard, but she could not let in the sun." Her entire appearance, writes Rölvaag, "seemed to reflect a never-ending struggle with unreality."

Per Hansa, on the other hand, meets the prairie with hope and a keen appreciation of its terms for survival. He grows "even louder in his optimism. . . . There were moments, even, when he felt confident that he would live to see the day when most of the land of the prairie would be taken up; in such moods, there was something fascinating about him; bright emanations of creative force seemed to issue out. . . . [W]henever he spoke a tone of deep joy rang in his words." But for Beret the future is only grim: "they would all become wild beasts if they remained here much longer. Everything human in them would gradually be blotted out. . . . They saw nothing, learned nothing. . . . Couldn't he understand that if the Lord God had intended these infinities to be peopled, He would not have left them desolate down through the ages?"

Each failure, each onslaught, extracts a bit more of Beret's sanity. Per Hansa, however, not only holds on to his high spirits but finds them sustained and generated anew by the cycles of nature: "the power to create a new life over this Endless Wilderness, and transform it into a habitable land for human beings. Wasn't it wonderful? . . . As the mild spring weather set in, a feverish restlessness seized him. . . . He walked so lightly; everything that had life he touched with a gentle hand, but talk to it he must; his voice sounded low, yet it thrilled with a vibrant energy. . . . [Beret] felt a force that made her tremble, emanating from him. . . . Per Hansa became more restless, but it only seemed to fill him with greater joy."

Raw optimism is a defining element not only in Per Hansa's life, but in Rölvaag's portrayal of pioneer life in general:

> It was as if nothing affected people in those days. They threw themselves blindly into the Impossible, and accomplished the Unbelievable. If anyone succumbed in the struggle—and that happened often—another would come and take his place. Youth was in the race; the unknown, the

> untried, the unheard-of, was in the air; people caught it, were intoxicated by it, threw themselves away, and laughed at the cost. Of course it was possible—everything was possible out here. There was no such thing as the Impossible anymore. The human race had not known such faith and such self-confidence since history began. . . . And so had been the Spirit since the day the first settlers landed on the eastern shores; it would rise and fall at intervals, would swell and surge on again with every new wave of settlers that rolled westward into the unbroken solitude.

Himself an immigrant and a pioneer, Albert Einstein once said that America is more capable of enthusiasm than any other country. It is certainly the case that America seems particularly to value exuberance and optimism. Not surprisingly, perhaps—in a country that gave birth to Walt Whitman and John Philip Sousa; invented jazz, square dancing, and rock and roll; gave the world Chuck Yeager, Ted Turner, and P. T. Barnum; created *Oklahoma!;* and glories in Louis Armstrong and Theodore Roosevelt—Americans see enthusiasm as an advantage. When asked which emotions they most like to experience, Americans are far more likely than people from other countries to say that they favor enthusiasm. They are also far more likely to say that they believe enthusiasm is a useful and constructive emotion in their lives. (Optimism is a related and defining American trait. The results of a Pew Carter poll conducted in 2002 of 38,000 people in forty-four countries found that more Americans [65 percent] than respondents from any other country disagreed with the statement "Success in life is pretty much determined by forces outside our control.")

Interestingly, high rates of manic-depressive illness have been observed in American immigrant groups, which suggests the possibility that individuals with mild forms of the illness, or temperamental variants of high energy and exuberant mood, may have been selected for. Individuals who sought the new, who took risks that others would not, or who rebelled against repressive social systems may have been more likely to immigrate to America and, once there, to succeed.

Exuberance is a part of the national vision, as well as its character. Whitman proclaimed, "My ties and ballasts leave me, my elbows rest in sea-gaps, / I skirt sierras, my palms cover continents, / I am afoot with my vision," and Lindbergh took this exultant notion to the skies. He weighed risk against adventure and chose freedom with a vengeance. "I began to feel that I lived on a higher plane than the skeptics of the ground," he wrote, "one that was richer because of its very association with the element of danger they dreaded, because it was freer of the earth to which they were bound. In flying, I tasted a wine of the gods of which they could know nothing. Who valued life more highly, the aviators who spent it on the art they loved, or these misers who doled it out like pennies through their antlike days? I decided that if I could fly for ten years before I was killed in a crash, it would be a worthwhile trade for an ordinary lifetime."

What happens, though, when the wine of the gods disappears, or if nothing matters enough to stake one's life and dreams on? What happens when enthusiasts become jaded? A passion for life is essential to the renewal of life. If passion is lost, the future itself is diminished. Scott Fitzgerald made this point well. There are no second acts in American lives, he said, but he made an ironic exception for New York in the boom days of the 1920s. The first act had been dazzling. "New York had all the iridescence of the beginning of the world"; it was an age of miracle and promise. But it was also, glaringly, an age of excess. The city and its inhabitants had become restless and careless: "The buildings were higher, the morals were looser and the liquor was cheaper," he wrote. The city was "bloated, gutted, stupid with cake and circuses."

By 1927, "a widespread neurosis began to be evident, faintly signalled, like a nervous beating of the feet." (There was a brief burst of hope in the midst of the decay, but it proved ephemeral. "Something bright and alien flashed across the sky," Fitzgerald wrote. "A young Minnesotan who seemed to have had nothing to do with his generation did a heroic thing, and for a moment people set down their glasses in country clubs and speak-easies and thought of their old best dreams. Maybe there was a way out by flying, maybe our restless blood could find frontiers in the illimitable air. But by that time we were all pretty well committed; and the Jazz Age continued; we would all have one more.")

The Lost City's second act, when it came, proved a tragedy. "We were somewhere in North Africa," Fitzgerald recalled, "when we heard a dull distant crash which echoed to the farthest wastes of the desert." The 1929 collapse of the stock market capped a decade of hollow euphoria and overextension, years that had drawn from the city far more than they ever gave back in hope or vitality. Exuberance, ginned up to such an unnatural level, was brittle and could not last. Fitzgerald's contemporaries were jaded and doomed, he wrote, and "had begun to disappear into the dark maw of violence." A classmate killed his wife and himself, "another tumbled 'accidentally' from a skyscraper in Philadelphia, another purposely from a skyscraper in New York . . . still another had his skull crushed by a maniac's axe in an insane asylum where he was confined."

Fitzgerald's own moods swung with the city's. In his autobiographical essays, *The Crack-up,* which he wrote in the decade following the crash, he described the toll. "I began to realize that for two years my life had been a drawing on resources that I did not possess, but I had been mortgaging myself physically and spiritually up to the hilt." Like the desperate jubilation of the Jazz Age, his mind, as he had known it, could not last. He was paying the piper for "an over-extension of the flank, a burning of the candle at both ends; a call upon physical resources that I did not command, like a man over-drawing at his bank. In its impact this blow was more violent [than his earlier psychological crises] but it was the same in kind—a feeling that I was standing at twilight on a deserted range, with an empty rifle in my hands and the targets down." Before that time, his happiness, he said, had "often approached such an ecstasy that I could not share it even with the person dearest to me but had to walk it away in quiet streets and lanes with only fragments of it to distil into little lines in books—and I think that my happiness, or talent for self-delusion or what you will, was an exception. It was not the natural thing but the unnatural—unnatural as the Boom; and my recent experience parallels the wave of despair that swept the nation when the Boom was over."

Exuberance is an assailable thing, as is the hope that rides with it. Exultant mood often forewarns a harrowing fall. Champagne will go flat, passion burn itself out, and optimism be trimmed by experience. The ecstasy that beguiles, even as it ascends into madness, will plunge, shattering its bearer. "It's not much fun writing about these breakdowns after they themselves have broken," observed Robert Lowell in the aftermath of a manic attack. "One stands stickily splattered with patches of the momentary bubble." He took the image and emotion into verse:

It takes just a moment
for the string of the gas balloon
to tug itself loose from the hand.

If its string could only be caught in time
it could still be brought down
become once more a gay toy
safely tethered in the warm nursery world
of games, and tears, and routine.

But once let loose out of doors
being gas-filled the balloon can do nothing but rise

although the children who are left on the ground may cry
seeing it bobbing out of human reach.

On its long cold journey up to the sky
the lost balloon might seem to have the freedom of a bird.
But it can fly only as a slave
obeying the pull to rise which it cannot feel.

Having flown too high to have any more use as a plaything
who will care if it pays back its debt and explodes
returning its useless little pocket of air
to an uncaring air it has never been able to breathe.

Savingly, nature teaches that joy can be replenished, life can succeed death, and joy find its way out of sorrow. Nature gives of its exuberance in remarkable ways, at extraordinary times. In 1918 a scientist described the flowers emerging from the Somme battlefield, a place where death had so recently been all-dominant: "In all the woods where the fighting was most severe not a tree is left alive, and the trunks which still stand are riddled with shrapnel and bullets and torn by fragments of shell, while here and there unexploded shells may still be seen embedded in the stems. Aveluy Wood, however, affords [an] example of the effort being made by Nature to beautify the general scene of desolation. Here some of the trees are still alive, though badly broken, but the ground beneath is covered with a dense growth of the rose-bay willow herb (*Epilobium augustifolium*) extending over several acres. Seen from across the valley, this great sheet of rosy-pink was a most striking object, and the shattered and broken trees rising out of it looked less forlorn than elsewhere."

Peter Ackroyd writes of a similar beauty and life force in the wake of the German bombings of London during World War II.

"It was the invisible and intangible spirit or presence of London that survived and somehow flourished," he says.

> London itself would rise again. There was even a natural analogy. Air damage to the herbarium in the Natural History Museum meant that certain seeds became damp, including mimosa brought from China in 1793. After their trance of 147 years, they began to grow again.
>
> Yet there was also a curious interval when the natural world was reaffirmed in another sense. One contemporary has described how "many acres of the most famous city in the world have changed from the feverish hum and activity of man into a desolate area grown over with brightly coloured flowers and mysterious with wildlife." . . . [The streets] bloomed [with] ragwort, lilies of the valley, white and mauve lilac. "Quiet lanes lead to patches of wild flowers and undergrowth not seen in these parts since the days of Henry VIII." . . . This earth had been covered with buildings for more than seven centuries, and yet its natural fertility was revived. It is indirect testimony, perhaps, to the force and power of London which kept this "fertility" at bay. The power of the city and the power of nature had fought an unequal battle, until the city was injured; then the plants, and the birds, returned.

For some who lose hope and vitality, nature will act on its own, as it did in London and the Somme, to reinfuse life. However dreadful the circumstances—death or madness, war, betrayal—the passion for life will surge back. For these individuals, it is an innate and irrepressible force; they are, in every true sense, exuberant by nature. For others, joy and laughter seep back in more slowly. They are less resilient; their healing is more hesitant and perhaps less complete. By whatever means joy comes back—however naturally or however haltingly—it is an amazing thing that it does. It is a gift of grace that allows us to move on, to seek, to love again.

The love for life returns in a profusion of ways. Biologist Joyce Poole, for example, writes that only Africa's natural beauty could heal the terrible pain she knew following the slaughter of the elephants she had studied and loved. Great matriarchs had been butchered by ivory poachers and entire family groups destroyed. By the time Poole left Amboseli, she writes, "I could no longer take any pleasure from the sights and smells and feelings that had once been so evocative." She watched on television as Kenya burned tons of ivory: "As the flames consumed the remains of so many elephants' lives, I cried for the thousands of violent deaths those elephants had suffered and for the hundreds of orphans still running bewildered through the Tsavo bush. . . . My dreams went up in smoke with the ivory, and I was left shattered and brittle like so many sharp-edged pieces of the burnt-to-blue ivory. . . . [T]he remaining magic had gone, and my bush life filled me with a terrible emptiness."

Only Africa itself could restore life. Poole recounts an early-morning flight she took with Richard Leakey long after she had watched the ivory burn:

> There were no roads here, and only the narrow cattle tracks crisscrossed the dry plains. A few Maasai settlements dotted the otherwise flat, desolate landscape. I looked ahead as Richard pointed out the two mountains that bordered the southern end of Lake Natron: *Gelai* and *Lenkai*. As we rounded the slope of *Oldoinyo Gelai*, one of the most breathtakingly beautiful sights in the world met my eyes, a kaleidoscope of colors, patterns, and movement, and changing horizons: the intense pinks of the lake, the soft blues of

> the mountains, and the millions of wing beats of the flamingos. As I watched the shifting light and colors, I found myself speechless, witnessing the exquisite beauty of Africa in juxtaposition with so many bittersweet memories. It had been so many years since I had allowed myself really to feel the beauty in anything. . . . I had been trying to discover my passion for life again and could not seem to find it.

Even under the most unforgivable circumstances, some joy and defiance exist. In 1850, Frederick Douglass said, "I admit that the slave *does* sometimes sing, dance, and appear to be merry. But what does this prove? It only proves to my mind, that though slavery is armed with a thousand strings, it is not able entirely to kill the elastic spirit of the bondman. That spirit will rise and walk abroad, despite of whips and chains, and extract from the cup of nature, occasional drops of joy and gladness. No thanks to the slaveholder, nor to slavery, that the vivacious captive may sometimes dance in his chains, his very mirth in such circumstances, stands before God, as an accusing angel against his enslaver." Even drops of joy abet defiance and make suffering more endurable.

Exuberance defies in strange and powerful ways; it asserts a future that others contrive to deny. Philippe Petit, a French juggler and acrobat, rigged a cable between the Twin Towers of New York City's World Trade Center in 1974 and made eight crossings a quarter of a mile above the ground. He was exultant. "I sit down on the wire, balancing pole on my lap. Leaning against the steel corner, I offer to myself, for a throne, the highest tower ever built by man; for a ceremonial carpet, the most savagely gigantic city of the Americas; for my dominion, a tray of seas wetting my forehead; while the folds of my wind-sculpted cape surround me with majestically mortal whirls. I rise, standing up on the wire. . . . I start walking. And walking, and walking." It is a sacred expedition, he chants to himself, a mythological journey. He calls out to the "gods of the billion constellations": "Watch closely. You're not going to believe your zillion eyes. . . . Standing up again, I recognize I am at the top of the world, with all of New York City at my feet! How not to laugh with joy? I laugh with joy—and conclude the crossing with ecstasy instead of oxygen in my lungs."

After the destruction of the World Trade Center in 2001, Philippe Petit offered up his song again. Defy, he said. Fight terror with what is great in the human spirit. Build:

> Let us print WE SHALL NOT BE DOOMED and paste the message high in the sky, for all in the world to read aloud.
>
> Let us rebuild the twin towers.
>
> We need the fuel of time and money, the mortar of ideas. . . .
>
> Bring yours.
>
> Here is mine. . . . Architects, please make them more magnificent—try a twist, a quarter turn along their longitudinal axes. Make them higher—yes, one more floor, so they reach 111 stories high. . . .
>
> When the towers again twin-tickle the clouds, I offer to walk again, to be the expression of the builders' collective voice. Together, we will rejoice in an aerial song of victory. I will carry my life across the wire, as your life, as all our lives, past, present, and future—the lives lost, the lives welcomed since.
>
> We can overcome.

The future, to act in its fullest, needs all the exuberance it can call in. It is true, as President William Jefferson Clinton said, that we are a questing people, but questing—and the energy and enthusiasm to fuel it—must be kept topped up. To explore requires vitality and curiosity; complaisance is death to discovery and to its attendant joy. If, as individuals or as a country, we stop pursuing new frontiers, become glutted on our past rather than drawn to the future, we will cease to be explorers. Exuberance is the headwater of motion, as it is of resilience; to lose our joy is to lose our ability to fight back and to advance. We have gone west and to the moon, but we need to bring a like passion to exploring the sea and the brain, to chasing comets and tracking down the first light in the universe, to writing symphonies and seeking social justice.

In 1962, President Kennedy said, "The United States was not built by those who waited and rested and wished to look behind them. This country was conquered by those who moved forward." This is a theme returned to time and again, and with great eloquence, by Michael Collins, the *Apollo 11* astronaut who orbited the moon as Neil Armstrong and Buzz Aldrin walked it. He writes, in *Carrying the Fire*, that

> our nation's strength has always derived from our youthful pioneers. . . . Some people were never content to huddle in protective little clumps along the East Coast, but pushed westward as boldly as circumstances permitted. When horizontal exploration met its limits, it was time to try the vertical, and thus has it been since, ever higher and faster.
>
> Now we have the capability to leave the planet, and I think we should give careful consideration to taking that option. Man has always gone where he has been able to go, it is a basic satisfaction of his inquisitive nature, and I think we all lose a little bit if we choose to turn our backs on further exploration. Exploration produces a mood in people, a widening of interest, a stimulation of the thought process, and I hate to see it wither.

Collins spoke to the same point in his address to the Congress not long after *Apollo 11* had returned to earth. "Man has always gone where he has been able to go," he said. "It's that simple. He will continue pushing back his frontier, no matter how far it may carry him from his homeland." The *Apollo 12* astronaut Alan Bean said much the same thing nearly thirty-five years later. Nothing, he declared, is going to change the "inexorable motion of human beings off this planet and out into the universe." For those of us who remember that magical July night in 1969 when Armstrong and Aldrin walked upon the moon as Collins circled it: we hope so.

Michael Collins describes his thoughts and the immensity of his solitude while orbiting the moon: "I am alone now, truly alone, and absolutely isolated from any known life. I am it. If a count were taken, the score would be three billion plus two over on the other side of the moon, and one plus God only knows what on this side. I feel this powerfully—not as fear or loneliness—but as awareness, anticipation, satisfaction, confidence, almost exultation. I like the feeling." He starts to turn off the lights in *Columbia* so he can get some sleep and, as he looks around the spacecraft, he is taken back to his childhood days:

> As I scurry about, blocking off the windows with metal plates and dousing the lights, I have almost the same feeling I used to have years ago when, as an altar boy, I snuffed out the candles one by one at the end of a long service. Come to think of it, with the center couch removed, *Columbia*'s floor plan is not unlike that of the National Cathedral, where I used to serve. Certainly it is cruciform, with the tunnel up above where the bell tower would be, and the navigation instruments at the altar. The main instrument panels span the north and south transepts, while the nave is where the

center couch used to be. If not a miniature cathedral, then at least it is a happy home, and I have no hesitation about leaving its care to God and Houston.

Five years later, a special service of commemoration and thanksgiving was held at the National Cathedral in Washington. A new stained-glass window, depicting swirling stars and orbiting planets floating in deep space, was dedicated. A narrow white line traces round the planets, representing the trajectory of a spacecraft. Carved near the window is a print of an astronaut's boot on the surface of the moon, and embedded in the window itself is a piece of lunar rock collected by the *Apollo 11* astronauts. Underneath the stars, at the bottom of the window in scarcely noticeable lettering, is an inscription taken from the book of Job: "Is not God in the height of heaven?" It is a question Hipparchus must have asked and Cecilia Payne-Gaposchkin surely did.

We have, in echo of this, always sung up great hymns of praise and wonder to the heavens. We have joined our exuberance to that of nature and hoped, in some measure, that ours will be as generous. We recognize that the joyous need not only sanction but shield, that the possibility of renewal brings joy, and that renewal is in turn made more likely by joy. In the words of the nineteenth-century hymn, an exultant song rings triumphant over despair:

My life flows on in endless song;
Above Earth's lamentation
I hear the sweet tho' far-off hymn
That hails a new creation;
Through all the tumult and the strife,
I hear the music ringing;
It finds an echo in my soul—
How can I keep from singing?

A passion for life is life's ultimate affirmation. To ask the question is to know this to be so; it is to know that exuberance is a god within:

How can I keep from singing?

Notes

Chapter 1: "Incapable of Being Indifferent"

3 "Shield your joyous ones": Variations of this prayer appear in the Anglican *Book of Common Prayer, The Book of Common Prayer* of the U.S. Episcopal church, and the Church of Scotland's "An Order of a Service of Healing."

4 "Under every grief & pine": William Blake, "Auguries of Innocence," lines 61–62, in *The Complete Poetry and Selected Prose of John Donne and the Complete Poetry of William Blake* (New York: Random House, 1941), p. 598.

5 "The Greeks understood": quoted in R. J. Dubos, *Louis Pasteur* (Boston: Little, Brown, 1950), p. 391.

5 as the psalm promises: "Weeping may endure for a night, but joy cometh in the morning," Psalms 30:5.

6 "Why should man want to fly at all?": Charles A. Lindbergh, *The Spirit of St. Louis* (1953; New York: Scribners, 2003), p. 269.

6 "Our earliest records": Charles A. Lindbergh, *Autobiography of Values* (1976; San Diego: Harcourt Brace Jovanovich, 1992), p. 352.

7 "create infectious enthusiasm": Lou Dobbs said of Ted Turner, "He is a natural-born leader. I once asked him his definition of a leader. He said, 'A leader has the ability to create infectious enthusiasm.' " Quoted in Ken Auletta, "The Lost Tycoon," *The New Yorker*, April 23 and 30, 2001, p. 148.

7 Life for Theodore Roosevelt: In addition to the specific works cited, the following general works, among others, were consulted for the section on Theodore Roosevelt: Corrinne Roosevelt Robinson, *My Brother Theodore Roosevelt* (New York: Scribners, 1921); Carleton Putnam, *Theodore Roosevelt: The Formative Years, 1858–1886* (New York: Scribners, 1958); William H. Harbaugh, *The Life and Times of Theodore Roosevelt* (New York: Oxford University Press, 1975); Edmund Morris, *The Rise of Roosevelt* (New York: Coward, McCann & Geoghegan, 1979); David McCullough, *Mornings on Horseback* (New York: Simon & Schuster, 1981); H. W. Brands, *TR: The Last Romantic* (New York: Basic Books, 1997); Edmund Morris, *Theodore Rex* (New York: Random House, 2001).

7 "unpacking of endless Christmas stockings": Mrs. Winthrop Chanler, *Roman Spring: Memoirs* (Boston: Little, Brown, 1934), p. 195.

7 "literally delirious joy": Theodore Roosevelt, *An Autobiography* (1913; New York: Da Capo, 1985), p. 7.

7 "who knows the great enthusiasms": Theodore Roosevelt, *The Works of Theodore Roosevelt*, 20 vols. (New York: Scribners, 1928), vol. 13, pp. 506–29.

8 "I never knew any one": Roosevelt, *Autobiography*, p. 9.

8 "went by in a round": ibid., p. 7.

8 "What an excitement": Letter from Theodore Roosevelt to his mother, April 28, 1868, in *The Letters of Theodore Roosevelt*, 8 vols., ed. E. E. Morison, J. M. Blum, and J. J. Buckley (Cambridge, Mass.: Harvard University Press, 1951–54), vol. 1, p. 3.

8 One debutante said: Nathan Miller, *Theodore Roosevelt: A Life* (New York: William Morrow, 1992), quoted on p. 72.

8 "unquenchable gaiety": ibid., pp. 61–62.

8 "I should almost perish": Theodore Roosevelt, diary entry, February 12, 1878, in *Theodore Roosevelt's Diaries of Boyhood and Youth* (New York: Scribners, 1928).

8 "Sometimes, when I fully realize my loss": Miller, *Theodore Roosevelt*, p. 81.

9 "He'll kill himself": quoted in Edmund Morris, *The Rise of Theodore Roosevelt* (1979; New York: Modern Library, 2001), p. 75.

9 "I am of a very buoyant temper": letter from Theodore Roosevelt to his sister, March 3, 1878, *Letters*, vol. 1, p. 32.

9 "rose like a rocket": letter to Theodore Roosevelt, Jr., October 20, 1903, *Letters*, vol. 3, p. 635.

9 "You could not talk to him": Miller, *Theodore Roosevelt*, p. 157.

9 "The light has gone out": Roosevelt, diary entry, February 14, 1884, in *Diaries*.

9 "black care rarely sits": Theodore Roosevelt, *Ranch Life and the Hunting Trail*, in *The Works of Theodore Roosevelt*, National Edition (New York: Scribners, 1926), vol. 1, p. 329; first published in 1888. He wrote, "These long, swift rides in the glorious spring mornings are not soon to be forgotten. The sweet, fresh air, with a touch of sharpness thus early in the day, and the rapid motion of the fiery little horses combine to make a man's blood thrill and leap with the sheer buoyant lightheartedness and eager, exultant pleasure in the boldness and freedom of the life he is leading" (ibid.).

9 "We felt the beat": Roosevelt, *Autobiography*, p. 95.

9 "I enjoyed life to the full": ibid., p. 96.

10 "wanted to put an end": Miller, *Theodore Roosevelt*, p. 205.

10 "I curled up on the seat": quoted in Morris, *Rise of Theodore Roosevelt*, p. 493.

10 "energy and enthusiasm": Richard Harding Davis, quoted in Miller, *Theodore Roosevelt*, p. 284.

10 "bully," "the great day": Roosevelt's account of the war is given in Theodore Roosevelt, *The Rough Riders*, in *Works*, National Edition, vol. 11, p. 81; first published in 1899.

10 "The President goes from one to another": William Bayard Hale, *A Week in the White House with Theodore Roosevelt* (New York: Putnam, 1908).

11 "You go into Roosevelt's presence": Mark Sullivan, *Our Times: 1900–1925*, 6 vols. (New York: Scribners, 1926–35), vol. 3, p. 81.

11 "You must always remember": Cecil Spring-Rice was Roosevelt's best man when he married Edith Carow. His comment about Roosevelt is quoted in Miller, *Theodore Roosevelt*, p. 50.

11 "surrounded him as a kind of nimbus": Lawrence F. Abbott, *Impressions of Theodore Roosevelt* (New York: Doubleday, Page, 1919), p. 267. Abbott also said that no individual "in modern times touched so many and so varied fields of activity in human life with such zest and vitality" (p. 266).

11 "fully intended to make science my life-work": Roosevelt, *Autobiography*, p. 26.

12 Native bison herds were decimated: Paul Russell Cutright, *Theodore Roosevelt: The Naturalist* (New York: Harper & Brothers, 1958), p. 2.

12 "Ever since man": Theodore Roosevelt, "The Conservation of Wild Life," *The Outlook*, January 20, 1915, in *Works*, National Edition, vol. 12, p. 424.

12 "There can be no greater issue": Theodore Roosevelt, "A Confession of Faith," address to the National Convention of the Progressive Party, August 6, 1912, in *Works*, National Edition, vol. 17, pp. 293–94.

12 "He is doubtless the most vital man": John Burroughs, *Camping and Tramping with Roosevelt* (Boston and New York: Houghton Mifflin, 1907), pp. 60–61. Another friend, Senator Henry Cabot Lodge, said shortly after Roosevelt died: "He touched a subject and it suddenly began to glow as when the high-power electric current touches the metal and the white light starts forth and dazzles the onlooking eyes. We know the air played by the Pied Piper of Hamelin no better than we know why Theodore Roosevelt thus drew the interest of men after him. We only know they followed wherever his insatiable activity of mind invited them." Address of Senator Henry Cabot Lodge of Massachusetts in Honor of Theodore Roosevelt, Before the Congress of the United States, February 9, 1919 (Washington, D.C.: U.S. Government Printing Office, 1919), p. 44.

13 "there would be little ground left": quoted in Miller, *Theodore Roosevelt*, p. 470.

13 "During the seven and a half years": Roosevelt, *Autobiography*, pp. 434–35.

13 "Wild beasts and birds": Theodore Roosevelt, "The Conservation of Wild Life," in *Works*, National Edition, vol. 12, pp. 423–31; quote on p. 425.

13 "A grove of giant redwoods": Theodore Roosevelt, *A Book-Lover's Holidays in the Open* (New York: Scribners, 1923); published in *Works*, vol. 4, p. 227.

13 "It is not the critic who counts": Roosevelt, *Works*, vol. 13, pp. 506–29.

14 "When I was a boy in Scotland": John Muir, *The Story of My Boyhood and Youth,* in *The Wilderness Journeys* (1913; Edinburgh: Canongate, 1996), pp. 1, 23.

15 "flying to the woods": ibid., p. 130.

15 "My eyes never closed": ibid.

15 "University of the Wilderness": ibid., p. 132.

15 "glorious botanical and geological excursion": ibid. Muir said that the excursion, looking back on it, "lasted nearly fifty years and is not yet completed, always happy and free . . . urged on and on through endless, inspiring, Godful beauty" (ibid.).

15 "glowing with Heaven's unquenchable enthusiasm": John Muir, *My First Summer in the Sierra,* in *The Wilderness Journeys* (1911; Edinburgh: Canongate, 1996), pp. 59, 63.

15 "Our camp grove fills": ibid., p. 71.

15 "joyful, wonderful, enchanting": ibid., p. 90.

15 "I shouted and gesticulated": ibid., p. 66.

15 "Exhilarated with the mountain air": ibid., p. 51.

16 "rocking and swirling": John Muir, "A Wind-Storm in the Forest," in John Muir, *The Mountains of California* (New York: Modern Library, 2001), p. 183; first published as "A Wind Storm in the Forest of the Yuba," *Scribner's Monthly,* November 1878.

16 "so noble an exhilaration of motion": ibid.

16 "Muir at once went wild": Samuel Hall Young, *Alaska Days with John Muir,* in *John Muir: His Life and Letters and Other Writings,* ed. T. Gifford (London: Bâton Wicks, 1996), p. 627. *Alaska Days* was first published in 1915.

16 "I feel as if driven": Letter from John Muir to Sarah Muir Galloway, February 26, 1875, in Gifford, *Life and Letters,* pp. 215–16.

16 "Every summer my gains": Letter from John Muir to Louie Wanda Strentzel, October 1879, ibid., p. 249.

17 "Do behold the King": Letter from John Muir to Mrs. Ezra S. Carr, n.d. [probably 1870], ibid., pp. 139–40.

17 "There is a balm": ibid., p. 140.

17 "He sung the glory of nature": Robert Underwood Johnson, quoted ibid., p. 873. ("John Muir as I Knew Him," talk given before the American Academy of Arts and Letters in New York, January 6, 1916.)

18 "Muir was always discovering": Young, *Alaska Days,* p. 647.

18 "How often have I longed for": ibid., p. 678.

18 "To have explored with Muir": Charles Keeler, "Recollections of John Muir," in Gifford, *Life and Letters,* p. 880.

18 Muir's was the most original mind: Ralph Waldo Emerson, quoted in Graham White's introduction to *The Wilderness Journeys*, p. vi.

18 "spell of fire and enthusiasm": Marion Randall Parsons, "John Muir and the Alaska Book," *Sierra Club Bulletin,* 10: 33–34 (1916). Theodore Roosevelt also commented on Muir's verbal persuasiveness: "John Muir talked even better than he wrote. His greatest influence was always upon those who were brought into close personal contact with him." Quoted in Stephen Fox, *John Muir and His Legacy* (Boston: Little, Brown, 1981), p. 126.

18 "I write to you personally": Letter from Theodore Roosevelt to John Muir, 1903, quoted in Paul Russell Cutright, *Theodore Roosevelt: The Making of a Conservationist* (Urbana and Chicago: University of Illinois Press, 1985), p. 247.

19 "I had a perfectly glorious time": Letter from John Muir to his wife, in William Frederic Badè, *The Life and Letters of John Muir,* 2 vols. (Boston: Houghton Mifflin, 1924), vol. 2, p. 412.

19 "I fairly fell in love with him": Letter from John Muir to C. Hart Merriam, ibid.

19 "I trust I need not tell you": quoted in Cutright, *Making of a Conservationist,* pp. 115–16.

19 an additional sense of urgency: On Muir's death, Roosevelt said "he was also—what few nature-lovers are—a man able to influence contemporary thought and action on the subjects to which he had devoted his life. He was a great factor in influencing the thought of California and the thought of the entire country so as to secure the preservation of those great natural phenomena—wonderful canyons, giant trees, slopes of flower-spangled hillsides. . . . [O]ur generation owes much to John Muir" (January 6, 1915). Roosevelt, *Works,* vol. 12. p. 566.

19 "I have just come from": Badè, *Life and Letters of John Muir,* p. 376.

19 "Any fool can destroy trees": Gifford, *Life and Letters,* pp. 372–73.

20 wilderness was a necessity: John Muir, quoted in Gretel Ehrlich, *John Muir: Nature's Visionary* (Washington, D.C.: National Geographic Society, 2000), p. 131. "Thousands of tired, nerve-shaken, over-civilized people are beginning to find out that going to the mountains is going home; that wilderness is a necessity."

20 "lies the hope of the world": Quoted in Frederick Turner, *John Muir: Rediscovering America* (New York: Viking Penguin, 1985), p. 290. Thoreau had said "in Wildness is the preservation of the World. Every tree sends its fibres forth in search of the Wild. The cities import it at any price. Men plow and sail for it. From the forest and wilderness come the tonics and barks which brace mankind." Henry David Thoreau, "Walking," in *The Essays of Henry D. Thoreau* (New York: North Point Press, 2002), p. 162; essay first published in 1862.

20 "The galling harness": Turner, *John Muir,* p. 290.

21 "All of us who give service": Roosevelt, *Works*, vol. 11, p. 267.
21 "I only went out for a walk": Turner, *John Muir*, p. 350.

Chapter 2: "This Wonderful Loveliness"

22 One pair of poppies: The figures for poppies and spiders come from an outstanding educational exhibit at the Natural History Museum in London, Origins of Species Gallery, June 2000.

22 The fertility and diversity of nature: For an excellent discussion, see Edward O. Wilson, *The Diversity of Life* (Cambridge, Mass: Belknap Press, 1992).

"The extravagant gesture is the very stuff of creation," writes Annie Dillard. "After the one extravagant gesture of creation in the first place, the universe has continued to deal exclusively in extravagances, flinging intricacies and colossi down aeons of emptiness, heaping profusion on profligacies with ever-fresh vigor. The whole show has been on fire from the word go." Annie Dillard, *Pilgrim at Tinker Creek* (New York: Harper & Row, 1974), p. 9.

22 1,500 species of butterfly: Wilson, *Diversity of Life*, pp. 185–86.

22 Lichens, among nature's oldest: For information about lichens, I relied upon two excellent books: Oliver Gilbert, *Lichens* (London: HarperCollins, 2000); and Irwin M. Brodo, Sylvia Duran Sharnoff, and Stephen Sharnoff, *Lichens of North America* (New Haven, Conn.: Yale University Press, 2001).

23 "endless forms most beautiful": Charles Darwin, *The Origin of Species* (1859; New York: Random House, 1993), p. 649. "There is grandeur in this view of life," wrote Darwin, "with its several powers, having been originally breathed by the Creator into a few forms or into one; and that, whilst this planet has gone cycling on according to the fixed law of gravity, from so simple a beginning endless forms most beautiful and most wonderful have been, and are being evolved" (pp. 648–49).

23 a million and a half species of fungi: Global Biodiversity Assessment, United Nations Environment Program, 2002. The structure of underlying genetic material is likewise diverse in number and complexity. The human genome has 3 billion DNA base pairs, for instance, but a trumpet lily has 90 billion and an amoeba 670 billion. Clearly the number of base pairs alone does not determine functional complexity, but the range of numbers and the structural diversity of life forms is a reminder of the variety of life evolved by nature. See Jonathan Knight, "All Genomes Great and Small," *Nature*, 417: 374–76 (2002).

23 more than a million species of bacteria: Recently, the molecular biologist Craig Venter discovered at least 1,800 new species of microbes and more than a million previously unknown genes while sampling water from the Sargasso Sea.

Andrew Pollack, "Groundbreaking Gene Scientist Is Taking His Craft to the Oceans," *New York Times*, March 5, 2004.

23 "green cathedral": Wilson, *Diversity of Life*, p. 184.

23 "How deeply with beauty": John Muir, *My First Summer in the Sierra*, in *The Wilderness Journeys* (1913; Edinburgh: Canongate, 1996), p. 74.

23 the thickness of the starfields: God, said Milton, had "sowed with stars the Heaven thick as a field." John Milton, *Paradise Lost*, Book VII.

24 Astronomers live among numbers: Astronomers reported at a January 2004 meeting of the American Astronomical Society that 3 billion years after the Big Bang a string of galaxies 300 million light-years long and 50 million light-years wide—a string that would be 2,000 billion billion miles long—had already formed. Kenneth Chang, "New-found Old Galaxies Upsetting Astronomers' Long-held Theories on the Big Bang," *New York Times*, January 8, 2004.

24 at least 10^{21} stars: The estimate for the number of stars is relatively consistent and is backed up by data from the Hubble Space Telescope (Stephen Telliet, Lunar and Planetary Institute in Houston, personal communication). Estimates of the number of galaxies range more widely, but recent data from the Hubble Space Telescope give a number of 125 billion. (The Hubble has allowed astronomers to see farther into space and thus locate more galaxies than previously thought.)

24 The Milky Way alone: A white dwarf star fifty light-years from Earth is estimated to contain 10 billion trillion trillion carats. T. S. Metcalfe, M. H. Montgomery, A. Kanaan, "Testing White Dwarf Crystallization Theory with Asteroseismology of the Massive Pulsating DA Star BPM 37093," paper submitted to *Astrophysical Journal Letters*, February 2, 2004; Guy Gugliotta, "White Dwarf Star Is a Girl's Best Friend," *Washington Post*, February 14, 2004.

24 10^{41} . . . grams of diamond dust: *Science*, 296: 1397 (2002).

24 "Of mingled blossoms": James Thomson, *The Seasons* (1726–30; New York: Frederick A. Stokes & Brother, 1889), pp. 21, 25.

25 "central theater of life": Edward Hoagland, introduction to Gavin Maxwell, *Ring of Bright Water* (New York: Penguin, 1996), p. v.

25 "bens and glens of stars": Robert Crawford, "From the Top," *The Tip of My Tongue* (London: Jonathan Cape, 2003), p. 37.

26 "Can you bind the beautiful Pleiades?": Job 38: 31–32.

26 "Who publishes the sheet-music": Turner, *John Muir*, p. 233.

27 "In the presence of nature": Ralph Waldo Emerson, "Nature," in *Nature* and *Walking* (Boston: Beacon Press, 1991), pp. 7–8. "Nature" first published in 1836.

27 landmark study of ecstasy: Marghanita Laski, *Ecstasy: A Study of Some Secular and Religious Experiences* (London: Cresset Press, 1961).

28 "fashioned for himself": James Frazer, *The Golden Bough: A Study in Magic and Religion* (New York: Penguin, 1996), p. 465; abridgment, first published in 1922.

28 "Nothing is so beautiful as Spring": Gerard Manley Hopkins, "Spring," in *The Poems of Gerard Manley Hopkins*, ed. W. H. Gardner and N. H. MacKenzie (London: Oxford University Press, 1967), p. 67.

28 Our vitalities change: Oyano Atotsugi has written:

> *No matter how tight*
> *the stitches are pulled*
> *they are rent asunder*
> *as the plum bursts into bloom*
> *and the warbler bursts into song.*

Quoted in Charlotte van Rappard-Boon, *Poetry and Image in Japanese Prints* (Leiden: Hotei Publishing, 2000), p. 61.

29 "The brooks sing carols": Henry David Thoreau, *Walden; or, Life in the Woods* (Boston: Beacon Press, 1997), p. 290. First published in Boston by Ticknor & Fields, 1854.

29 "Walden was dead": ibid., p. 291. "So our human life but dies down to its root," wrote Thoreau, "and still puts forth its green blade to eternity" (ibid.).

29 "I am the great Sun": John Heath-Stubbs, "Canticle of the Sun: Dancing on Easter Morning," in *Collected Poems: 1943–1987* (Manchester, U.K.: Carcanet, 1988), p. 333. Sir John Suckling, in his "Ballad Upon a Wedding," had written more than three hundred years earlier: "But oh, she dances such a way, / No sun upon an Easter day / Is half so fine a sight."

30 darkness covered the land: Matthew 27:45, "Now from the sixth hour there was darkness over the land unto the ninth hour."

30 "There was only—spring itself": Willa Cather, *My Ántonia* (1918; New York: Signet, 1994), p. 116.

31 "The earth / Puts forth": Langston Hughes, "In Time of Silver Rain," in *Selected Poems of Langston Hughes* (New York: Vintage, 1959), p. 56.

31 "With the sunshine": F. Scott Fitzgerald, *The Great Gatsby* (1926; London: Penguin, 1950), pp. 9–10.

31 "the dark night wakes": Phillips Brooks, "O Little Town of Bethlehem," in *The English Hymnal* (London: Oxford University Press, 1906), p. 14.

31 "When we try to pick out anything": Muir, *First Summer*, p. 91.

32 "The vastness of the heavens": Richard Feynman, M. L. Sands, and R. B. Leighton, *The Feynman Lectures on Physics*, vol. 1 (New York: Addison-Wesley, 1994).

33 "Amazed and thrilled": Wilson A. Bentley, "Forty Years' Study of Snow Crystals," *Monthly Weather Review*, 52: 530–32 (1924).

33 "The deeper one enters": Wilson A. Bentley, "The Latest Designs in Snow and Frost Architecture," *The American Annual of Photography,* 20: 166–70 (1906).

33 "it was the snowflakes that fascinated me": Mary B. Mullet, "Snowflake Bentley," *The American Magazine,* February 1925.

33 "I found that snowflakes": ibid. The study of the structure of snow crystals goes back at least to the second century before Christ. The oldest record of Chinese investigations of snowflakes was made by Han Ying about 135 B.C.: "Flowers of plants and trees are generally five-pointed, but those of snow . . . are always six-pointed." Joseph Needham and Lu Gwei-Djen, "The Earliest Snow Crystal Observations, *Weather,* 16: 312–27 (1961).

33 "When a snowflake melted": Mullet, "Snowflake Bentley."

33 "a wonderful little splinter of ice": ibid. Bentley's interviewer reports: " 'That was a tragedy!' he said, shaking his head mournfully. 'In spite of my carefulness, the crystal was broken in transferring it to the slide.' His voice actually shook with emotion. 'It makes me almost cry, even now,' he said, as if he were speaking of the death of a friend."

34 "great desire to show": ibid.

34 "Perhaps they come to us": Wilson A. Bentley, "The Wonders and Beauties of Snow," *Christian Herald,* March 2, 1904.

34 "How full of the creative genius": Henry David Thoreau, *H. D. Thoreau: A Writer's Journal,* selected and ed. Laurence Stapleton (New York: Dover, 1960), p. 134.

35 he who is *too* much a master: Interestingly, Matthew Cobb suggests in an essay in *Nature* that scientists have become too objective in recent times: "Science is, after all, about communication. Would the objectivity and precision of the modern scientific article really suffer if we were to express just a fragment of our feelings about our work?" Matthew Cobb, "Wondrous Order," *Nature,* 413: 779 (2001).

35 Bentley used the words "beauty" or "beautiful": Wilson A. Bentley, "Studies Among the Snow Crystals During the Winter in 1901–2," *Monthly Weather Review,* 30: 607–16 (1902).

36 "I assume that the configurations": ibid.

36 the definitive biography: Duncan C. Blanchard, *The Snowflake Man: A Biography of Wilson A. Bentley* (Blacksburg, Va.: McDonald & Woodard, 1998).

36 Sir Galahad's for the Holy Grail: "Bentley returned again and again to this idea of 'the one preeminently beautiful snow crystal.' It haunted him long after his search for scientific understanding had diminished." Personal correspondence from Duncan Blanchard to the author, September 12, 2000.

36 "preeminently beautiful" snow crystal: Bentley, "Studies Among the Snow Crystals." "It is extremely improbable that anyone has as yet found, or, indeed,

ever will find, the one preeminently beautiful and symmetrical snow crystal that nature has probably fashioned when in her most artistic mood" (p. 616).

37 "The beautiful branching one": Wilson A. Bentley, "Some Recent Treasures of the Snow," *Monthly Weather Review,* 55: 358–59 (1927).

37 "wonderfully brilliant closing": ibid.

37 No two will be alike: "New and beautiful designs seem to be as numerous now as when I began the work 40 years ago. While many of them are very similar to one another, I have, as yet, found no exact duplicates." Bentley, "Forty Years' Study," p. 532.

37 "considering all the ways those molecules": Fred Hapgood, "When Ice Crystals Fall from the Sky Art Meets Science," *Smithsonian,* 6: 66–73 (1976), p. 71.

37 "it could snow day and night": Kenneth Libbrecht, *The Snowflake: Winter's Secret Beauty,* photographs by Patricia Rasmussen (Stillwater, Minn.: Voyageur Press, 2003), p. 102.

38 six hundred auroras: Blanchard, in his 1998 biography of Bentley, discusses these and other scientific contributions in detail.

38 a little cracked: Bentley was aware of his neighbors' concerns. In an interview published after his death, he was quoted as saying, "A fool, and probably crazy, is what some folks call me, and they mostly demand to know what good it does to get all these pictures of just snow! I don't argue with them." Vrest Orton, " 'Snowflake' Bentley," *Vermont Life,* 2: 11–13 (1948).

39 "John Ruskin declared": "Bentley's Contribution," *Burlington Free Press,* December 28, 1931.

39 "So long as eyes shall see": Bentley, "Wonders and Beauties," p. 191.

39 "insistent ardor of the lover": W. A. Bentley and W. J. Humphreys, *Snow Crystals* (New York: McGraw-Hill, 1931), p. 2.

39 "Cold west wind afternoon": December 7, 1931, entry in weather notebook, quoted in Blanchard, *Snowflake Man,* p. 214.

Chapter 3: "Playing Fields of the Mind"

40 a stuff which will not endure: "Youth's a stuff will not endure," William Shakespeare, *Twelfth Night,* Act II, scene 3, line 53.

40 "In the sun that is young": Dylan Thomas, "Fern Hill," in *The Collected Poems of Dylan Thomas* (New York: New Directions, 1953), p. 178.

41 "Natural selection": Karl Groos, *The Play of Animals,* trans. E. L. Baldwin (New York: D. Appleton and Company, 1898), p. xx.

41 *Plein,* meaning, in Middle Dutch: John Ayto, *Dictionary of Word Origins* (New York: Arcade, 1991).

42 "soup of behavior": S. Miller, "Ends, Means, and Galumphing: Some Leitmotifs of Play," *American Anthropologist,* 75: 87–98 (1973).

43 "went simply galumphing about": Lewis Carroll, *The Hunting of the Snark: An Agony in Eight Fits* (1876; New York: Pantheon, 1966), p. 26.

43 "One, two! One, two!": Lewis Carroll, *Through the Looking-Glass,* in *Alice's Adventures in Wonderland & Through the Looking-Glass* (1871; New York: Signet, 2000), p. 138.

43 Rhesus monkeys running to play: D. S. Sade, "An Ethogram for Rhesus Monkeys: I. Antithetical Contrasts in Posture and Movement," *American Journal of Physical Anthropology,* 38: 537–42 (1973).

43 facial expressions of black bear cubs: J. D. Henry and S. M. Herrero, "Social Play in the American Black Bear: Its Similarity to Canid Social Play and an Examination of Its Identifying Characteristics," *American Zoologist,* 14: 371–89 (1974).

43 "gives the whole world": Groos, *Play of Animals,* p. 326.

44 "It exploded with joy": George B. Schaller, *The Last Panda* (Chicago: University of Chicago Press, 1993), p. 66.

44 "Wombat play": Barbara Triggs, *The Wombat: Common Wombats in Australia* (Sydney: University of New South Wales, 1996), pp. 79–80.

45 "hair-trigger mousetraps": Carolyn King, *The Natural History of Weasels and Stoats* (Ithaca, N.Y.: Cornell University Press, 1989), p. 4.

45 "From whichever retreat": quoted ibid., p. 6, from P. Drabble, *A Weasel in My Meatsafe* (London: Michael Joseph, 1977).

45 "extremely bad at doing nothing": Gavin Maxwell, *Ring of Bright Water* (1960; New York: Penguin, 1996), p. 92.

45 "what soon became": ibid., p. 102.

46 "would set out from the house": ibid., p. 138.

46 Marc Bekoff, a biologist: M. Bekoff, "The Development of Social Interaction, Play, and Metacommunication in Mammals: An Ethological Perspective," *Quarterly Review of Biology,* 47: 412–34 (1972). Among the general sources I consulted about animal play—in addition to those listed elsewhere in these notes—were F. A. Beach, "Current Concepts of Play in Animals," *American Naturalist,* 79: 523–41 (1945); D. F. Lancy, "Play in Species Adaptation," *Annual Review of Anthropology,* 9: 471–95 (1980); Paul D. MacLean, *The Triune Brain: Role in Paleocerebral Functions* (New York: Plenum, 1990); Marc Bekoff and John A. Byers, *Animal Play: Evolutionary, Comparative, and Ecological Perspectives* (Cambridge, U.K.: Cambridge University Press, 1998); David F. Bjorklund and Anthony D. Pellegrini, *The Origins of Human Nature: Evolutionary Developmental Psychology* (Washington, D.C.: American Psychological Association Press, 2002).

46 "He won't say": Annie Dillard, *Teaching a Stone to Talk: Expeditions and Encounters* (New York: Harper Colophon, 1982), p. 15.

47 "Young bulls love to chase things": Katy Payne, *Silent Thunder: In the Presence of Elephants* (New York: Penguin, 1998), pp. 72–73.

47 Young sea lions: Marianne Riedman, *The Pinnipeds: Seals, Sea Lions, and Walruses* (Berkeley: University of California Press, 1990), p. 338.

47 Harbor seals of all ages: D. Renouf, "Play in a Captive Breeding Colony of Harbour Seals (*Phoca vitulina*): Constrained by Time or by Energy?" *Journal of Zoology,* 231: 351–63 (1993).

48 Biologists studying harbor seals: ibid.

48 "One animal, probably a subadult": George B. Schaller, Hu Jinchu, Pan Wenshi, and Zhu Jing, *The Giant Pandas of Wolong* (Chicago: University of Chicago Press, 1985), p. 150.

48 Japanese macaque monkeys: G. Eaton, "Snowball Construction by a Feral Troop of Japanese Macaques (*Macaca fuscata*) Living Under Seminatural Conditions," *Primates,* 13: 411–14 (1972).

48 young wolf's first experience: John Fentress, "Animal Emotions: Wolves," paper presented at conference on animal emotions held at the Smithsonian Institution, Washington, D.C., October 28, 2000.

48 ravens pushing themselves: Bernd Heinrich, *Mind of the Raven* (New York: Cliff Street Books, 1999).

49 dancelike behaviors: Payne, *Silent Thunder,* p. 63.

49 "He'd whirl in a circle": Ronald Rood, *How Do You Spank a Porcupine?* (Shelburne, Vt.: New England Press, 1969), p. 61.

49 The porcupines, in solitary play: A. R. Shadle, "The Play of American Porcupines," *Journal of Comparative Psychology,* 37: 145–49 (1944), p. 147.

50 Young weasels and stoats: King, *Natural History of Weasels and Stoats.*

50 "The random high spirits of youth": P. Chalmers Mitchell, *The Childhood of Animals* (London: William Heinemann, 1912), p. 242.

51 "in order that they may have surplus energy": ibid., p. 245.

51 facilitate an animal's ability to move: J. Byers and C. Walker, "Refining the Motor Training Hypothesis for the Evolution of Play," *American Nature,* 146: 25–40 (1995).

51 Very young cheetah cubs: T. M. Caro, "Short-Term Costs and Correlates of Play in Cheetahs," *Animal Behaviour,* 49: 333–45 (1995).

51 Serengeti lion cubs: George B. Schaller, *The Serengeti Lion: A Study of Predator-Prey Relations* (Chicago: University of Chicago Press, 1972).

52 even domestic cattle will play: A. Brownlee, "Play in Domestic Cattle in Britain: An Analysis of Its Nature," *British Veterinary Journal,* 110: 48–68 (1954).

52 Sea lions and seals play: Reidman, *Pinnipeds.*

52 Dolphins have been seen: Rachel Smolker, *To Touch a Wild Dolphin* (New York: Doubleday, 2001).

52 "stretch their necks above the surface": Hope Ryden, *Lily Pond* (New York: Lyons & Burford, 1989), pp. 215–16.

52 He smelled, touched, and tasted: Joyce Poole, *Coming of Age with Elephants* (New York: Hyperion, 1996), p. 157.

53 Birds also play: W. H. Thorpe, "Ritualization in Ontogeny: I. Animal Play," *Philosophical Transactions of the Royal Society of London,* series B, 251: 311–19 (1966); Heinrich, *Mind of the Raven.*

53 Play is conspicuously rare in reptiles: Play has been observed only rarely in reptiles, presumably due to their limited energy and thermal resources. An interesting exception is a Nile soft-shelled turtle at the National Zoo in Washington who played with several objects added to his tank (including a rubber hoop, sticks, and a basketball). Unlike reptiles in the wild, however, this turtle had his metabolic and food needs met by others. G. M. Burghardt, B. Ward, and R. Rosscoe, "Problem of Reptile Play: Environmental Enrichment and Play Behavior in a Captive Nile Soft-Shelled Turtle, *Trionyx triunguis,*" *Zoo Biology,* 15: 223–38 (1996).

53 "The young of the Komodo dragon": P. D. MacLean, "Brain Evolution Relating to Family, Play, and the Separation Call," *Archives of General Psychiatry,* 42: 402–17 (1985).

54 communal den of spotted hyenas: C. M. Drea, J. E. Hawk, and S. E. Glickman, "Aggression Decreases as Play Emerges in Infant Spotted Hyaenas: Preparation for Joining the Clan," *Animal Behaviour,* 51: 1323–36 (1996).

55 "Not a single night had passed": Ryden, *Lily Pond,* pp. 87, 104.

55 Wolves, who form close packs: Bekoff, "Development of Social Interaction."

55 Common seals engage: S. Wilson, "Juvenile Play of the Common Seal *Phoca vitulina* with Comparative Notes on the Grey Seal *Halichoerus grypus,*" *Behaviour,* 48: 37–60 (1971).

55 "merely nuzzled each other": Personal communication from Desmond Morris to S. Wilson, quoted ibid.

56 "Something was obviously amiss": Benjamin Kilham and Ed Gray, *Among the Bears: Raising Orphan Cubs in the Wild* (New York: Henry Holt, 2002), pp. 254–55.

56 "almost delirious with excitement": Cynthia Moss, "Animal Emotions: Elephants," paper presented at conference on animal emotions held at the Smithsonian Institution, Washington, D.C., October 28, 2000.

56 "rush together, heads high": Joyce Poole, "Family Reunions," in *The Smile of a Dolphin,* ed. Marc Bekoff (New York: Discovery Books, 2000), p. 122.

57 "Protected, comforted, cooed over": Joyce Poole, keynote address to the 22nd Annual Elephant Managers Workshop, presented by Disney's Animal Kingdom, Orlando, Fla., November 9–12, 2001.

57 "elephantine joy": Moss, Smithsonian lecture, 2000.

57 "The bird suddenly stretches": William J. L. Sladen, "Social Structure Among Penguins," in *Group Processes,* ed. B. Schaffner (New York: Josiah Macy, Jr., Foundation, 1956), p. 45.

57 African wild dogs: R. D. Estes and J. Goodard, "Prey Selection and Hunting Behavior of the African Wild Dog," *Journal of Wildlife Management,* 31: 52–70 (1967), p. 57.

58 powerfully reinforcing opioid systems: A. P. Humphreys and D. F. Einon, "Play as a Reinforcer for Maze-Learning in Juvenile Rats," *Animal Behaviour,* 29: 259–70 (1981); W. W. Beatty and K. B. Costello, "Naloxone and Play Fighting in Juvenile Rats," *Pharmacological and Biochemical Behavior,* 17: 905–7 (1982); J. Panksepp, J. E. Jalowiec, F. G. DeEskenazi, and P. Bishop, "Opiates and Play Dominance in Juvenile Rats," *Behavioral Neuroscience,* 99: 441–53 (1985); L.J.M.J. Vanderschuren, R.J.M. Niesink, B. M. Spruijt, and J. M. Van Ree, "Effects of Morphine on Different Aspects of Social Play in Juvenile Rats," *Psychopharmacology,* 117: 225–31 (1995); L.J.M.J. Vanderschuren, E. A. Stein, V. M. Wiegant, and J. M. Van Ree, "Social Play Alters Regional Brain Opioid Receptor Binding in Juvenile Rats," *Brain Research,* 680: 148–56 (1995); Jaak Panksepp, *Affective Neuroscience* (New York: Oxford University Press, 1998), p. 248.

58 "brain source of joy": Panksepp, *Affective Neuroscience,* p. 280.

58 play probably increases gene expression: ibid., p. 291.

59 Trout raised in hatcheries: Research findings presented by Michael Marchetti and Gabrielle Nevitt to the Ecological Society of America in August 2000, *Science,* 289 (August 25, 2000).

59 "the brains of domestic rabbits": Charles Darwin, *The Descent of Man,* in *The Works of Charles Darwin,* vol. 21 (London: Pickering & Chatto, 1989), p. 59; first published in 1871.

59 "learns during play": Jane Goodall, *In the Shadow of Man* (Boston: Houghton Mifflin, 1988), p. 156.

60 "Inasmuch as new objects": William James, *The Principles of Psychology,* vol. II (1890; New York: Dover, 1950), p. 429.

60 "susceptibility for being excited": ibid.

60 "It's cat and monkey spirit": Eugene Walter, as told to Katherine Clark, *Milking the Moon: A Southerner's Story of Life on This Planet* (New York: Crown, 2001).

60 The systems in the brain: Panksepp, *Affective Neuroscience.*

61 "There was a child went forth": Walt Whitman, "There Was a Child Went

Forth," in *Leaves of Grass,* ed. Sculley Bradley and Harold W. Blodgett (New York: Norton, 1965), p. 364.

61 "an acting out of options": Heinrich, *Mind of the Raven,* p. 294.

61 "One by one": Quoted in Cynthia Asquith, *Portrait of Barrie* (London: Greenwood Press, 1954), p. 220.

61 Highly creative children: E. P. Torrance, "Priming Creative Thinking in the Primary Grades," *Elementary School Journal,* 62: 139–45 (1961); J. W. Getzels and P. W. Jackson, *Creativity and Intelligence* (New York: Wiley, 1962); M. A. Wallach and N. Kogan, *Modes of Thinking in Young Children* (New York: Holt, 1965).

Playfulness in childhood is also associated with increased creativity in adulthood: J. L. Singer, *The Child's World of Make-Believe: Experimental Studies of Imaginative Play* (New York: Academic Press, 1973); L. R. Sherrod and J. L. Singer, "The Development of Make-Believe Play," in *Sports, Games, and Play,* ed. J. Goldstein (Northvale, N.J.: Jason Aronson, 1989), pp. 1–15.

62 children's ability to produce: J. N. Lieberman, "Playfulness and Divergent Thinking: An Investigation of Their Relationship at the Kindergarten Level," *Journal of General Psychology,* 107: 219–24 (1965); J. L. Dansky and I. W. Silverman, "Effect of Play on Associative Fluency in Preschool-Aged Children," *Developmental Psychology,* 9: 38–43 (1973); J. S. Bruner, A. Jolly, and K. Sylva, eds., *Play* (New York: Basic Books, 1976); J. N. Lieberman, *Playfulness: Its Relationship to Imagination and Creativity* (New York: Academic Press, 1977); J. Singer, "Affect and Imagination in Play and Fantasy," in *Emotions in Personality and Psychopathology,* ed. C. E. Izard (New York: Plenum, 1979), pp. 13–34.

62 The level of elation: Singer, *Child's World of Make-Believe;* Singer, "Affect and Imagination."

62 two dimensions of play: L. A. Barnett, "Playfulness: Definition, Design and Measurement," *Play and Culture,* 3: 319–36 (1990); L. Barnett, "Characterizing Playfulness Correlates with Individual Attributes and Personal Traits," *Play and Culture,* 4: 371–93 (1991).

62 first and most strikingly apparent: Communication with author, 2000; Ellen Winner, *Gifted Children: Myth and Realities* (New York: Basic Books, 1996). Earlier studies also have shown the importance of high energy levels in creative individuals: V. Goertzel and M. Goertzel, *Cradles of Eminence* (Boston: Little, Brown, 1962); J. Bergman, "Energy Levels: An Important Factor in Identifying and Facilitating the Development of Giftedness in Young Children," *Creative Child and Adult Quarterly,* 4: 181–88 (1979).

63 "It is better to have a broken bone": Lady Allen of Hurtwood, cited in J. Scott, "When Child's Play Is Too Simple," *New York Times,* July 15, 2000.

63 "I am interested": Margaret Mead, quoted in Schaffner, *Group Processes,* p. 21.

64 It was said of John Muir: Samuel Hall Young, *Alaska Days with John Muir,* in *John Muir: His Life and Letters and Other Writings,* ed. T. Gifford (London: Bâton Wicks, 1996), p. 678.

64 "child heart": Charles Keeler, "Recollections of John Muir," in Gifford, *Life and Letters,* p. 878.

64 "It's my last chance": quoted in Nathan Miller, *Theodore Roosevelt: A Life* (New York: William Morrow, 1992), p. 535.

64 "the boy in him had died": quoted ibid., p. 562.

64 " 'I am cherry alive' ": Delmore Schwartz, " 'I Am Cherry Alive,' the Little Girl Sang," in *Selected Poems (1938–1958): Summer Knowledge* (New York: New Directions, 1959), p. 161.

Chapter 4. "The Glowing Hours"

67 "flashing from one end": Robert Louis Stevenson, "Crabbed Age and Youth," in *The Lantern Bearers and Other Essays* (New York: Farrar, Straus and Giroux, 1988), p. 65; essay first published in 1877.

68 "stripped off himself": Max Beerbohm, "The Child Barrie," *Saturday Review,* January 7, 1905, pp. 13–14.

68 "I think one remains": James Matthew Barrie, from his dedication to *Peter Pan* (play) (New York: Dover, 2000), p. x; first performed in London in 1904.

68 "Perhaps we do change": ibid., p. xi.

68 "I can still remember everything": quoted in Humphrey Carpenter, *Secret Gardens: A Study of the Golden Age of Children's Literature* (Boston: Houghton Mifflin, 1985), p. 119.

69 "*Could* you ask your friend": A. A. Milne, *The House at Pooh Corner* (1928; New York: Puffin, 1992), p. 31.

69 "with a way of saying": ibid., p. 61.

69 "who was always in front": ibid., p. 75.

69 "But whatever his weight in pounds": ibid., p. 32.

70 "I don't think they ought to be there": ibid., p. 33.

70 "They wanted to come in": ibid.

70 "Stornry good flyers": ibid., p. 62.

70 "Can they climb trees": ibid., p. 63.

70 "Climbing trees is what they do best": ibid.

71 "Once, it was nothing but sailing": Kenneth Grahame, *The Wind in the Willows* (1908; New York: St. Martin's Griffin, 1994), p. 29.

71 "There you are!": ibid., pp. 42, 44.

72 "The poetry of motion!": ibid., p. 51.

72 "What are we going to do": ibid., p. 52.

72 "We can't all": A. A. Milne, *Winnie-the-Pooh* (1926; New York: Puffin, 1992), p. 74.
72 "an animal of tilled field": Grahame, *Wind in the Willows*, p. 93.
73 " 'O, Mole!' cries the Rat": ibid., p. 143.
74 "It's time we taught him a lesson": Milne, *House at Pooh Corner*, p. 109.
74 "there's too much of him": ibid., p. 111.
74 "Taking people by surprise": ibid., p. 104.
74 "He just *is* bouncy": ibid., p. 105.
74 "Piglet settled it all": ibid., p. 111.
74 "a different Tigger altogether": ibid., pp. 112–13.
75 "If we can make Tigger": ibid., p. 113.
75 "Tigger was tearing around": ibid., pp. 126–27.
75 "We'll take Toad seriously in hand": Grahame, *Wind in the Willows*, p. 81.
75 "I'm *not* sorry": ibid., p. 123.
76 "It's for your own good": ibid., p. 124.
76 "We'll take great care": ibid., p. 124.
76 "They descended the stair": ibid., pp. 125–26.
77 "Toad once more": ibid., p. 133.
77 "It was too late": ibid., p. 238.
79 "I'm a cheerful sort of man": P. L. Travers, *Mary Poppins* (1934; New York: Dell, 1991), p. 32.
79 "I become so filled": ibid., p. 33.
79 "rolling over and over": ibid., p. 34.
79 "growing lighter and lighter": ibid., p. 35.
79 "The thought that they would have to go home": ibid., p. 45.
80 "Where and How and When": P. L. Travers, *Mary Poppins Opens the Door* (1944; London: HarperCollins, 1994), pp. 255–56.
80 "interrupted poetry": Umberto Eco, foreword to *Arriva Charlie Brown!* (Milan, Italy: Cartonato Milano Libri, 1963).
80 "simplicity and depth change": Art Spiegelman, "Abstract Art Is a Warm Puppy," *The New Yorker*, February 14, 2000, p. 61.
80 "gold standard": Garry Trudeau, " 'I Hate Charlie Brown': An Appreciation," *Washington Post*, December 16, 1999.
80 "The most terrifying loneliness": Charles M. Schulz, Introduction to *"Peanuts" 35th Anniversary Collection* (New York: Holt, Rinehart & Winston, 1985).
81 "He was the wildest": Charles M. Schulz, interview with Barnaby Conrad, *New York Times Magazine*, April 16, 1967, p. 49.
81 "I wonder why Snoopy": Charles M. Schulz, *Around the World in 45 Years* (Kansas City: Andrews McMeel, 1994), p. 27.
82 "has to retreat into": Charles M. Schulz, interview with Gary Groth, in *Charles*

M. Schulz: Conversations (Jackson: University Press of Mississippi, 2000), p. 221. Originally published in *Comics Journal*, 200: 3–48 (1997).

83 "Life for Snoopy": author's interview with Judy Sladky, September 22, 2000.

84 "Snoopy has the freedom to express": Correspondence from Jean Schulz to author, September 21, 2000.

85 "On with the dance!": George Gordon, Lord Byron, *Childe Harold's Pilgrimage*, Canto the Third, lines 192–95, in *Lord Byron: The Complete Poetical Works*, vol. II, ed. Jerome J. McGann (Oxford: Clarendon, 1980), p. 84; first published in 1816.

86 "The man's true life": Robert Louis Stevenson, "The Lantern Bearers," in *The Lantern Bearers and Other Essays*, pp. 233–34; essay first published in 1887.

86 "All children, except one": James Matthew Barrie, *Peter Pan* (book) (New York: Scribners, 1980), p. 1; first published as *Peter and Wendy* in 1911.

86 "I ran away the day I was born": ibid., p. 26.

87 "You just think lovely wonderful thoughts": ibid., p. 33.

87 "Second to the right": ibid., p. 36.

87 "He would come back laughing": ibid., p. 39.

87 "If he forgets them so quickly": ibid.

87 "all the four seasons": Barrie, *Peter Pan* (play), p. 21.

88 "There are zigzag lines on it": Barrie, *Peter Pan* (book), p. 6.

88 "He was fond of variety": ibid., p. 37.

88 "had seen many tragedies": ibid., p. 81.

89 "Fancy your forgetting": Barrie, *Peter Pan* (play), p. 70.

89 "which look like black candles": ibid., p. 23. Because Barrie had portrayed Captain Hook as a graduate of Eton, he was invited to Eton to discuss the Provost's contention that "James Hook, the pirate captain, was a great Etonian but not a good one." Barrie suggested to the assembled audience that Hook's disreputable ways might have been avoided had he not fallen in with bad companions—that is to say, Harrow graduates—while at Oxford. Barrie also noted that Cook's position in life gave proof to the long-held belief that "the Etonian is a natural leader of men." J. M. Barrie, "Captain Hook at Eton," in *M'Connachie and J.M.B.: Speeches by J. M. Barrie* (London: Peter Davies, 1938), pp. 115–29.

89 "is not wholly evil": Barrie, *Peter Pan* (play), p. 51.

89 "Pan, who and what art thou?": ibid., p. 61.

89 "saw that he was higher": Barrie, *Peter Pan* (book), pp. 87–88.

89 "its true meaning came to me": quoted in Carpenter, *Secret Gardens*, p. 187.

90 "wherever they go": Milne, *House at Pooh Corner*, p. 180.

90 "I remember my youth": Joseph Conrad, *Youth and Two Other Stories* (New York: Doubleday, 1931), pp. 36–37.

90 "The regret we have": Robert Louis Stevenson, "Child's Play," in R. L. Stevenson, *Essays and Poems,* ed. Claire Harman (London: J. M. Dent, 1992), p. 53; essay first published in 1878.

Chapter 5: "The Champagne of Moods"

91 Improbably, the English: Patrick Forbes, *Champagne: The Wine, the Land and the People* (New York: William Morrow, 1967); Tom Stevenson, *Christie's World Encyclopedia of Champagne and Sparkling Wine* (San Francisco: Wine Appreciation Guild, 1998).

91 Christopher Merret described: Christopher Merret, "Some Observations Concerning the Ordering of Wines," paper presented to the Royal Society in London, December 17, 1662.

92 Champagne historian Tom Stevenson: Stevenson, *Christie's World Encyclopedia,* pp. 9–10.

92 attempting to annihilate them: Hugh Johnson writes that Dom Pérignon, the cellar master of the Abbey of Hautvillers at the end of the seventeenth century, "took every precaution to avoid bubbles." Hugh Johnson, *The World Atlas of Wine,* 4th ed. (New York: Simon & Schuster, 1994), p. 77. Robert Joseph concurs: "the last thing the monk was aiming to make was fizzy wine." Robert Joseph, *The Ultimate Encyclopedia of Wine* (London: Carlton, 1996), p. 126.

92 seven bottles of Champagne: *Hachette Atlas of French Wines & Vineyards,* gen. ed. Pascal Ribéreau-Gayon (New York: Viking, 2000), p. 126.

92 250 million bubbles: Stevenson, *Christie's World Encyclopedia,* p. 47. What Champagne would Stevenson recommend? "If money and rarity were no obstacle," he writes, "I would crack a bottle of Heidsieck & Co. 1907 Goût Americain, which has spent the last 80 years on the bottom of the Baltic courtesy of a German U-Boat in 1916. The bottles are incredibly consistent. With no trace of seawater penetration, the constant 2°C has put this wine through suspended animation and the result is a Champagne which is profoundly and succulently sweet, with impeccably balanced fruit of amazing purity and freshness" (ibid., p. 311).

92 "men and girls came and went": F. Scott Fitzgerald, *The Great Gatsby* (1926; London: Penguin, 1990), p. 41.

92 "Hardly did it appear": quoted in Forbes, *Champagne,* p. 131.

92 "Champagne should laugh at you": ibid., p. 352. Forbes is marvelous on the energy and colors of Champagne: "Is it snow-white?" he asks. "Does its agitation, its anxiety to vanish into thin air, convey an impression of force? If so, excellent." He goes on: "Good Champagnes vary in colour through a range of yellows which extend from straw to primrose and buttercup to bright gold and bronze. A hint of green is exciting; a tinge of brown is a danger signal" (ibid.).

93 "O the joy": Walt Whitman, "A Song of Joys," in *Leaves of Grass*, ed. Sculley Bradley and Harold W. Blodgett (New York: Norton, 1973), p. 177.

93 "I must say": Winston Churchill, *Painting as a Pastime* (London: Penguin, 1964), p. 29.

94 "fly to the sky": Alan Jay Lerner and Frederick Loewe, *Gigi*, Warner Brothers, 1958.

94 psychology textbooks have devoted: E. R. Carlson, "The Affective Tone of Psychology," *Journal of General Psychology*, 75: 65–78 (1966).

94 For every hundred journal articles: Martin Seligman, *Authentic Happiness* (New York: Free Press, 2001), p. 14.

94 Cross-cultural analyses: James R. Averill, "On the Paucity of Positive Emotions," in *Advances in the Study of Communication and Affect*, ed. K. R. Blankstein, P. Pliner, and J. Polivy, vol. 6: *Assessment and Modification of Emotional Behavior* (New York: Plenum, 1980), pp. 7–45; P. C. Ellsworth and C. A. Smith, "Shades of Joy: Patterns of Appraisal Differentiating Pleasant Emotions," *Cognition and Emotion*, 2: 301–31 (1988).

95 Survival is made more likely: For an excellent review, see R. F. Baumeister, E. Bratslavsky, C. Finkenauer, and K. D. Vohs, "Bad Is Stronger Than Good," *Review of General Psychology*, 5: 323–70 (2001).

95 brain imaging studies: S. Paradiso, D. L. Johnson, N. C. Andreasen, D. S. O'Leary, G. L. Watkins, L. L. Boles Ponto, and R. D. Hichwa, "Cerebral Blood Flow Changes Associated with Attribution of Emotional Valence to Pleasant, Unpleasant, and Neutral Visual Stimuli in a PET Study of Normal Subjects," *American Journal of Psychiatry*, 156: 1618–29 (1999); S. B. Hamann, T. D. Ely, J. M. Hoffman, and C. D. Kilts, "Ecstasy and Agony: Activation of the Human Amygdala in Positive and Negative Emotions," *Psychological Science*, 13: 135–41 (2002).

95 positive experiences associated with mania: K. R. Jamison, R. H. Gerner, C. Hammen, and C. Padesky, "Clouds and Silver Linings: Positive Experiences Associated with Primary Affective Disorders," *American Journal of Psychiatry*, 137: 198–202 (1980).

96 "Bright Bubbles": Samuel Taylor Coleridge, *Collected Letters of Samuel Taylor Coleridge*, 6 vols., ed. E. L. Griggs (Oxford, U.K.: Oxford University Press, 1956–71), vol. 1, p. 209.

96 an entire issue of *American Psychologist:* special issue on Happiness, Excellence, and Optimal Human Functioning, guest editors Martin E. P. Seligman and Mihaly Csikszentmihalyi, *American Psychologist*, vol. 55 (January 2000).

97 "Our message": ibid., p. 7.

97 "The trouble with the emotions": William James, *The Principles of Psychology*, vol. II (1890; New York: Dover, 1950), p. 449.

98 high on activation: Exuberance is importantly different from Mihaly Czikszentmihalyi's concept of flow (Mihaly Csikszentmihalyi and Isabella Selega Csikszentmihalyi, *Optimal Experience: Psychological Studies of Flow in Consciousness* [Cambridge, U.K.: Cambridge University Press, 1988]), most particularly in the intensity and energy of the high mood state of exuberance. As Martin Seligman points out, "There is no positive emotion on the list of [flow's] essential components. While positive emotions like pleasure, exhilaration, and ecstasy are occasionally mentioned, typically in retrospect, they are not usually felt. In fact it is the absence of emotion, of any kind of consciousness, that is at the heart of flow. Consciousness and emotion are there to correct your trajectory; when what you are doing is seamlessly perfect, you don't need them." Martin Seligman, *Authentic Happiness* (New York: Free Press, 2002), pp. 115–16.

98 anyone who experiences joy: C. S. Lewis, *Surprised by Joy: The Shape of My Early Life* (San Diego: Harvest, 1955), p. 18.

98 Captive foxes: paper presented by Samantha Bremner to the British Ecological Society, December 2001, and reported in *New Scientist*, 22/29 (December 2001).

98 Guppies, even: E. W. Warren and S. Callaghan, "Individual Differences in Response to an Open Field Test by the Guppy—*Poecilia reticulata* (Peters)," *Journal of Fish Biology*, 7: 105–13 (1975); S. V. Budaev, " 'Personality' in the Guppy (*Poecilia reticulata*): A Correlational Study of Exploratory Behavior and Social Tendency," *Journal of Comparative Psychology*, 111: 399–411 (1997).

99 "This fundamental duality": Antonio Damasio, *The Feeling of What Happens: Body and Emotion in the Making of Consciousness* (San Diego: Harvest, 2000), pp. 78–79.

100 "class of 'raw material' ": Gordon Allport, *Pattern and Growth in Personality* (New York: Holt, Rinehart & Winston, 1961), pp. 33–34.

101 "bubbles over happily": Karl Jaspers, *General Psychopathology* (London: Manchester University Press, 1949), p. 440.

101 the concept of "hyperthymia": H. S. Akiskal and G. Mallya, "Criteria for the 'Soft' Bipolar Spectrum: Treatment Implications," *Psychopharmacology Bulletin*, 23: 68–73 (1987); H. S. Akiskal, "Delineating Irritable and Hyperthymic Variants of the Cyclothymic Temperament," *Journal of Personality Disorders*, 6: 326–42 (1992).

102 The extravert, as defined: J. A. Gray, "The Psychophysiological Basis of Introversion-Extraversion," *Behavior Research & Therapy*, 8: 249–66 (1970); S.B.J. Eysenck and H. J. Eysenck, *Manual of the Eysenck Personality Questionnaire* (London: Hodder & Stoughton, 1975); C. R. Cloninger, D. M. Svrakic, and T. Przybeck, "A Psychobiological Model of Temperament and Character," *Archives of General Psychiatry*, 50: 975–90 (1993); D. Watson and L. A. Clark,

"Extraversion and Its Positive Emotional Core," in *Handbook of Personality Psychology*, ed. R. Hogan, L. Johnson, and S. Briggs (San Diego: Academic Press, 1997), pp. 767–93; O. P. John and S. Srivastava, "The Big Five Trait Taxonomy: History, Measurement, and Theoretical Perspectives," in *Handbook of Personality Theory and Research*, ed. L. A. Pervin and O. P. John (New York: Guilford, 1999); R. E. Lucas, E. Diener, A. Grob, E. M. Suh, and L. Shao, "Cross-Cultural Evidence for the Fundamental Features of Extraversion," *Journal of Personality and Social Psychology*, 79: 452–68 (2000).

102 exquisitely alert and sensitive: R. J. Larsen and T. Ketelaer, "Extraversion, Neuroticism and Susceptibility to Positive and Negative Mood Induction Procedures," *Personality and Individual Differences*, 10: 1221–28 (1989); L. A. Clark, D. Watson, and S. Mineka, "Temperament, Personality, and the Mood and Anxiety Disorders," *Journal of Abnormal Psychology*, 103: 103–16 (1994); Lucas et al., "Cross-Cultural Evidence."

102 The state of one's mood: H. Berenbaum and T. F. Ottmanns, "Emotional Experience and Expression in Schizophrenia and Depression," *Journal of Abnormal Psychology*, 101: 37–44 (1992); B. E. Wexler, L. Levenson, S. Warrenburg, and L. H. Price, "Decreased Perceptual Sensitivity to Emotion-Evoking Stimuli in Depression," *Psychiatry Research*, 51: 127–38 (1994); D. M. Sloan, M. E. Strauss, S. W. Quirk, and M. Sajatovic, "Subjective and Expressive Emotional Responses in Depression," *Journal of Affective Disorders*, 46: 135–41 (1997); N. B. Allen, J. Trinder, and C. Brennen, "Affective Startle Modulation in Clinical Depression: Preliminary Findings," *Biological Psychiatry*, 46: 542–50 (1999); J. B. Henriques and R. J. Davidson, "Decreased Responsiveness to Reward in Depression," *Cognition and Emotion*, 14: 711–24 (2000); D. M. Sloan, M. E. Strauss, and K. L. Wisner, "Diminished Response to Pleasant Stimuli by Depressed Women," *Journal of Abnormal Psychology*, 110: 488–93 (2001); J. Rottenberg, K. L. Kasch, J. J. Gross, and I. H. Gotlib, "Sadness and Amusement Reactivity Differentially Predict Concurrent and Prospective Functioning in Major Depressive Disorder," *Journal of Abnormal Psychology*, 111: 302–12 (2002); L. K. Murray, T. J. Wheeldon, I. C. Reid, D. A. Rowland, D. M. Burt, and D. I. Perrett, "Depression and Facial Expression Sensitivity: Exploratory Studies: Facial Expression Sensitivity in Depression," submitted for publication.

102 cross-species review: S. D. Gosling and O. P. John, "Personality Dimensions in Nonhuman Animals: A Cross-Species Review," *Current Directions in Psychological Science*, 8: 69–75 (1999).

103 significant differences between octopuses: J. A. Mather and R. C. Anderson, "Personalities of Octopuses (*Octopus rubescens*)," *Journal of Comparative Psy-

chology, 107: 336–40 (1993); D. L. Sinn, N. Perrin, J. A. Mather, and R. C. Anderson, "Early Temperamental Traits in an Octopus (*Octopus bimaculoides*)," *Journal of Comparative Psychology,* 115: 351–64 (2001).

103 species with the most diverse diets: A. S. Clarke and S. Boinski, "Temperament in Nonhuman Primates," *American Journal of Psychiatry,* 37: 103–25 (1995).

103 one house cat in seven: R. E. Adamec, "Anxious Personality in the Cat," in *Psychopathology and the Brain,* ed. B. J. Carroll and J. E. Barrett (New York: Raven, 1991), pp. 153–68.

104 One in five young rhesus monkeys: There have been many studies of individual differences in personality and temperament in nonhuman primates, including: R. Bolig, C. S. Price, P. L. O'Neill, and S. J. Suomi, "Subjective Assessment of Reactivity Level and Personality Traits of Rhesus Monkeys," *International Journal of Primatology,* 13: 287–306 (1992); M. T. McGuire, M. J. Raleigh, and D. B. Pollack, "Personality Features in Vervet Monkeys: The Effects of Sex, Age, Social Status, and Group Composition," *American Journal of Primatology,* 33: 1–14 (1994); G. Byrne and S. J. Suomi, "Development of Activity Patterns, Social Interactions, and Exploratory Behavior in Infant Tufted Capuchins (*Cebus apella*)," *American Journal of Primatology,* 35: 255–70 (1995); S. L. Watson and J. P. Ward, "Temperament and Problem Solving in the Small-Eared Bushbaby (*Otolemur garnettii*)," *Journal of Comparative Psychology,* 110: 377–85 (1996); D. M. Dutton, R. A. Clark, and D. W. Dickins, "Personality in Captive Chimpanzees: Use of a Novel Rating Procedure," *International Journal of Primatology,* 18: 539–52 (1997); A. Weiss, J. E. King, and A. J. Figueredo, "The Heritability of Personality Factors in Chimpanzees (*Pan troglodytes*)," *Behavior Genetics,* 30: 213–21 (2000); L. A. Fairbanks, "Individual Differences in Response to a Stranger: Social Impulsivity as a Dimension of Temperament in Vervet Monkeys (*Cercopithecus aethiops sabaeus*)," *Journal of Comparative Psychology,* 115: 22–28 (2001).

104 Among zebra finches and pumpkinseed sunfish: D. S. Wilson, K. Coleman, A. B. Clark, and L. Biederman, "Shy–Bold Continuum in Pumpkinseed Sunfish (*Lepomis gibbosus*): An Ecological Study of a Psychological Trait," *Journal of Comparative Psychology,* 107: 250–60 (1993); G. Beauchamp, "Individual Differences in Activity and Exploration Influence Leadership in Pairs of Foraging Zebra Finches," *Behaviour,* 137: 301–14 (2000).

104 Melvin Konner: Melvin Konner, *Why the Reckless Survive* (New York: Viking, 1990). See also Marvin Zuckerman, *Behavioral Expressions and Biosocial Bases of Sensation Seeking* (Cambridge, U.K.: Cambridge University Press, 1994).

104 "just plain boring": letter from Joyce Poole to the author, September 27, 2000. All quotes in this paragraph come from Poole's letter.

104 study of brown bears: R. Fagen and J. M. Fagen, "Individual Distinctiveness in Brown Bears, *Ursus arctos* L.," *Ethology,* 102: 212–26 (1996).

105 "We find it interesting": ibid., p. 222.

105 a related personality trait: Bolig et al., "Subjective Assessment of Reactivity Level."

105 pleasant affect, and extraversion: P. T. Costa and R. R. McCrae, "Influence of Extraversion and Neuroticism on Subjective Well-Being: Happy and Unhappy People," *Journal of Personality and Social Psychology,* 38: 668–78 (1980); G. J. Meyer and J. R. Shack, "Structural Convergence of Mood and Personality: Evidence for Old and New Directions," *Journal of Personality and Social Psychology,* 57: 691–706 (1989); D. Watson and L. A. Clark, "On Traits and Temperament: General and Specific Factors of Emotional Experience and Their Relation to the Five Factor Model," *Journal of Personality,* 60: 441–76 (1992); D. Watson and L. A. Clark, "Extraversion and Its Positive Emotional Core," in *Handbook of Personality Psychology,* ed. R. Hogan, J. Johnson, and S. Briggs (San Diego: Academic Press, 1997), pp. 767–93; E. Diener and R. E. Lucas, "Personality and Subjective Well-being," in *Well-being: The Foundations of Hedonic Psychology,* ed. D. Kahneman, E. Diener, and N. Schwarz (New York: Russell Sage, 1999), pp. 214–29; R. E. Lucas and F. Fujita, "Factors Influencing the Relation Between Extraversion and Pleasant Affect," *Journal of Personality and Social Psychology,* 79: 1039–56 (2000); W. Fleeson, A. B. Malanos, and N. M. Achille, "An Intraindividual Process Approach to the Relationship Between Extraversion and Positive Affect: Is Acting Extraverted as 'Good' as Being Extraverted?" *Journal of Personality and Social Psychology,* 83: 1409–22 (2002).

105 the correlation between the two traits: A correlation coefficient, which ranges between −1.00 and 1.00, provides an estimate of the relatedness of two variables. If they are completely and negatively related, the correlation coefficient will be −1.00; if there is no correlation at all it will be 0.00; and if they are perfectly correlated it will be 1.00. A correlation of 0.80 indicates that the degree of relatedness of extraversion and positive emotions is very high.

106 tend to be happier: Costa and McCrae, "Influence of Extraversion"; R. A. Emmons and E. Diener, "Personality Correlates of Subjective Well-being," *Personality and Social Psychology Bulletin,* 11: 89–97 (1985); B. Heady and A. Wearing, "Personality, Life Events, and Subjective Well-being: Towards a Dynamic Equilibrium Model," *Journal of Personality and Social Psychology,* 57: 731–39 (1989); W. Pavot, E. Diener, and F. Fujita, "Extraversion and Happiness," *Personality and Individual Differences,* 11: 1299–1306 (1990); J. A. Averill and T. A. More, "Happiness," in M. Lewis and J. M. Haviland-Jones, *Handbook of Emotions* (New York: Guilford, 2000), pp. 663–76; E. Diener and M. Seligman, "Very Happy People," *Psychological Science,* 13: 81–84 (2002).

106 greater intensity in such moods: E. Diener, R. J. Larsen, S. Levine, and R. A. Emmons, "Intensity and Frequency: Dimensions Underlying Positive and Negative Affect," *Journal of Personality and Social Psychology,* 48: 1253–65 (1985).

106 "We are the ones who gallop": Eugene Walter, as told to Katherine Clark, *Milking the Moon: A Southerner's Story of Life on This Planet* (New York: Crown, 2001), pp. 3–4.

107 An analysis of 24,000 twins: J. C. Loehlin, *Genes and Environment in Personality Development* (Newberry Park, Calif.: Sage, 1992); R. Plomin and A. Caspi, "Behavioral Genetics and Personality," in *Handbook of Personality: Theory and Research,* ed. L. A. Pervin and O. P. John (New York: Guilford, 2000), pp. 251–76.

107 Thomas Bouchard and his colleagues: T. Bouchard and M. McGue, "Genetic and Rearing Environmental Influences on Adult Personality: An Analysis of Adopted Twins Raised Apart," *Journal of Personality,* 68: 263–82 (1990); T. J. Bouchard, "Genes, Environment, and Personality," *Science,* 264: 1700–1701 (1994); T. J. Bouchard and Y.-M. Hur, "Genetic and Environmental Influences on the Continuous Scales of the Myers-Briggs Type Indicator: An Analysis Based on Twins Raised Apart," *Journal of Personality,* 66: 135–49 (1998).

107 "Joy, good cheer": A. Tellegen, D. T. Lykken, T. J. Bouchard, K. J. Wilcox, N. L. Segal, and S. Rich, "Personality Similarity in Twins Reared Apart and Together," *Journal of Personality and Social Psychology,* 54: 1031–39 (1988).

107 Studies of young children: A. Matheny, "Developmental Behavior Genetics: The Louisville Study," in *Developmental Behavior Genetics: Neural Biometrical and Evolutionary Approaches,* ed. M. E. Hahn, J. K. Hewitt, N. D. Henderson, and R. Benno (New York: Oxford University Press, 1990), pp. 25–38; J. L. Robinson, J. Kagan, J. S. Reznick, and R. Corley, "The Heritability of Inhibited and Uninhibited Behavior," *Developmental Psychology,* 28: 1030–37 (1992).

108 All breeds of dog: John Paul Scott and John L. Fuller, *Genetics and the Social Behavior of the Dog* (Chicago: University of Chicago Press, 1965); T. W. Draper, "Canine Analogs of Human Personality Factors," *Journal of General Psychology,* 122: 241–52 (1995); James Serpell, *The Domestic Dog: Its Evolution, Behaviour, and Interactions with People* (Cambridge, U.K.: Cambridge University Press, 1995).

108 Swedish researchers studied behavior: K. Svartberg and B. Forkman, "Personality Traits in the Domestic Dog (*Canis familiaris*)," *Applied Animal Behaviour Science,* 79: 133–55 (2002).

108 nine-week-old wolf pups: K. MacDonald, "Stability of Individual Differences in Behavior in a Litter of Wolf Cubs (*Canis lupus*)," *Journal of Comparative Psychology,* 97: 99–106 (1983).

108 little difference in playfulness: B. J. Hart, "Analyzing Breed and Gender Differences in Behaviour," in Serpell, *The Domestic Dog,* pp. 65–77.

108 boys are more likely: L. Barnett, "Characterizing Playfulness: Correlates with Individual Attributes and Personal Traits," *Play and Culture*, 4: 371–93 (1991); Jerome Kagan, *Galen's Prophecy: Temperament in Human Nature* (New York: Basic Books, 1994).

108 men are more likely: For the proposition that men have higher rates of hyperthymic temperament, see G. Perugi, E. Simonini, L. Musetti, F. Piagentini, G. B. Cassano, and H. S. Akiskal, "Gender-Mediated Clinical Features of Depressive Illness: The Importance of Temperamental Differences," *British Journal of Psychiatry*, 157: 835–41 (1990); G. B. Cassano, H. S. Akiskal, G. Perugi, L. Musetti, and M. Savino, "The Importance of Measures of Affective Temperaments in Genetic Studies of Mood Disorders," *Journal of Psychiatric Research*, 26: 257–68 (1992). Temperaments with depressive features are more common in women, and those with hypomanic features more common in men: R. Depue, J. F. Slater, and H. Wolfstetter-Kausch, "A Behavioral Paradigm for Identifying Persons at Risk for Bipolar Depressive Disorder: A Conceptual Framework and Five Validating Studies," *Journal of Abnormal Psychology*, 90: 381–437 (1981); M. Eckblad and L. J. Chapman, "Development and Validation of a Scale of Hypomanic Personality," *Journal of Abnormal Psychology*, 95: 214–22 (1986).

109 In a landmark series: J. Kagan, J. S. Resnick, and N. Snidman, "Biological Basis of Childhood Shyness," *Science*, 240: 167–71 (1988); J. Kagan, "Temperamental Contributions to Social Behavior," *American Psychologist*, 44: 688–74 (1989); J. Kagan, J. S. Resnick, and N. Snidman, "The Temperamental Qualities of Inhibition and Lack of Inhibition," in *Handbook of Developmental Psychopathology*, ed. M. Lewis and M. Miller (New York: Plenum, 1990), pp. 219–26; J. Kagan and N. Snidman, "Infant Predictors of Inhibited and Uninhibited Profiles," *Psychological Science*, 2: 40–44 (1991); J. Kagan, N. Snidman, and D. M. Arcus, "Initial Reactions to Unfamiliarity," *Current Directions in Psychological Science*, 1: 171–74 (1992).

109 "difficult to name": Kagan, *Galen's Prophecy*, p. 266.

110 Characterized by the researchers: N. A. Fox, H. A. Henderson, K. H. Rubin, S. D. Calkins, and L. A. Schmidt, "Continuity and Discontinuity of Behavioral Inhibition and Exuberance: Psychophysiological and Behavioral Influences Across the First Four Years of Life," *Child Development*, 72: 1–21 (2001).

110 "From an early age": letter to the author from Ellen Winner, April 2002.

111 Infants who gaze more: A review of fifteen samples of infants and young children found a significant association between early preference for novelty and later intelligence; see J. Fagan, "The Intelligent Infant," *Intelligence*, 8: 1–9 (1984).

111 Likewise in our primate cousins: S. L. Watson and J. P. Ward, "Temperament

and Problem Solving in the Small-Eared Bushbaby (*Otolemur garnetti*)," *Journal of Comparative Psychology*, 110: 377–85 (1996).

111 Children who scored high on stimulation-seeking: A. Raine, C. Reynolds, P. H. Venables, and S. A. Mednick, "Stimulation Seeking and Intelligence: A Prospective Longitudinal Study," *Journal of Personality and Social Psychology*, 82: 663–74 (2002).

112 curious, enthusiastic, and cheerful children: R. Bell and L. Harper, *Child Effects on Adults* (Lincoln: University of Nebraska Press, 1977); B. J. Breitmayer and H. N. Ricciuti, "The Effect of Neonatal Temperament on Caregiver Behavior in the Newborn Nursery," *Infant Mental Health Journal*, 9: 158–72 (1988); S. Scarr, "Developmental Theories for the 1990s: Development and Individual Differences," *Child Development*, 63: 1–19 (1992).

112 "Nature versus nurture": Matt Ridley, *Nature via Nurture: Genes, Experience, and What Makes Us Human* (London: HarperCollins, 2003).

112 Mice and rats, we know: R. Paylor, S. K. Morrison, J. W. Rudy, L. T. Waltrip, and J. M. Wehner, "Brief Exposure to an Enriched Environment Improves Performance on the Morris Water Task and Increases Hippocampal Cytosolic Protein Kinase C Activity in Young Rats," *Behavior Brain Research*, 52: 49–59 (1992); A. Fernandez-Teruel, R. M. Escorihuela, B. Castellano, B. Gonzalez, and A. Tobeña, "Neonatal Handling and Environmental Enrichment Effects on Emotionality, Novelty/Reward Seeking, and Age-Related Cognitive and Hippocampal Impairments: Focus on the Roman Rat Lines," *Behavior Genetics*, 27: 513–26 (1997); G. Kempermann, H. G. Kuhn, and F. H. Gage, "More Hippocampal Neurons in Adult Mice Living in an Enriched Environment," *Nature*, 386: 493–95 (1997).

112 Rhesus monkey infants: M. L. Schneider, C. F. Moore, S. J. Suomi, and M. Champoux, "Laboratory Assessment of Temperament and Environmental Enrichment in Rhesus Monkey Infants (*Macaca mulatta*)," *American Journal of Primatology*, 25: 137–55 (1991).

113 Dopamine does many things: R. A. Depue, M. Luciana, P. Arbisi, P. Collins, and A. Leon, "Dopamine and the Structure of Personality: Relation of Agonist-Induced Dopamine Activity to Positive Emotionality," *Journal of Personality and Social Psychology*, 67: 485–98 (1994); R. A. Depue and P. F. Collins, "Neurobiology of the Structure of Personality: Dopamine, Facilitation of Incentive Motivation, and Extraversion," *Behavioral and Brain Sciences*, 22: 491–569 (1999).

113 Brain imaging studies conducted: A. J. Blood and R. J. Zatorre, "Intensely Pleasurable Responses to Music Correlate with Activity in Brain Regions Implicated in Reward and Emotion," *Proceedings of the National Academy of Sciences*, 98: 11818–23 (2001).

113 brain's "pleasure center": J. Olds and P. Milner, "Positive Reinforcement Produced by Electrical Stimulation of Septal Area and Other Regions of Rat Brain," *Journal of Comparative Physiology and Psychology*, 47: 419–27 (1954).

114 brain's sensitivity to dopamine: Depue and Collins, "Neurobiology of the Structure of Personality."

114 a drug that increases dopamine transmission: S. Florin, C. Suaudeau, J. C. Meunier, and J. Cosentin, "Nociceptin Stimulates Locomotion and Exploratory Behaviour in Mice," *European Journal of Pharmacology*, 12: 9–13 (1996). For an excellent review of the role of dopamine in behavior, see Depue and Collins, "Neurobiology of the Structure of Personality."

114 A mouse born without the genes: J. B. Eells, "The Control of Dopamine Neuron Development, Function and Survival: Insights from Transgenic Mice and the Relevance to Human Disease," *Current Medicinal Chemistry*, 10: 857–70 (2003); R. E. Nally, F. N. McNamara, J. J. Clifford, A. Kinsella, O. Tighe, D. T. Croke, A. A. Fienberg, P. Greengard, and J. L. Waddington, "Topographical Assessment of Ethological and Dopamine Receptor Agonist-Induced Behavioral Phenotype in Mutants with Congenic DARPP-32 'Knockout,' " *Neuropsychopharmacology*, 28: 2055–63 (2003).

114 extraverts are exquisitely sensitive: Larsen and Ketelaar, "Extraversion, Neuroticism and Susceptibility"; Clark, Watson, and Mineka, "Temperament, Personality"; Lucas et al., "Cross-Cultural Evidence."

114 thirty-nine countries: Lucas et al., "Cross-Cultural Evidence."

115 most pathological manifestation: R. A. Depue and W. G. Iacono, "Neurobehavioral Aspects of Affective Disorders," *Annual Review of Psychology*, 40: 457–92 (1989); R. A. Depue, M. Luciana, P. Arbisi, P. Collins, and A. Leon, "Dopamine and the Structure of Personality: Relation of Agonist-Induced Dopamine Activity to Positive Emotionality," *Journal of Personality and Social Psychology*, 67: 485–98 (1994).

115 Amphetamines promote the release of dopamine: D. Jacobs and T. Silverstone, "Dextroamphetamine-Induced Arousal in Human Subjects as a Model for Mania," *Psychological Medicine*, 16: 323–29 (1986).

115 dopamine precursor L-dopa: F. K. Goodwin, D. L. Murphy, H. K. Brodie, and W. E. Bunney, "L-dopa, Catecholamines, and Behavior: A Clinical and Biochemical Study in Depressed Patients," *Biological Psychiatry*, 2: 341–66 (1970); D. L. Murphy, H. K. Brodie, F. K. Goodwin, and W. E. Bunney, "Regular Induction of Hypomania by L-dopa in 'Bipolar' Manic-Depressive Patients," *Nature*, 229: 135–36 (1971); H. M. Van Praag and J. Korf, "Endogenous Depression With and Without Disturbances in 5-hydroxytryptamine Metabolism: A Biochemical Classification?" *Psychopharmacologia*, 19: 148–52 (1971).

115 an antidepressant effect: R. H. Gerner, R. M. Post, and W. E. Bunney, "A

Dopaminergic Mechanism in Mania," *American Journal of Psychiatry*, 133: 1177–80 (1976); T. Silverstone, "Response to Bromocriptine Distinguishes Bipolar from Unipolar Depression," *Lancet*, 1: 903–4 (1984).

115 therapeutic effect against mania: Summarized in F. K. Goodwin and K. R. Jamison, *Manic-Depressive Illness* (New York: Oxford University Press, 1990), pp. 419–21, 578–79, 622–23.

115 Greater activation in the left frontal area: Paradiso et al., "Cerebral Blood Flow Changes"; D. H. Zald, D. L. Mattson, and J. V. Pardo, "Brain Activity in Ventromedial Prefrontal Cortex Correlates with Individual Differences in Negative Affect," *Proceedings of the National Academy of Sciences*, 99: 2450–54 (2002).

115 activate the left amygdala: Hammann et al., "Ecstasy and Agony." In nonhuman studies, the amygdala has been implicated in conditioning and addiction to reward stimuli; see M. Gallagher and P. C. Holland, "Understanding the Function of the Central Nucleus: Is Simple Conditioning Enough?" in *The Amygdala: Neurobiological Aspects of Emotion, Memory, and Mental Dysfunction*, ed. J. P. Aggleton (New York: Wiley-Liss, 1992), pp. 307–21; B. J. Everitt, J. A. Parkinson, M. C. Olmstead, M. Aroyo, P. Robledo, and T. W. Robbins, "Associative Processes in Addiction and Reward: The Role of Amygdala-Ventral Striatal Subsystems," in J. F. McGinty, ed., *Annals of the New York Academy of Sciences: Advancing from the Ventral Striatum to the Extended Amygdala*, vol. 877: 412–38 (1999).

Amygdalar activation in response to viewing happy faces is significantly correlated with levels of extraversion. The activation was located within the left hemisphere, the one associated with positive emotions and approach behavior: T. Canli, H. Sivers, S. L. Whitfield, I. H. Gotlib, and J. D. E. Gabrieli, "Amygdala Response to Happy Faces as a Function of Extraversion," *Science*, 296: 2191 (2002).

115 damage in the left frontal areas: Goodwin and Jamison, *Manic-Depressive Illness*, pp. 503–40.

115 right frontal region: ibid.

116 reduction in gray-matter volume: W. C. Drevets, J. L. Price, J. R. Simpson, R. D. Todd, T. Reich, M. Vannier, and M. E. Raichle, "Subgenual Prefrontal Cortex Abnormalities in Mood Disorders," *Nature*, 386: 824–27 (1997); V. Sharma, R. Menon, T. J. Carr, M. Densmore, D. Mazmanian, and P. C. Williamson, "An MRI Study of Subgenual Prefrontal Cortex in Patients with Familial and Non-familial Bipolar I Disorder," *Journal of Affective Disorders*, 77: 167–71 (2003).

116 Carl Schwartz, Jerome Kagan, and their colleagues: C. E. Schwartz, C. J. Wright, L. M. Shin, J. Kagan, and S. L. Rauch, "Inhibited and Uninhibited

Infants 'Grown Up': Adult Amygdalar Response to Novelty," *Science*, 300: 1952–53 (2003).

116 the amygdala is primarily responsive: letter from Jerome Kagan to the author, October 17, 2003.

116 "A merry heart": Proverbs 17: 22.

116 "purgeth the blood": Robert Burton quoting Vives in *The Anatomy of Melancholy*, ed. Holbrook Jackson (New York: New York Review Books, 2001), pt. 2, p. 119; first published in 1621.

117 induced a high-arousal: B. L. Fredrickson and R. W. Levenson, "Positive Emotions Speed Recovery from the Cardiovascular Sequelae of Negative Emotions," *Cognition and Emotion*, 12: 191–220 (1998).

117 positive attitudes such as optimism: S. E. Taylor and J. D. Brown, "Positive Illusions and Well-being Revisited: Separating Fact from Fiction," *Psychological Bulletin*, 116: 21–27 (1994); S. E. Taylor, R. L. Repetti, and T. L. Seeman, "Health Psychology: And How Does It Get Under the Skin?" *Annual Review of Psychology*, 48: 411–47 (1997); S. E. Taylor, M. E. Kemeny, G. M. Reed, J. E. Bower, and T. L. Gruenewald, "Psychological Resources, Positive Illusions, and Health," *American Psychologist*, 55: 99–109 (2000).

See also: Lionel Tiger, *Optimism: Biology of Hope* (New York: Simon & Schuster, 1979); Richard S. Lazarus, *Emotion and Adaptation* (New York: Oxford University Press, 1991); G. Affleck and H. Tennen, "Construing Benefits from Adversity: Adaptational Significance and Dispositional Underpinnings," *Journal of Personality*, 64: 899–922 (1996); P. Salovey, A. J. Rothman, J. B. Detweiler, and W. T. Steward, "Emotional States and Physical Health," *American Psychologist*, 55: 110–21 (2000); B. L. Fredrickson and T. Joiner, "Positive Emotions Trigger Upward Spirals Toward Emotional Well-being," *Psychological Science*, 13: 172–75 (2002).

117 A study of 180 nuns: D. D. Danner, D. A. Snowdon, and W. V. Friesen, "Positive Emotions in Early Life and Longevity: Findings from the Nun Study," *Journal of Personality and Social Psychology*, 80: 804–13 (2001).

A study of 839 patients referred to the Mayo Clinic obtained similar results. Patients who were assessed as optimistic by a series of psychological and physical tests had a 19 percent increase in their expected life span when compared with patients who were classified as pessimists. See T. Maruta, R. Colligan, M. Malinchoc, and K. Offord, "Optimists vs. Pessimists: Survival Rates Among Medical Patients over a 30-Year Period," *Mayo Clinic Proceedings*, 75: 140–43 (2000).

117 "passion, imagination, self-will": William Hazlitt, "On the Love of Life," in *The Collected Works of William Hazlitt*, ed. A. R. Waller and Arnold Glover (London: J. M. Dent & Co., 1902), vol. I, p. 3.

117 "Lively passions": David Hume, *Treatise of Human Nature* (1739; Amherst, N.Y.: Prometheus, 1992), p. 427.

118 more likely to make decisions: D. Aderman, "Elation, Depression, and Helping Behavior," *Journal of Personality and Social Psychology*, 24: 91–101 (1972); A. M. Isen and B. Means, "The Influence of Positive Affect on Decision-Making Strategy," *Social Cognition*, 2: 18–31 (1983); M. Carlson, V. Charlin, and N. Miller, "Positive Mood and Helping Behavior: A Test of Six Hypotheses," *Journal of Personality and Social Psychology*, 55: 211–29 (1988); R. A. Baron and J. Thomley, "A Whiff of Reality: Positive Affect as a Potential Mediator of Pleasant Fragrances on Task Performance and Helping," *Environment and Behavior*, 26: 766–84 (1994); A. M. Isen, in M. Lewis and J. M. Haviland-Jones, *Handbook of Emotions* (New York: Guilford, 2000), pp. 417–35.

118 more actively explore: N. H. Frijda, *The Emotions* (Cambridge, U.K.: Cambridge University Press, 1986); B. L. Fredrickson, "The Broaden-and-Build Theory of Positive Emotions," *American Psychologist*, 56: 216–26 (2001).

118 a larger number of responses: A. M. Isen and K. A. Daubman, "The Influence of Affect on Categorization," *Journal of Personality and Social Psychology*, 47: 1206–17 (1984); A. M. Isen, M. S. Johnson, E. Mertz, and G. F. Robinson, "The Influence of Positive Affect on the Unusualness of Word Associations," *Journal of Personality and Social Psychology*, 48: 1413–26 (1985); A. M. Isen, K. A. Daubman, and G. P. Nowicki, "Positive Affect Facilitates Creative Problem Solving," *Journal of Personality and Social Psychology*, 52: 1122–31 (1987); A. M. Isen, "On Creative Problem Solving," in *Affect, Creative Experience, and Psychological Adjustment*, ed. S. Russ (Philadelphia: Taylor and Francis, 1999), pp. 3–17.

118 in a global way: A. M. Isen, "Positive Affect, Cognitive Process, and Social Behavior," in *Advances in Experimental Social Psychology*, ed. L. Berkowitz (San Diego: Academic Press, 1987), pp. 203–53; H. Bless and K. Fiedler, "Affective States and Knowledge," *Personality and Social Psychology Bulletin*, 21: 766–78 (1995); M. R. Basso, B. K. Schefft, M. D. Ris, and W. N. Dember, "Mood and Global-Local-Visual Processing," *Journal of the International Neuropsychological Society*, 2: 249–55 (1996); G. L. Clore, R. S. Wyer, B. Dienes, K. Gasper, C. Gohm, and L. Isbell, "Affective Feelings as Feedback: Some Cognitive Consequences," in *Theories of Mood and Cognitive*, ed. L. L. Martin and G. L. Clore (Mahwah, N.J.: Erlbaum, 2001), pp. 27–62; K. Gasper and G. L. Clore, "Attending to Local Processing of Visual Information," *Psychological Science*, 13: 34–40 (2002).

119 "Not by constraint": Henry David Thoreau, June 23, 1840, journal entry, in *Journal*, vol. 1: 1837–1844, gen. ed. J. C. Broderick, ed. E. H. Witherell, W. L. Howarth, R. Sattelmeyer, and T. Blanding (Princeton, N.J.: Princeton University Press, 1981), p. 140.

119 In a typical study: T. R. Greene and H. Noice, "Influence of Positive Affect upon Creative Thinking and Problem Solving in Children," *Psychological Reports*, 63: 895–98 (1988).

119 incompatible with anxiety: G. Mandler, "Stress and Thought Processes," in *Handbook of Stress: Theoretical and Clinical Aspects*, ed. L. Goldberger and S. Breznitz (New York: Free Press, 1982), pp. 88–104; G. Keinan, N. Friedland, and Y. Ben-Porath, "Decision Making Under Stress: Scanning of Alternatives Under Physical Threat," *Acta Psychologia*, 64: 219–28 (1987); D. Derryberry and M. A. Reed, "Anxiety and Attentional Focusing: Trait, State and Hemispheric Influences," *Personality and Individual Differences*, 25: 745–61 (1998).

119 the way in which cognitive material is organized: F. G. Ashby, A. M. Isen, and K. Turken, "A Neuropsychological Theory of Positive Affect and Its Influence on Cognition," *Psychological Review*, 106: 529–50 (1999).

120 Surges in dopamine: ibid.

120 nor is it inconsistent across studies: Not all investigators find that positive mood necessarily improves cognitive functioning; for example, G. Kaufman and S. K. Vosburg, " 'Paradoxical' Mood Effects on Creative Problem Solving," *Cognition and Emotion*, 11: 151–70 (1997); L. Clark, S. D. Iverson, and G. M. Goodwin, "The Influence of Positive and Negative Mood States on Risk Taking, Verbal Fluency, and Salivary Cortisol," *Journal of Affective Disorders*, 63: 179–87 (2001).

121 It is relatively common: The lifetime prevalence for the severe form of bipolar illness is about 1 percent, but estimates for the milder forms range from 5 to nearly 9 percent. L. N. Robins and D. A. Regier, eds., *Psychiatric Disorders in America: The Epidemiologic Catchment Area Study* (New York: Free Press, 1991); R. C. Kessler, D. R. Rubinow, C. Holmes, J. M. Abelson, and S. Zhao, "The Epidemiology of DSM-III-R Bipolar I Disorder in a General Population Survey," *Psychological Medicine*, 29: 1079–89 (1997); J. Angst, "The Emerging Epidemiology of Hypomania and Bipolar II Disorder," *Journal of Affective Disorders*, 50: 143–51 (1998); L. L. Judd and H. S. Akiskal, "The Prevalence and Disability of Bipolar Spectrum Disorders in the U.S. Population: Re-analysis of the ECA Database Taking into Account Sub-threshold Cases," *Journal of Affective Disorders*, 73: 123–31 (2003).

121 "the blood becomes changed": quoted in F. Walker, *Hugo Wolf: A Biography* (London: J. M. Dent & Sons, 1968), p. 359.

122 "unrestrained, merry": Emil Kraepelin, *Manic-Depressive Insanity and Paranoia* (1921; New York: Arno Press, 1976), p. 63.

122 "they show off in public": Aretaeus of Cappadocia, quoted in G. Roccatagliata, *A History of Ancient Psychiatry* (New York: Greenwood Press, 1986), pp. 230–31.

122 mood and energy generally soar: A review of 14 clinical studies comprising a total of nearly 800 manic patients found that 71 percent exhibited euphoria (a similar percentage showed irritability and/or depression; many had symptoms of both depressed mood and euphoria): Goodwin and Jamison, *Manic-Depressive Illness,* p. 31. As early as Aretaeus of Cappadocia in A.D. 150, observers of mania and depression observed a close link between the two states, often regarding them as different forms of the same clinical condition.

122 "like a man with air balloons": Benjamin Haydon quoted in Walter Jackson Bate, *John Keats* (Cambridge, Mass.: Belknap Press, 1963), p. 98.

122 more colorful language: N.J.C. Andreasen and B. Pfohl, "Linguistic Analysis of Speech in Affective Disorders," *Archives of General Psychiatry,* 33: 1361–67 (1976).

123 Rhyming and sound associations: Kraepelin refers to at least five earlier word-association experiments in his 1921 monograph *Manic-Depressive Insanity;* G. Murphy, "Types of Word-Association in Dementia Praecox, Manic-Depressives, and Normal Persons," *American Journal of Psychiatry,* 79: 539–71 (1923); L. Pons, J. I. Nurnberger, and D. L. Murphy, "Mood-Independent Aberrancies in Associative Processes in Bipolar Affective Disorder: An Apparent Stabilizing Effect of Lithium," *Psychiatry Research,* 14: 315–22 (1985).

123 "There are moments": Vincent van Gogh, letter to his brother Theo, February 3, 1889, in *The Complete Letters of Vincent van Gogh* (Boston: New York Graphic Society, published by Little, Brown, 1985), vol. 3, p. 134.

123 Artwork produced during mania: The studies of artistic expression during mania and depression are summarized and reviewed in Goodwin and Jamison, *Manic-Depressive Illness* (tabularized on p. 288).

123 "flight of ideas": A review of nine clinical studies comprising a total of more than 600 manic patients found that 71 percent reported a flight of ideas or racing thoughts: ibid.

123 "First and foremost": John Custance, *Wisdom, Madness, and Folly: The Philosophy of a Lunatic* (New York: Farrar, Straus and Giroux, 1952), p. 30.

124 "As I sit here": ibid., pp. 33–34.

124 "All the problems of the universe": E. Reiss, *Konstitutionelle Verstimmung und Manisch-Depressives Irresein* (Berlin: J. Springer, 1910); quoted in Goodwin and Jamison, *Manic-Depressive Illness,* pp. 26–27.

125 "I roll on like a ball": quoted in John Rosenberg, *The Darkening Glass: A Portrait of Ruskin's Genius* (New York: Columbia University Press, 1986), p. 151.

125 "bustle along like a Surinam toad": quoted in W. Jackson Bate, *Coleridge* (Cambridge, Mass.: Harvard University Press, 1987), p. 211.

126 "mania is a sickness": Robert Lowell, "A Conversation with Ian Hamilton," in

Robert Lowell: Collected Prose, ed. Robert Giroux (New York: Farrar, Straus and Giroux, 1987), p. 286.

126 "I must record everything": Morag Coate, *Beyond All Reason* (London: Constable & Co., 1964), pp. 84–85.

126 More than twenty studies: In addition to the more than twenty studies discussed in Kay Redfield Jamison, *Touched with Fire: Manic-Depressive Illness and the Artistic Temperament* (New York: Free Press, 1993), see F. Post, "Creativity and Psychopathology: A Study of 291 World-Famous Men," *British Journal of Psychiatry*, 165: 22–34 (1994); J. J. Schildkraut, A. J. Hirshfeld, and J. M. Murphy, "Mind and Mood in Modern Art: II. Depressive Disorders, Spirituality, and Early Deaths in the Abstract Expressionist Artists of the New York School," *American Journal of Psychiatry*, 151: 482–88 (1994); A. M. Ludwig, *The Price of Greatness: Resolving the Creativity and Madness Controversy* (New York: Guilford Press, 1995); F. Post, "Verbal Creativity, Depression, and Alcoholism: An Investigation of One Hundred American and British Writers," *British Journal of Psychiatry*, 168: 545–55 (1996); E. M. Fodor, "Subclinical Inclination Toward Manic-Depression and Creative Performance on the Remote Associates Test," *Personality and Individual Differences*, 27: 1273–83 (1999); D. Schuldberg, "Six Subclinical Spectrum Traits in Normal Creativity," *Creativity Research Journal*, 13: 5–16 (2000–2001); C. M. Strong, C. M. Santosa, N. Sachs, C. M. Rennicke, P. W. Wang, A. Hier, and T. A. Ketter, "Relationships Between Creativity and Temperament in Bipolar Disorder Patients and Healthy Volunteers," paper presented to the American Psychiatric Association in May 2000; B. Das, C. M. Strong, N. Sachs, M. Eng, J. Mongolcheep, and T. A. Ketter, "Creativity Enhancement in Bipolar Patients More Specific Than in Creative Subjects," paper presented to the American Psychiatric Association in May 2001.

127 "The *thinking* of the manic is flighty": Eugen Bleuler, *Textbook of Psychiatry*, Eng. ed. A. A. Brill (New York: Macmillan, 1924), p. 466.

127 Both individuals who are manic: N. Andreasen and P. Powers, "Creativity and Psychosis: An Examination of Conceptual Style," *Archives of General Psychiatry*, 32: 70–73 (1975).

127 researchers at the Payne Whitney Psychiatric Clinic: L. Welch, O. Diethelm, and L. Long, "Measurement of Hyper-Associative Activity During Elation," *Journal of Psychology*, 21: 113–26 (1946).

128 Verbal associations increase: Pons, Nurnberger, and Murphy, "Mood-Independent Aberrancies"; M. R. Solovay, M. E. Shenton, and P. S. Holzman, "Comparative Studies of Thought Disorders: I. Mania and Schizophrenia," *Archives of General Psychiatry*, 44: 13–20 (1987); J. Levine, K. Schild, R. Kimhi,

and G. Schreiber, "Word Association Production in Affective Versus Schizophrenic Psychoses," *Psychopathology*, 29: 7–13 (1996).

128 study of eminent writers and artists: K. R. Jamison, "Mood Disorders and Patterns of Creativity in British Writers and Artists," *Psychiatry*, 52: 125–34 (1989).

129 Harvard study of manic-depression: R. L. Richards, D. K. Kinney, I. Lunde, and M. Bent, "Creativity in Manic-Depressives, Cyclothymes, and Their Normal First-Degree Relatives: A Preliminary Report," *Journal of Abnormal Psychology*, 97: 281–88 (1988).

129 students vulnerable to manic-depressive illness: Fodor, "Subclinical Inclination Toward Manic-Depression."

130 "The slightest forms of the disorder": Kraepelin, *Manic-Depressive Insanity*, pp. 129–30.

130 "a link in the long chain": ibid., p. 130.

130 certain temperaments, including hyperthymia: H. S. Akiskal and G. Mallya, "Criteria for the 'Soft' Bipolar Spectrum: Treatment Implications," *Psychopharmacology Bulletin*, 23: 68–73 (1987); H. S. Akiskal, "Delineating Irritable and Hyperthymic Variants of the Cyclothymic Temperament," *Journal of Personality Disorders*, 6: 326–42 (1992).

130 Hypomanic Personality Scale: M. Eckblad and L. J. Chapman, "Development and Validation of a Scale for Hypomanic Personality," *Journal of Abnormal Psychology*, 95: 214–22 (1986); T. D. Meyer, "The Hypomanic Personality Scale, the Big Five, and Their Relationship to Depression and Mania," *Personality and Individual Differences*, 32: 649–60 (2002); T. D. Meyer and M. Hautzinger, "Screening for Bipolar Disorders Using the Hypomanic Personality Scale," *Journal of Affective Disorders*, 75: 149–54 (2003).

131 thirteen-year follow-up study of students: T. R. Kwapil, M. B. Miller, M. C. Zinser, L. J. Chapman, J. Chapman, and M. Eckblad, "A Longitudinal Study of High Scorers on the Hypomanic Personality Scale," *Journal of Abnormal Psychology*, 109: 222–26 (2000).

131 study of American and Italian students: G. F. Placidi, S. Signoretta, A. Liguori, R. Gervasi, I. Maremmani, and H. S. Akiskal, "The Semi-Structured Affective Temperament Interview (TEMPS-I): Reliability and Psychometric Properties in 1010 14–26-Year-Old Students," *Journal of Affective Disorders*, 47: 1–10 (1998); Kagan, *Galen's Prophecy;* Fox et al., "Continuity and Discontinuity."

Chapter 6: "Throwing Up Sky-Rockets"

133 People *like* to be humbugged: P. T. Barnum, quoted in A. H. Saxon, *P. T. Barnum: The Legend and the Man* (New York: Columbia University Press, 1989), p. 335.

133 "throw up sky-rockets": quoted ibid.

134 "blessed with a vigor": P. T. Barnum, *Barnum's Own Story: The Autobiography of P. T. Barnum* (1855; Gloucester, Mass.: Peter Smith, 1972), p. 452.

134 "utterly fruitless": ibid., p. 401.

134 "We can hear it": Leon Edel, "The Madness of Art," *American Journal of Psychiatry*, 132: 1005–12 (1975), p. 1008.

136 "The colors of my life": "The Colors of My Life," lyrics by Michael Stewart, from the musical *Barnum* by Michael Stewart and Cy Coleman, first performed in 1980; Warner Brothers, 1980.

136 "I have lived so long": P. T. Barnum, letter to Mrs. Abel C. Thomas, May 22, 1874, in *Selected Letters of P. T. Barnum*, ed. A. H. Saxon (New York: Columbia University Press, 1983), p. 180.

137 "Through a night as dark as space": "The Prince of Humbug," lyrics by Michael Stewart, from the musical *Barnum* by Michael Stewart and Cy Coleman; Warner Brothers 1980.

138 Malcolm Gladwell argues: Malcolm Gladwell, *The Tipping Point* (Boston: Little, Brown, 2000.)

138 "How I long for": John Osborne, *Look Back in Anger: A Play in Three Acts* (London: Faber and Faber, 1960), p. 15.

138 "Good things as well as bad": C. S. Lewis, *Mere Christianity* (1952; New York: HarperCollins, 2001), p. 176.

139 "he had fun": Katharine Graham, obituary for Russ Wiggins, *Washington Post*, November 20, 2000.

139 "powerful senders": R. Buck, R. E. Miller, and W. F. Caul, "Sex, Personality and Physiological Variables in the Communication of Emotion via Facial Expression," *Journal of Personality and Social Psychology*, 30: 587–96 (1974).

139 scored high on measures of extraversion: ibid.; H. S. Friedman and R. E. Riggio, "Effect of Individual Differences in Nonverbal Expressiveness on Transmission of Emotion," *Journal of Nonverbal Behavior*, 6: 96–101 (1981); Elaine Hatfield, John Cacioppo, and Richard Rapson, *Emotional Contagion* (Cambridge, U.K.: Cambridge University Press, 1994).

139 "visible and convincing appearance": Carl Jung, *Psychological Types* (London: Kegan Paul, 1933), p. 407.

139 "kindles no flame of enthusiasm": ibid., p. 408.

140 The Affective Communication Test: H. S. Friedman, L. M. Prince, R. E. Riggio, and M. R. Di Matteo, "Understanding and Assessing Nonverbal Expressiveness: The Affective Communication Test," *Journal of Personality and Social Psychology*, 39: 333–51 (1980).

140 People who score high on this test: ibid.

140 Expressive individuals strongly influence: Friedman and Riggio, "Effect of Individual Differences"; E. S. Sullins, "Emotional Contagion Revisited: Effects of Social Comparisons and Expressive Style on Mood Convergence," *Personality and Social Psychology Bulletin*, 17: 166–74 (1991).

140 more emotional information: D. Newton, "Attribution and the Unit of Perception of Ongoing Behavior," *Journal of Personality and Social Psychology*, 28: 28–38 (1973); R. Buck, R. Baron, N. Goodman, and B. Shapiro, "Utilization of Spontaneous Nonverbal Behavior in the Study of Emotion Communication," *Journal of Personality and Social Psychology*, 39: 522–29 (1980); L. Z. McArthur, "What Grabs You? The Role of Attention in Impression Formation and Causal Attribution," in *Social Cognition*, ed. E. T. Higgins, C. P. Herman, and M. P. Zanna (Hillsdale, N.J.: Erlbaum, 1981), pp. 201–41; R. Buck, R. Baron, and D. Barrette, "Temporal Organization of Spontaneous Emotional Expression: A Segmentation Analysis," *Journal of Personality and Social Psychology*, 42: 506–17 (1982).

140 Women, although they in general score: Friedman et al., "Understanding and Assessing Nonverbal Expressiveness"; R. Buck, *The Communication of Emotion* (New York: Guilford, 1984); Judith Hall, *Nonverbal Sex Differences: Communication Accuracy and Expressive Style* (Baltimore: Johns Hopkins University Press, 1984).

140 Barbara Wild and her colleagues: B. Wild, M. Erb, and M. Bartels, "Are Emotions Contagious? Evoked Emotions While Viewing Emotionally Expressive Faces: Quality, Quantity, Time Course and Gender Differences," *Psychiatry Research*, 102: 109–24 (2001).

141 the most accurately communicated of the emotions: H. L. Wagner, C. J. MacDonald, and A.S.R. Manstead, "Communication of Individual Emotions by Spontaneous Facial Expressions," *Journal of Personality and Social Psychology*, 50: 737–43 (1986).

141 Negative emotions, although less accurately transmitted: E. S. Sullins, "Emotional Contagion Revisited: Effects of Social Comparisons and Expressive Style on Mood Convergence," *Personality and Social Psychology Bulletin*, 17: 166–74 (1991).

141 many types of negative stimuli: F. Pratto, "Automatic Vigilance: The Attention-Grabbing Power of Negative Social Information," *Journal of Personality and Social Psychology*, 61: 380–91 (1991); J.M.G. Williams, A. Matthews, and C. MacLeod, "The Emotional Stroop Task and Psychopathology," *Psychological Bulletin*, 120: 3–24 (1996); D. Wentura, K. Rothermund, and P. Bak, "Automatic Vigilance: The Attention-Grabbing Power of Approach- and Avoidance-Related Social Information," *Journal of Personality and Social Psychology*, 78:

1024–37 (2000); A. Dijksterhuis and H. Aarts, "On Wildebeests and Humans: The Preferential Detection of Negative Stimuli," *Psychological Science*, 14: 14–18 (2003).

141 Mice, for example: M. Luo, M. S. Fee, and L. C. Katz, "Encoding Pheromonal Signals in the Accessory Olfactory Bulb of Behaving Micc," *Science*, 299: 1196–1201 (2003).

141 Young male Asian elephants: L.E.L. Rasmussen, H. S. Riddle, and V. Krishnamurthy, "Mellifluous Matures to Malodorous in Musth," *Nature*, 415: 975–76 (2002).

142 "gather sweetness from the temples": ibid., p. 975.

142 Emotional Contagion Scale: Elaine Hatfield, John Cacioppo, and Richard Rapson, *Emotional Contagion* (Cambridge, U.K.: Cambridge University Press, 1994).

142 Gender is also a factor: J. M. Haviland and C. Z. Malatesta, "The Development of Sex Differences in Nonverbal Signals: Fallacies, Facts, and Fantasies," in *Gender and Nonverbal Behavior*, ed. C. Mayo and N. M. Henley (New York: Springer Verlag, 1981), pp. 183–208.

143 women were far more susceptible: Hatfield et al., *Emotional Contagion*.

143 people who are themselves happy: J. A. Easterbrook, "The Effect of Emotion on Cue-Utilization and the Organization of Behavior," *Psychological Review*, 66: 183–201 (1959); C. K. Hsee, E. Hatfield, and C. Chemtob, "Assessment of the Emotional States of Others: Conscious Judgments versus Emotional Contagion," *Journal of Social and Clinical Psychology*, 11: 119–28 (1991); K. Otley and J. M. Jenkins, "Human Emotions: Function and Dysfunction," *Annual Review of Psychology*, 43: 55–85 (1992); C. Sedikides, "Mood as a Determinant of Attentional Focus," *Cognition and Emotion*, 6: 129–48 (1992).

143 Depressed individuals: H. Berenbaum and T. F. Ottmanns, "Emotional Experience and Expression in Schizophrenia and Depression," *Journal of Abnormal Psychology*, 101: 37–44 (1992); B. E. Wexler, L. Levenson, S. Warrenburg, and L. H. Price, "Decreased Perceptual Sensitivity to Emotion-Evoking Stimuli in Depression," *Psychiatry Research*, 51: 127–38 (1994); D. M. Sloan, M. E. Strauss, S. W. Quirk, and M. Sajatovic, "Subjective and Expressive Emotional Responses in Depression," *Journal of Affective Disorders*, 46: 135–41 (1997); N. B. Allen, J. Trinder, and C. Brennen, "Affective Startle Modulation in Clinical Depression: Preliminary Findings," *Biological Psychiatry*, 46: 542–50 (1999); J. B. Henriques and R. J. Davidson, "Decreased Responsiveness to Reward in Depression," *Cognition and Emotion*, 14: 711–24 (2000); D. M. Sloan, M. E. Strauss, and K. L. Wisner, "Diminished Response to Pleasant Stimuli by Depressed Women," *Journal of Abnormal Psychology*, 110: 488–93 (2001); J. Rottenberg, K. L. Kasch, J. J. Gross, and I. H. Gotlib, "Sadness and Amusement

Reactivity Differentially Predict Concurrent and Prospective Functioning in Major Depressive Disorder," *Journal of Abnormal Psychology,* 111: 302–12 (2002); L. K. Murray, T. J. Wheeldon, I. C. Reid, D. A. Rowland, D. M. Burt, and D. I. Perrett, "Depression and Facial Expression Sensitivity: Exploratory Studies: Facial Expression Sensitivity in Depression," submitted for publication.

143 Nine-month-old infants: N. T. Termine and C. E. Izard, "Infants' Responses to Their Mothers' Expressions of Joy and Sadness," *Developmental Psychology,* 24: 223–29 (1988).

143 One-year-olds: D. L. Mumme and A. Fernald, "The Infant as Onlooker: Learning from Emotional Reactions Observed in a Television Scenario," *Child Development,* 74: 221–37 (2003).

143 Adults, too, when interacting: J. K. Hietanen, V. Surakka, and I. Linnankoski, "Facial Electromyographic Response to Vocal Affect Expressions," *Psychophysiology,* 35: 530–36 (1998); V. Surakka and J. K. Hietanen, "Facial and Emotional Reactions to Duchenne and Non-Duchenne Smiles," *International Journal of Psychophysiology,* 29: 23–33 (1998).

143 Darwin believed that laughter: Charles Darwin, *The Expression of the Emotions in Man and Animals,* 3d ed. (New York: Oxford University Press, 1998), p. 195; first edition published in 1872.

144 Chimpanzees and pygmy chimpanzees: M. J. Owren and J. Bachorowski, "The Evolution of Emotional Expression: A 'Selfish-Gene' Account of Smiling and Laughter in Early Hominids and Humans," in *Emotions: Current Issues and Future Directions,* ed. M. T. Mayne and G. A. Bonanno (New York: Guilford, 2001), pp. 152–91; J.A.R.A.M. van Hooff and S. Preuschoft, "Laughter and Smiling: The Intertwining of Nature and Culture," in *Animal Social Complexity: Intelligence, Culture, and Individualized Societies,* ed. F.B.M. de Waal and P. L. Tyack (Cambridge, Mass.: Harvard University Press, 2003), pp. 260–87.

144 Tickling, according to Roger Fouts: Roger Fouts, with Stephen Tukel Mills, *Next of Kin: What Chimpanzees Have Taught Me About Who We Are* (New York: William Morrow, 1997).

145 smiling and laughter had very different origins: van Hooff and Preuschoft, "Laughter and Smiling."

145 "learnt of friends": Rupert Brooke, "The Soldier," in *1914 & Other Poems* (London: Sidgwick & Jackson, 1919), p. 15.

145 "From quiet homes": Hilaire Belloc, "Dedicatory Ode," in *Complete Verse* (London: Duckworth, 1970), p. 60.

146 Epidemics of contagious laughter: Robert R. Provine, *Laughter: A Scientific Investigation* (New York: Viking, 2000), pp. 129–33.

146 "this plague of laughter": ibid., pp. 130–31; A. M. Rankin and P. J. Philip, "An Epidemic of Laughing in the Bukoba District of Tanganyika," *Central African*

Journal of Medicine, 9: 167–70 (1963); G. J. Ebrahim, "Mass Hysteria in School Children: Notes on Three Outbreaks in East Africa," *Clinical Pediatrics*, 7: 437–38 (1968).

146 Women laugh more often: Provine, *Laughter*, pp. 27–29.

146 chimpanzees and college students: R. R. Provine and K. R. Fischer, "Laughing, Smiling, and Talking: Relation to Sleeping and Social Context in Humans," *Ethology*, 83: 295–305 (1989); van Hooff and Preuschoft, "Laughter and Smiling."

146 students were thirty times more likely to laugh: Provine and Fischer, "Laughing, Smiling, and Talking."

147 "Niagara of laughter": quoted in Provine, *Laughter*, p. 136.

147 The long-term physical benefits: R. Holden, *Laughter: The Best Medicine* (New York: HarperCollins, 1993); P. Martin, *The Sickening Brain: Brain, Behavior, Immunity and Disease* (New York: HarperCollins, 1997); R. A. Martin, "Humor, Laughter, and Physical Health: Methodological Issues and Research Findings," *Psychological Bulletin*, 127: 504–19 (2001); R. A. Martin, "Is Laughter the Best Medicine? Humor, Laughter, and Physical Health," *Psychological Science*, 11: 216–20 (2002).

147 extraverts laugh more often: W. Ruch, "Exhilaration and Humor," in *Handbook of Emotions*, ed. M. Lewis and J. M. Haviland (New York: Guilford, 1993), pp. 605–16.

147 Rats that chirp: B. Knutson, J. Burgdorf, and J. Panksepp, "Anticipation of Play Elicits High-Frequency Ultrasonic Vocalizations in Young Rats," *Journal of Comparative Psychology*, 112: 1–9 (1998); J. Panksepp and J. Burgdorf, "Laughing Rats? Playful Tickling Arouses 50KHz Ultrasonic Chirping in Rats," *Society for Neuroscience Abstracts*, 24: 691 (1998); J. Panksepp and J. Burgdorf, "Laughing Rats? Playful Tickling Arouses High-Frequency Ultrasonic Chirping in Young Rodents," in *Toward a Science of Consciousness III*, ed. S. Hameroff, D. Chalmers, and A. Kazniak (Cambridge, Mass.: MIT Press, 1999), pp. 231–44.

147 Brain scans taken: V. Goel and R. Dolan, "The Functional Anatomy of Humor: Segregating Cognitive and Affective Components," *Nature Neuroscience*, 4: 237–38 (2001).

147 The funnier the joke: ibid.

147 Scientists believe that the reward: ibid.

148 "Cold Cape Cod clams": Cole Porter, "Let's Do It, Let's Fall in Love," from *Paris*, first performed in 1928; Warner Brothers Publications, Miami, Fla.

148 "Hot blood begets hot thoughts": William Shakespeare, *Troilus and Cressida*, Act III, scene 1, lines 126–27. He wrote as well that "affection is a coal that must be cooled, / Else, suffered, it will set the heart on fire. / The sea hath bounds, but deep desire hath none" (*Venus and Adonis*, lines 387–89).

149 "Yea, to such rashness": Thomas Hardy, "Lines, To a Movement in Mozart's E-Flat Symphony," in *The Complete Poems of Thomas Hardy*, ed. James Gibson (London: Macmillan, 1976), p. 459.

149 "The simple accident": Robert Louis Stevenson, "On Falling in Love," in *The Lantern Bearers and Other Essays*, ed. J. Treglown (New York: Farrar, Straus and Giroux, 1988), p. 45; essay first published in 1876.

149 "sets the whole world to a new tune": William James, *The Varieties of Religious Experience* (1902; New York: Penguin, 1982), p. 150.

149 Love had been his tutor: W. S. Gilbert and Arthur Sullivan, *H.M.S. Pinafore*, in *The Complete Annotated Gilbert and Sullivan* (Oxford: Oxford University Press, 1996), Act I, p. 125; first performed in 1878.

149 This marriage of affinity and joy: T. R. Insel, J. T. Winslow, Z.-X. Wang, L. Young, and T. J. Hulihan, "Oxytocin and the Molecular Basis of Monogamy," *Advances in Experimental Medical Biology*, 395: 227–34 (1996); T. R. Insel and Larry J. Young, "The Neurobiology of Attachment," *Nature Reviews: Neuroscience*, 2: 1–8 (2001); T. R. Insel, "Is Attachment an Addictive Disorder?" *Physiology and Behavior*, 79: 351–57 (2003).

150 When a chemical that blocks oxytocin receptors: J. R. Williams, T. R. Insel, C. R. Harbaugh, and C. S. Carter, "Oxytocin Administered Centrally Facilitates Formation of a Partner Preference in Female Prairie Voles (*Microtus ochrogaster*)," *Journal of Neuroendocrinology*, 6: 247–50 (1994); T. R. Insel and T. J. Hulihan, "A Gender-Specific Mechanism for Pair Bonding: Oxytocin and Partner Preference Formation in Monogamous Voles," *Behavioral Neuroscience*, 109: 782–89 (1995).

150 mammal species that are not monogamous: D. G. Kleiman, "Monogamy in Mammals," *Quarterly Review of Biology*, 52: 39–69 (1977).

150 Examination of the montane vole's brain: T. R. Insel and L. E. Shapiro, "Oxytocin Receptor Distribution Reflects Social Organization in Monogamous and Polygamous Voles," *Proceedings of the National Academy of Sciences*, 89: 5981–85 (1992).

150 brain activation patterns: H. Breiter, R. L. Golub, R. M. Weisskoff, D. N. Kennedy, N. Makris, J. D. Berke, J. M. Goodman, H. L. Kantor, D. R. Gastfriend, J. P. Riorden, R. T. Mathew, B. R. Rosen, and S. E. Hyman, "Acute Effects of Cocaine on Human Brain Activity and Emotion," *Neuron*, 19: 591–611 (1997).

150 "Passions are the only orators": François de La Rochefoucauld, *Maxims*, trans. Stuart D. Warner and Stéphane Douard (South Bend, Ind.: St. Augustine's Press, 2001), p. 4.

151 repeated bouts of depression: Churchill's physician, Lord Moran, wrote extensively about Churchill's depressions and noted, as Churchill had, that melancholy permeated the Marlborough line. Churchill's depression lasted at times

for only hours and at other times for months. Lord Moran, *Churchill: Taken from the Diaries of Lord Moran. The Struggle for Survival 1940–1965* (Boston: Houghton Mifflin, 1966).

151 "conspicuously shared": Isaiah Berlin, *The Proper Study of Mankind: An Anthology of Essays* (New York: Farrar, Straus and Giroux, 1998), p. 621.

151 "with all his buoyant sparkle": quoted in Jon Meacham's excellent book, *Franklin and Winston: An Intimate Portrait of an Epic Friendship* (New York: Random House, 2003), p. xiv.

151 "At a time of weakness": Berlin, *Proper Study of Mankind*, p. 629.

151 "by his astonishing appetite for life": ibid., p. 615.

152 "I can see to this day": Wilson Brown, quoted in Geoffrey C. Ward, *A First-Class Temperament: The Emergence of Franklin Roosevelt* (New York: Harper Perennial, 1989), p. 221.

152 "showed immediately that he was at home": Ward, *First-Class Temperament*, p. 222.

152 "I'm nearly dead": FDR to Frances Perkins, quoted in Meacham, *Franklin and Winston*, p. 237.

152 "he seemed to have been endowed": Violet Bonham Carter, obituary of Churchill, *The Times* (London), January 26, 1965.

152 "We are all worms": Violet Bonham Carter, "Winston Churchill—As I Know Him," in *Winston Spencer Churchill: Servant of Crown and Commonwealth*, ed. James Marchant (London: Cassell, 1954), p. 149.

153 "tearing spirits": Brendan Bracken, quoted in Moran, *Churchill*, p. 795.

153 "was our hope": C. P. Snow, *Variety of Men* (New York: Scribners, 1966), p. 149.

153 "Churchill had a very powerful mind": ibid., p. 167.

153 "The multitudes were swept forward": Winston Churchill, *Great Contemporaries* (1937; Safety Harbor, Fla.: Simon, 2001), p. 123.

153 "I remember early in the war": quoted in Moran, *Churchill*, p. 773.

154 "He was indeed made for the hour": Moran, *Churchill*, pp. 832–33.

154 "I was very glad": Winston Churchill, speech to Parliament, November 30, 1954, in *Never Give In!: The Best of Winston Churchill's Speeches*, selected by his grandson Winston S. Churchill (New York: Hyperion, 2003), p. 490.

154 "rain festival": Jane Goodall, *Africa in My Blood: An Autobiography in Letters: The Early Years*, ed. Dale Peterson (Boston: Houghton Mifflin, 2000), pp. 171–72.

155 Dance and music are an ancient part: Anthony Storr, *Music and the Mind* (London: HarperCollins, 1992); Robert Jourdain, *Music, the Brain, and Ecstasy* (New York: William Morrow, 1997); Nils L. Wallin, Björn Merker, and Steven

Brown, eds., *The Origins of Music* (Cambridge, Mass.: MIT Press, 2001); R. J. Zatorre and C. L. Krumhansi, "Mental Models and Musical Minds," *Science*, 298: 2138–39 (2002); P. Janata, J. L. Birk, J. D. Van Horn, M. Leman, B. Tillman, and J. J. Bharucha, "The Cortical Topography of Tonal Structures Underlying Western Music," *Science*, 298: 2167–70 (2002).

156 "To fling my arms wide": Langston Hughes, "Dream Variations," in *Selected Poems of Langston Hughes* (New York: Vintage, 1959), p. 14.

157 "The [river] boat was still far off": quoted in Laurence Bergreen, *Louis Armstrong: An Extravagant Life* (New York: Broadway Books, 1997), p. 148.

157 "It was a breakdown": Hoagy Carmichael and Stephen Longstreet, *Sometimes I Wonder* (New York: Farrar, Straus and Giroux, 1965), pp. 57–58.

157 "distinctly American brand of optimism": Bergreen, *Louis Armstrong*, pp. 5–6.

157 "his sound is both the most modern and the most ancient": quoted in Dick Russell, *Black Genius and the American Experience* (New York: Carroll & Graf, 1998), p. 32.

157 "The question in jazz": quoted ibid., p. 33.

158 "They all know I'm there": quoted ibid., p. 35. Armstrong continued, "Through all of the misfortunes, etc., I did not plan anything. Life was there for me and I accepted it. And life, whatever came out, has been beautiful to me, and I love everybody" (pp. 35–36).

158 Music evolved as a "play-space": quoted in Susan Milius, "Face the Music," *Natural History*, December 2001–January 2002, pp. 48–57.

158 "was game for anything": Bergreen, *Louis Armstrong*, p. 161.

159 the songs of humpback whales: R. Payne, "Whale Songs: Musicability or Mantra?" paper presented at BioMusic Symposium, American Association for the Advancement of Science Annual Meeting, 2000. See also P. M. Gray, B. Krause, J. Atema, C. Krumhansl, and L. Batista, "The Music of Nature and the Nature of Music," *Science*, 291: 52–54 (2001).

160 Music activates the same reward systems: A. J. Blood and R. J. Zatorre, "Intensely Pleasurable Responses to Music Correlate with Activity in Brain Regions Implicated in Reward and Emotion," *Proceedings of the National Academy of Sciences*, 98: 11818–23 (2001).

160 decreases activity in brain structures: ibid.

160 "Music exalts life": Storr, *Music and the Mind*, p. 188.

160 "Man's extremity": William James, *The Varieties of Religious Experience* (1902; London: Penguin, 1982), pp. 47–48.

161 The ecstasy associated with religious experiences: James, *The Varieties of Religious Experience;* Marghanita Laski, *Ecstasy: A Study of Some Secular and Religious Experiences* (London: Cresset, 1961).

161 "broke up in a single moment": C. S. Lewis: *Surprised by Joy: The Shape of My Early Life* (San Diego: Harvest, 1955), p. 72.

161 "I knew (with fatal knowledge)": ibid., p. 73.

162 "allows you to keep going": David Sloan Wilson, *Darwin's Cathedral: Evolution, Religion, and the Nature of Society* (Chicago: University of Chicago Press, 2002).

162 "denote the ravings of insanity": Henry Maudsley, *Natural Causes and Supernatural Seemings* (London: Kegan Paul, 1886), p. 221.

162 "The visitation [of Swedenborg's hallucinations]": ibid., pp. 241–42.

162 Neptune and Uranus: J. F. Nisbit, *The Insanity of Genius* (London: Grant Richards, 1900).

162 manic delusions and hallucinations: Emil Kraepelin, *Manic-Depressive Insanity and Paranoia,* trans. R. M. Barclay, ed. G. M. Robertson (Edinburgh: E. & S. Livingstone, 1921; reprinted New York: Arno Press, 1976); George Winokur, Paula Clayton, and Theodore Reich, *Manic-Depressive Illness* (St. Louis: C. V. Mosby, 1969); Frederick K. Goodwin and Kay R. Jamison, *Manic-Depressive Illness* (New York: Oxford University Press, 1990), pp. 262–68.

163 We are not the only species: Ronald K. Siegel, *Intoxication: Life in the Pursuit of Artificial Paradise* (New York: Pocket Books, 1989); Cindy Engel, *Wild Health* (Boston: Houghton Mifflin, 2002).

163 self-medication is also involved: R. K. Siegel and M. Brodie, "Alcohol Self-Administration by Elephants," *Bulletin of the Psychonomic Society,* 22: 49–52 (1984).

164 There are many nonchemical routes: Peter T. Furst, " 'High States' in Culture-Historical Perspectives," in N. E. Zinberg, *Alternate States of Consciousness* (New York: Free Press, 1977), pp. 53–88.

164 the ancient Greeks were the first: William J. Broad, "For Delphic Oracle, Fumes and Visions," *New York Times,* March 19, 2002.

165 "absolutely intoxicated me": letter from Sir Humphry Davy to Davies Giddy, April 10, 1799, quoted in John Ayrton Paris, *The Life of Sir Humphry Davy* (London: Colburn, 1831), pp. 79–80.

165 "Such a gas has Davy discovered!": Robert Southey, *The Life and Correspondence of Robert Southey,* 6 vols., ed. Charles Cuthbert Southey (London: Longman, Brown, Green & Longmans, 1849), vol. 2, pp. 21–22.

165 "united power of 700 instruments": Henry Wansey, quoted in James Hamilton, *Faraday: The Life* (London: HarperCollins, 2002), p. 68.

165 "Depth beyond depth": James, *Varieties of Religious Experience,* pp. 387–88.

165 "utterly what they are": William James, "The Psychology of Belief," *Mind,* 14: 321–52 (1889), p. 322.

166 "nine cases out of ten": Sir James Crichton-Browne, "The Cavendish Lecture on Dreamy Mental States," *Lancet,* July 13, 1895, 73–75, p. 73.

166 "A medical man": ibid., pp. 73–74.

167 by stimulating a part of the brain: G. F. Koob and E. Nestler, "The Neurobiology of Drug Addiction," *Journal of Neuropsychiatry and Clinical Neuroscience,* 9: 482–97 (1997); W. Schultz, P. Dayan, and P. R. Montaque, "A Neural Substrate of Prediction and Reward," *Science,* 275: 1593–99 (1997); C. W. Bradberry, R. L. Barrett-Larimore, P. Jatlow, and S. R. Rubino, "Impact of Self-Administered Cocaine and Cocaine Cues on Extracellular Dopamine in Mesolimbic and Sensorimotor Striatum in Rhesus Monkeys," *Journal of Neuroscience,* 20: 3874–83 (2000); P.E.M. Phillips, G. D. Stuber, M.L.A.V. Helen, R. M. Wightman, and R. M. Carelli, "Subsecond Dopamine Release Promotes Cocaine Seeking," *Nature,* 422: 614–18 (2003).

167 Prolonged cocaine use: J. M. Wilson, A. Levey, C. Bergeron, K. Kalasinsky, L. Ang, F. Peretti, V. I. Adams, J. Smialek, W. R. Anderson, K. Shannak, J. Deck, H. B. Niznik, and S. J. Kish, "Striatal Dopamine, Dopamine Transporter, and Vesicular Monoamine Transporter in Chronic Cocaine Users," *Annals of Neurology,* 40: 428–39 (1996); K. Y. Little, L. Zhang, T. Desmond, K. A. Frey, G. W. Dalack, and B. J. Cassin, "Striatal Dopaminergic Abnormalities in Human Cocaine Users," *American Journal of Psychiatry,* 156: 238–45 (1999); K. Y. Little, D. M. Krolewski, L. Zhang, and B. J. Cassin, "Loss of Striatal Vesicular Monoamine Transporter Protein (VMAT2) in Human Cocaine Users," *American Journal of Psychiatry,* 160: 47–55 (2003).

167 "Hashish spreads out": Charles Baudelaire, "The Poem of Hashish," in *The Essence of Laughter and Other Essays, Journals and Letters by Charles Baudelaire,* ed. Peter Quennell (New York: Meridian Books, 1956), p. 101.

168 "This will be deducted": Théophile Gautier, *Revue des Deux Mondes,* first published in 1846; reprinted from *The Drug Experience: First-Person Accounts of Addicts, Writers, Scientists and Others* (New York: Orion Press, 1961), pp. 6–15.

168 A love of festivities: William James, *The Principles of Psychology,* 2 vols. (1890; New York: Dover, 1950), vol. 1, p. 428.

168 As a bomber pilot: letter from Senator George McGovern to the author, February 29, 2000.

169 "It is difficult for me to imagine": ibid.

169 "I am apt to believe": John Adams, letter to Abigail Adams, July 3, 1776, in *The Adams Papers: Adams Family Correspondence,* ed. L. H. Butterfield, vol. 2: June 1776–March 1778 (Cambridge, Mass.: Belknap Press, 1963), p. 30.

170 "The endless crackling of torpedoes": Julia Ward Howe, quoted in Robert Haven Schauffler, *Independence Day* (New York: Dodd, Mead, 1927), p. 25.

Chapter 7: "Forces of Nature"

173 "The excitement of discovery": Richard Fortey, *Life: A Natural History of the First Four Billion Years of Life on Earth* (New York: Knopf, 1998), pp. 13–14.

174 "When you're the first person": Sue Hendrickson, interview with Robert Kurson, "Close to the Bone," *New York Times Magazine*, May 28, 2000.

174 "It's the thrill of discovery": quoted in Steve Fiffer, *Tyrannosaurus Sue: The Extraordinary Saga of the Largest, Most Fought Over T. Rex Ever Found* (New York: W. H. Freeman, 2000), p. 9.

174 "We were now getting into areas": Richard E. Byrd, *Skyward* (New York: G. P. Putnam's Sons, 1928), p. 176.

175 "an ecstasy induced not by drugs": Neil Campbell, *Biology*, 2d ed. (1990), cited in Nicola McGirr, *Nature's Connections: An Exploration of Natural History* (London: Natural History Museum, 2000), p. 75.

175 "there was no runway up there": Buzz Aldrin, "What It Feels Like to Walk on the Moon," *Esquire*, June 2001, p. 90.

175 "Then felt I like some watcher": John Keats, "On First Looking into Chapman's Homer," in *The Complete Poetical Works and Letters of John Keats* (Boston: Houghton Mifflin, 1899), p. 9.

176 "the heavens afford the most sublime": Mary Somerville, *Mechanism of the Heavens* (London: John Murray, 1831); quoted in Kathryn A. Neeley, *Mary Somerville: Science, Illumination and the Female Mind* (Cambridge, U.K.: Cambridge University Press, 2001), pp. 110–11.

176 crystallizing experiences: Howard Gardner, *Creating Minds* (New York: Basic Books, 1993), p. 32.

176 "an essential step in the scientific demonstration": Jesse L. Greenstein, "An Introduction to 'The dyer's hand,' " in *Cecilia Payne-Gaposchkin: An Autobiography and Other Recollections*, 2d ed., Katherine Haramundanis (Cambridge, U.K.: Cambridge University Press, 1996), p. 3.

177 "the bravery and adventure": ibid., p. 10.

177 fervor for science: The title of Payne-Gaposchkin's autobiography, *The dyer's hand*, reflects the intensity of her life of science. It is taken from Shakespeare's Sonnet 111: "My nature is subdued / To what it works in, like the dyer's hand."

177 heritage "dominated by women": Cecilia Payne-Gaposchkin, *The dyer's hand: An Autobiography*, in *Cecilia Payne-Gaposchkin: An Autobiography and Other Recollections*, p. 83.

177 "The Bee Orchis": ibid., p. 84.

177 "When I won a coveted prize": ibid., p. 102.

178 "I had, in a sense, converted": ibid., p. 99.

178 "was peopled with legendary figures": ibid., pp. 115–16.

178 "when I returned to my room": ibid., pp. 117–18. In her intense enthusiasm for work, she resembles another remarkable woman scientist, the crystallographer Dorothy Hodgkins, who elucidated the structure of Vitamin B_{12}, insulin, and penicillin and received the Nobel Prize in Chemistry in 1994. A friend is quoted as saying, "She arrived back just before the lecture [she was an undergraduate at Oxford] in one of her utterly irresponsible and delirious moods—leaping about on one foot and saying she had still two hours work to do that night. . . . Dorothy had had nothing since lunch to eat and was obviously in a state of nervous excitement. . . . she left the labs at 3 a.m. . . . She was in a state of excitement all [the next day]—" Quoted in Georgina Ferry, *Dorothy Hodgkins: A Life* (London: Granta, 1999), p. 70.

179 "lived largely on her enthusiasms": Quoted in Peggy A. Kidwell, "An Historical Introduction to 'The dyer's hand,' " in *Cecilia Payne-Gaposchkin*, p. 17.

179 "had opened the doors": Payne-Gaposchkin, *The dyer's hand*, p. 120.

179 "who walked with the stars": ibid., p. 124.

179 "in the heady atmosphere of New England": ibid., p. 136.

179 "the most brilliant Ph.D. thesis": In their history of twentieth-century astronomy, Otto Struve and Velta Zebergs write that Payne-Gaposchkin's dissertation was "undoubtedly the most brilliant Ph.D. thesis ever written in astronomy." O. Struve and V. Zebergs, *Astronomy in the 20th Century* (New York: Macmillan, 1962), p. 220.

179 "in a kind of ecstasy": Payne-Gaposchkin, *The dyer's hand*, p. 165.

179 "Being a woman": ibid., p. 227.

179 "Astronomers are incorrigible optimists": from the introduction to Cecilia Payne-Gaposchkin, *Stars in the Making* (Cambridge, Mass.: Harvard University Press, 1952), p. xi.

180 "is the emotional thrill": Cecilia Payne-Gaposchkin, Russell Prize Lecture, American Astronomical Society, Honolulu, Hawaii, January 17, 1977.

180 "Martin had one characteristic": Sinclair Lewis, *Arrowsmith* (1925; New York: Signet, 1998), p. 292.

181 "The joy I felt": quoted in F. D. Drewitt, *The Life of Edward Jenner* (London: Longmans, 1931), p. 53.

181 "passion for collecting": Charles Darwin, *The Autobiography of Charles Darwin and Selected Letters*, ed. Francis Darwin (1892; Mineola, N.Y.: Dover, 1958), p. 6.

181 "I had strong and diversified tastes": ibid., p. 9.

181 "to the utmost": ibid., p. 31.

181 "has been steady and ardent": ibid., p. 55.

182 "We are not looking into the universe from outside": George Wald, "The Ori-

gins of Life," *Proceedings of the National Academy of Sciences*, 52: 595–611 (1964), pp. 609–10.

183 "I had been attracted to aviation": Charles A. Lindbergh, *Autobiography of Values* (1976; San Diego: Harcourt Brace Jovanovich/Harvest, 1992), p. 310.

183 "I know nothing": Antoine de Saint-Exupéry, *Wind, Sand and Stars*, trans. Lewis Galantière (1939; San Diego: Harvest, 1992), p. 130. A classmate of Saint-Exupéry said of him: "He was above all a dreamer. I remember him, his chin resting on his hand, staring out the window at the cherry tree. . . . I recall an unassuming boy, an original, who was not bookish, and yet who was prone from time-to-time to certain explosions of joy, of exuberance." Quoted in Stacy Schiff's excellent biography, *Saint-Exupéry* (New York: Knopf, 1994), p. 54.

183 "There was a fierce kind of joy": quoted in Paul Hoffman, *Wings of Madness: Alberto Santos-Dumont and the Invention of Flight* (New York: Theia, 2003), p. 49.

183 "is with the wind": Saint-Exupéry, *Wind, Sand and Stars*, p. 166.

183 "I was neither hungry nor thirsty": ibid., p. 131.

184 "I have lifted my plane": Beryl Markham, *West with the Night* (London: Virago, 1984), p. 9.

184 "Of the gladdest moments in human life": quoted in James C. Simmons, *Star-Spangled Eden* (New York: Carroll & Graf, 2000), pp. 186–87.

This is a sentiment expressed as well by the scientist and explorer Alexander von Humboldt; in 1801 he wrote, "I was spurred on by an uncertain longing for what is distant and unknown, for whatever excited my fantasy: danger at sea, the desire for adventures, to be transported from a boring daily life to a marvellous world." Discontent was a sharp spur. "I despised anything to do with bourgeois life," he said, "that slow rhythm of home life and fine manners sickened me." Alexander von Humboldt, *Personal Narrative of a Journey to the Equinoctial Regions of the New Continent*, abridged and trans. Jason Wilson (London: Penguin, 1995), p. xxxv; first published in 3 vols., 1814–1825.

184 "I do not think there is any thrill": quoted in Margaret Cheney, *Tesla: Man Out of Time* (New York: Laurel, 1981), p. 107.

185 "The point of the search for comets": David Levy, "How to Catch a Comet: A Night Watchman's Journey," lecture given at the National Air and Space Museum, Washington, D.C., April 9, 2003.

185 "passion for dazzling pursuits": Thomas Jefferson, "Life of Captain Lewis," August 18, 1813, in M. Lewis, *The Lewis and Clark Expedition*, "The 1814 edition" (Philadelphia: J. B. Lippincott, 1961), vol. I, p. xvi.

185 "no season or circumstance": ibid.

185 "I miss the intensity": Alan Lightman, "Looking Back at Pure World of Theoretical Physics," *New York Times*, May 9, 2000.

186 "I couldn't believe": Glenn Seaborg, quoted in George Johnson, "Sometimes the March of Science Goes Backward," *New York Times*, July 23, 2002.

186 "Beginning with the discovery": Lewis Thomas, "Connections," in *The Fragile Species* (New York: Collier, 1993), p. 180.

187 "The book, the statue, the sonata": Robert Louis Stevenson, "Letter to a Young Gentleman Who Proposes to Embrace the Career of Art," in *The Lantern Bearers and Other Essays*, ed. Jeremy Treglown (New York: Cooper Square, 1999), p. 245; essay first published in 1888.

187 "Science is not everything": Thomas Morgan, "With Oppenheimer on an Autumn Day," *Look*, December 27, 1966, p. 63.

188 "The mathematician's patterns": G. H. Hardy, *A Mathematician's Apology* (Cambridge, U.K.: Cambridge University Press, 1984), p. 85.

188 "Bohr's atom seemed to me": Oliver Sacks, *Uncle Tungsten: Memories of a Chemical Boyhood* (New York: Knopf, 2001), p. 307.

188 "we had lunch": James D. Watson, *The Double Helix: A Personal Account of the Discovery of the Structure of DNA* (1968; New York: Mentor, 1969), p. 131.

188 "I have seen more than one speaker": Francis Crick, *What Mad Pursuit: A Personal View of Scientific Discovery* (New York: Basic Books, 1988), pp. 78–79.

189 *Physics of the Air:* W. J. Humphreys, *Physics of the Air*, 3d ed. (New York: McGraw-Hill, 1940). Humphreys was an early supporter of Wilson A. Bentley and wrote the text for their classic book of snow crystal photography, *Snow Crystals*.

190 "seldom proceeds in the straight-forward logical manner": Watson, *Double Helix*, p. ix.

190 "there remains general ignorance": ibid., pp. ix–x.

190 the greatest achievement of science: Peter Medawar, "Lucky Jim," *New York Review of Books*, March 28, 1968, 3–5, p. 4.

190 "towered over all that the rest of us had achieved": E. O. Wilson, *Naturalist* (Washington, D.C.: Island Press, 1994), p. 233.

190 "The great thing about their discovery": Medawar, "Lucky Jim," p. 3.

191 "adventure characterized": Watson, *Double Helix*, p. ix.

191 "communicates the spirit of science": Jacob Bronowski, "Honest Jim and the Tinker Toy Model," *The Nation*, March 18, 1968, 381–82, p. 382.

191 "wonderfully candid self-portrait": Robert Merton, "Making It Scientifically," *New York Times Book Review*, February 25, 1968.

191 "You are describing how science *is* done": letter from Richard Feynman to James Watson, February 10, 1967; quoted in James Gleick, *Genius: The Life and Science of Richard Feynman* (New York: Vintage, 1993), p. 386.

191 "has never been anything": Alex Comfort, "Two Cultures No More," *Manchester* (England) *Guardian*, May 16, 1968.

191 "The style is elated": ibid.

192 "no fewer than a dozen": editorial, "Professor Watson's Memoirs," *Nature* 217, March 25, 1968.

192 "bleak recitation of bickering": John Lear, "Heredity Transactions," in *The Double Helix*, Norton Critical Edition, ed. Gunther S. Stent (New York: Norton, 1980), p. 195; article first published in *Saturday Review*, March 16, 1968.

192 "Like geographical explorers of old": Sir Howard Florey, "Development of Modern Science," *Nature*, 200: 397–402 (1963), p. 397.

193 "What every scientist knows": Richard Lewontin, "Honest Jim Watson's Big Thriller About DNA," in *The Double Helix*, Norton Critical Edition, p. 186; article first published in the *Chicago Sunday Sun-Times*, February 25, 1968.

193 "He betrays in himself": Medawar, "Lucky Jim," p. 5.

193 "If, of course": editorial, *Nature*, March 25, 1968.

193 "What a much duller": Matt Ridley, foreword to John Inglis, Joseph Sambrook, and Jan Witkowski, eds., *Inspiring Science: Jim Watson and the Age of DNA* (Cold Spring Harbor, N.Y.: Cold Spring Harbor Laboratory Press, 2003), p. xv.

194 "Damn the men": Sinclair Lewis, *Arrowsmith*, p. 272.

194 "just wanted the answer": Crick, *What Mad Pursuit*, pp. 69–70.

194 "Maurice continually frustrated Francis": Watson, *The Double Helix*, pp. 19–20.

195 "Our characters were imperfect": James Watson, speech at Harvard University, March 11, 2002.

196 When I asked him: author's interview with James Watson, February 24, 2002.

196 "It is necessary to share it": ibid.

196 "Both young men are somewhat mad hatters": from Gerard Pomerat's diary, April 1, 1953; quoted in *Inspiring Science*, p. 66.

196 "would pop up from his chair": Watson, *Double Helix*, p. 127.

197 "winged into the Eagle": ibid., p. 126.

197 more innate than learned: I asked most of the individuals I interviewed for my book whether they thought exuberance was innate, learned, or a mixture of both. They responded as follows: Dr. Samuel Barondes—"Heredity plays a big role but environment does too. I think you can teach people to be a little more or less exuberant. But it remains difficult to make a silk purse from a sow's ear"; J. Carter Brown—"Primarily innate"; Dr. Andrew Cheng—"It's innate. In my experience, it can't really be learned. But it can be influenced"; Dr. Robert Farquhar—"It's in you"; Dr. Robert Gallo—"One's innate biology cannot be excluded as a major force dictating emotions, and exuberance must be regarded chiefly as part of the emotions not the intellect. I come back to favoring the innate because there are many ways one can select to avoid hurt [and I] seem prone to use exuberance"; Senator George McGovern—"Equally innate and learned"; Dr. James Watson—"Most likely of innate origin"; Senator Paul

Wellstone—"Innate. I was born bouncing"; Dr. Ellen Winner—"My guess is that it is temperamental and inborn. I think that the environment can kill it, but I don't think the environment can create it."

197 "I just like to know": quoted in Melvyn Bragg, with Ruth Gardiner, *On Giants' Shoulders: Great Scientists and Their Discoveries from Archimedes to DNA* (New York: John Wiley, 1998), p. 327.

197 "it's not the precious": quoted in J. D. McClatchy, "Braving the Elements," *The New Yorker,* March 27, 1995.

197 "Our various brains have been programmed": James Watson, "The Pursuit of Happiness," Liberty Medal Address, City of Philadelphia, July 4, 2000.

198 "always a student": quoted in Victor K. McElheny, *Watson and DNA: Making a Scientific Revolution* (Cambridge, Mass.: Perseus, 2003), pp. 193–94.

198 "It was not only in the lab": Lionel Crawford, "It Smells Right . . . ," in *Inspiring Science,* p. 179.

199 "very depressing, but it's not really": James Watson, remarks at Harvard University, May 24, 1998; quoted in McElheny, *Watson and DNA,* p. 275.

199 Passion in the service of reason: James Watson, lecture at the National Institutes of Health, Bethesda, Maryland, April 14, 2003.

200 "Have Fun and Stay Connected": James D. Watson, *A Passion for DNA* (Cold Spring Harbor, N.Y.: Cold Spring Harbor Laboratory Press, 2000), p. 125.

200 "I may not be religious": James D. Watson, with Andrew Berry, *DNA: The Secret of Life* (New York: Knopf, 2003), pp. 404–5.

200 When asked to define "exuberance": author's interview and correspondence with Robert Gallo, August 24 and September 6, 2000.

202 "One might anticipate": Office of Research Integrity, United States Department of Health and Human Services, November 1993. Accounts of the decision are given in *Nature,* 366: 191 (November 11, 1993) and in *Science,* 262: 1202–3 (November 19, 1993).

202 This critical issue of dispute: The issue of cell culture contamination is exceptionally complicated in the dispute over the discovery of the AIDS virus. The codiscoverers, Robert Gallo and Luc Montagnier, in discussing their related discoveries, have acknowledged the problem of cross-contamination in public interviews and in jointly published scientific writings, most recently R. C. Gallo and L. Montagnier, "The Discovery of HIV as the Cause of AIDS," *New England Journal of Medicine,* 349: 2283–85 (2003). Dr. Gallo has summarized the controversy and its resolution as follows, a summary that has been endorsed by Professor Montagnier (May 2004).

> Gallo and Montagnier each proved he independently isolated HIV from different patients. Montagnier reported the first virus isolate from a per-

son with enlarged lymph glands in 1983, a syndrome considered to be a prelude to AIDS. By early 1984, Gallo was able to report that he and his coworkers had obtained HIV in cell cultures from forty-eight different patients. This linkage was one of his demonstrations that this new virus was the cause of AIDS. He and his colleague Mika Popovic were also the first to succeed in propagating *some* strains of HIV continuously in laboratory cell cultures, a technique that would be crucial in their development of the HIV blood test. One of the HIV strains grew better than the rest, so Gallo selected it for use in the blood test.

Later it became clear that this HIV strain was one that Montagnier had sent to Gallo to examine as a part of their planned collaboration. It had accidentally cross-contaminated the Gallo-Popovic culture and would thus cause years of controversy. However, Montagnier had emphatically stated that his first HIV strain (BRU) could not replicate continuously in cell culture, and indeed Gallo and Popovic verified this the first time Montagnier sent them his virus sample. Not having adequate virus material, they asked Montagnier to send another sample to build their collaboration. Unbeknownst to Montagnier, by this time his original HIV isolate had been accidentally contaminated with HIV from a specimen from an AIDS patient he had just began to work with. When sent to Gallo, the new HIV strain also contaminated one of the Gallo-Popovic cultures—the one they had selected for the blood test. This very strain remains the best strain of HIV for growth in cell culture, and because of this property later contaminated the cultures of several other scientists.

Ironically, the HIV strain that was to cause so much political turmoil not only grew best in the laboratory but also saved many lives.

A similar account of the contaminated strains was recounted by Montagnier in "A History of HIV Discovery" (*Science*, 298: 1727–28 [2002]), and jointly by Gallo and Montagnier in their 2003 article in the *New England Journal of Medicine*, reflecting back on their discoveries of twenty years earlier. Two short but excellent journalistic accounts have been given, by Malcolm Gladwell ("NIH Vindicates Researcher Gallo in AIDS Virus Dispute," *Washington Post*, April 26, 1992) and Nicholas Wade ("Method and Madness: The Vindication of Robert Gallo," *New York Times Magazine*, December 26, 1993).

202 "He's incredibly resilient": quoted in Francis X. Clines, "For Besieged Scientist, New Start in New Lab," *New York Times*, March 11, 1997.

202 "We are not alike in our styles": Robert Gallo, *Virus Hunting* (New York: Basic Books, 1991), p. 169.

203 "It is highly competitive in science": "In Their Own Words: NIH Researchers Recall the Early Years of AIDS," oral history interview at the National Institutes of Health in Bethesda, Maryland, November 4, 1994. The interviewers were Dr. Victoria Harden and Dennis Rodrigues. Archived at http://aidshistory.nih.gov.

203 "He does an excellent job": Bernie Poiesz, quoted in "The Untold Story of HUT78," *Science*, 248: 1499–1507 (1990), p. 1501.

203 "No. Empathetic is a better word": quoted in Clines, "For Besieged Scientist."

203 "capacity to reemerge": Gallo interview with author.

204 "You know, the juices flow": quoted in "The Untold Story of HUT78," p. 1507.

204 "I have to say": Oral history interview at the National Institutes of Health in Bethesda, Maryland, August 25, 1994. The interviewers were Dr. Victoria Harden and Dennis Rodrigues. Archived at http://aidshistory.nih.gov.

204 "To Dr. Gallo for his tenacious": citation for the 1982 Albert Lasker Basic Medical Research Award.

205 "To a desperate moment": citation for the 1986 Albert Lasker Clinical Medical Research Award.

205 "I am here only to introduce": letter from Robert Gallo to the author, April 27, 2001.

205 "I saw science": Gallo, *Virus Hunting*, p. 19.

205 "I read the books": author's conversation with James Watson, June 18, 2003.

206 "values criticism almost higher than friendship": author's conversation with James Watson, August 6, 2000.

206 "temperament for laughter": D. Carleton Gajdusek, autobiographical preface to "Unconventional Viruses and the Origin and Disappearance of Kuru," Nobel Lecture, December 13, 1976, Stockholm, Sweden; reprinted in *Science*, 197: 943–60 (1977).

207 "As a boy of five": ibid.

207 "intellectual curiosity and playful enthusiasm": D. Carleton Gajdusek, "Early Inspiration," *Creativity Research Journal*, 7: 341–49 (1994), p. 341.

208 "Everything he possessed": quoted in Richard Rhodes, *Deadly Feasts: The "Prion" Controversy and the Public's Health* (New York: Touchstone, 1998), p. 28.

209 "extraordinarily fundamental advance": citation for the 1976 Nobel Prize in Physiology or Medicine.

209 "who combines the intelligence": quoted in Rhodes, *Deadly Feasts*, p. 55.

210 "exceeded the sacred forty-five-minute limit": George Klein, *Live Now* (New York: Prometheus, 1997), pp. 89–90.

210 When I asked Gajdusek: letters to the author from Carleton Gajdusek, April 5, 2000; November 16, 2001; July 24, 2002; October 30, 2003.

211 "I often radiate it": letter from Gajdusek to the author, November 16, 2001.
211 "Gajdusek is quite manically energetic": quoted in Rhodes, *Deadly Feasts*, p. 32.
211 "I play with ideas": Gajdusek, "Early Inspiration," p. 343.
211 "I look on my current joyful years": letter from Carleton Gajdusek to Richard Wyatt, April 5, 2000.
211 Samuel Barondes: letter from Samuel Barondes to the author, April 12, 2000.
213 "is a genius with celestial pinball": Don Yeomans, quoted in Oliver Morton, "The Art of Falling," *Wired*, December 1999.
213 "a classic case study": author's interview with Robert Farquhar, Johns Hopkins Applied Physics Laboratory, May 11, 2001.
215 Exuberance, Cheng makes clear: author's interview with Andrew Cheng, Johns Hopkins Applied Physics Laboratory, May 11, 2001.
219 "I remember my first encounter": Katy Payne, *Silent Thunder: In the Presence of Elephants* (New York: Penguin, 1998), pp. 38–39.
219 "the water tastes of the pipes": ibid., p. 10.
219 "The organ was alive": ibid., pp. 20–21.
220 "Happy fishermen": ibid., p. 109.
220 When I asked Payne: letter from Katy Payne to the author, August 8, 2000.
222 "I went into an intensely manic state": Joyce Poole, *Coming of Age with Elephants* (New York: Hyperion, 1996), p. 106.
222 "Cynthia and I spent the long days": ibid., p. 82.
222 "What do elephants think about?": Joyce Poole, keynote address to the 22nd Annual Elephant Managers Workshop, Disney's Animal Kingdom, Orlando, Fla., November 9–12, 2001.
222 "for giving me a life of meaning": Poole, *Coming of Age*, p. xi.
222 When I asked Poole: letter from Joyce Poole to the author, September 27, 2000.
224 "my response to what I see": letter from Hope Ryden to the author, April 22, 2000.
225 "Good science as a way of life": James Watson, speech at the Nobel Banquet in Stockholm, December 10, 1962, printed in *Les Prix Nobel* (Stockholm: Almquist and Wiksell, 1963).

Chapter 8: "Nothing Is Too Wonderful to Be True"

226 "just a little more cheerful": Hans Bethe, quoted in John Gribbin and Mary Gribbin, *Richard Feynman: A Life in Science* (London: Penguin, 1997), p. 102.
227 To teach is to show: The word goes back to the prehistoric Indo-European base *deik*, "show," which also produced the Greek *deiknunai*, "show." See John Ayto, *Dictionary of Word Origins* (New York: Arcade, 1991), p. 522.

228 "All vigor is contagious": Ralph Waldo Emerson, "Progress of Culture," in *The Works of Ralph Waldo Emerson: Letters and Social Aims* (Boston: Fireside, 1919), pp. 217–18; essay based on a lecture given by Emerson in 1867.

228 "My life was extacy": Henry David Thoreau, journal entry, July 16, 1851, in Henry D. Thoreau, *Journal,* vol. 3: 1848–1851, gen. ed. John C. Broderick, ed. Robert Sattelmeyer, Mark R. Patterson, and William Rossi (Princeton, N.J.: Princeton University Press, 1990), pp. 305–6.

228 "Summer was drunken": Henry Adams, *The Education of Henry Adams* (1907; Boston: Houghton Mifflin, 2000), p. 8.

The contrast between city and country and the lasting influence of nature on a young child were similarly observed by Robert Louis Stevenson: "I have never again been happy in the same way. For indeed, it was scarce a happiness of this world, as we conceive it when we are grown up, and was more akin to that of an animal than that of a man. The sense of sunshine, of green leaves, and of the singing of birds, seems never to have been so strong in me as in that place. The deodar upon the lawn, the laurel thickets, the mills, the river, the church bell, the sight of people ploughing . . . the sharp contrast between this place and the city where I spent the other portion of my time, all these took hold of me, and still remain upon my memory, with a peculiar sparkle and sensuous excitement" (*Memories and Portraits, Memoirs of Himself, Selections from His Notebook* [London: William Heinemann, 1924], p. 151).

228 "hot pine-woods": Adams, *Education,* p. 8.

229 "knew the taste of everything": ibid.

229 "the taste of A-B": ibid.

229 "the cumuli in a June afternoon sky": ibid.

229 "passed in summer": ibid., p. 39.

229 "The justification for a university": Alfred North Whitehead, "Universities and Their Function," in *The Aims of Education and Other Essays* (New York: Macmillan, 1929), p. 139

230 "so few went to hear Him": quoted in Richard S. Westfall, *Never at Rest: A Biography of Isaac Newton* (Cambridge, U.K.: Cambridge University Press, 1980), p. 209.

230 "When Newton saw an apple fall": George Gordon, Lord Byron, *Don Juan,* Canto X, in *Lord Byron: The Complete Poetical Works,* ed. Jerome J. McGann (Oxford, U.K.: Oxford University Press), vol. 5, p. 437.

231 "buckles flew, stays popped": James Hamilton, *Faraday: The Life* (New York: HarperCollins, 2002), p. 17.

231 "talks rapidly": quoted in J. G. Crowther, *British Scientists of the Nineteenth Century* (London: Routledge and Kegan Paul, 1935), p. 8.

231 "I have never witnessed": quoted in June Z. Fullmer, *Young Humphry Davy: The Making of an Experimental Chemist* (Philadelphia: American Philosophical Society, 2000), p. 148.

231 "naturally ardent and speculative": Sir Henry Holland, quoted in Gwendy Caroe, *The Royal Institution* (London: John Murray, 1985), p. 39.

231 "The temperament of Davy": J. A. Paris, *The Life of Sir Humphry Davy* (London: Colburn, 1831).

232 "minute globules of potassium": quoted in Crowther, *British Scientists*, pp. 48–49.

232 "are like blessings of heaven": Humphry Davy, *Fragmentary Remains, Literary and Scientific*, ed. John Davy (London: John Churchill, 1858), p. 59.

232 Samuel Taylor Coleridge: quoted in Crowther, *British Scientists*, p. 17.

233 "The sensation created": quoted in Anne Treneer, *The Mercurial Chemist: A Life of Sir Humphry Davy* (London: Methuen, 1963), p. 86.

233 "I was seized with the desire": from the memoirs of Humphry Davy, quoted in Fullmer, *Young Humphry Davy*, p. 181.

234 "The appearances of the greater number": quoted in Crowther, *British Scientists*, pp. 7–8.

235 "A lecturer should exert": quoted in Caroe, *Royal Institution*, p. 123.

235 "bangs, flashes, soap bubbles": Michael Faraday, *A Course of Six Lectures on the Chemical History of a Candle* (London: Chatto & Windus, 1861), p. 88.

235 "placed two vessels": ibid.

236 "The Christmas lectures": Hamilton, *Faraday*, p. 208.

236 "I claim the privilege": Faraday, *Chemical History of a Candle*, p. 10; a course of six lectures delivered at the Royal Institution during the Christmas holidays of 1860–1861.

236 "Nothing is too wonderful": diary entry for March 19, 1849, in *Faraday's Diary*, ed. Thomas Martin (London: Royal Institution, 1932–36), vol. 5, p. 152.

236 "*Wonderful* it is": All quotations in this paragraph are taken from Faraday's *Chemical History of a Candle*.

237 "Look at the way we see it": Richard Feynman, *The Meaning of It All* (London: Penguin, 1998), p. 10; lectures given at the University of Washington in April 1963. For an excellent biography of Feynman, see James Gleick's *Genius: The Life and Science of Richard Feynman* (New York: Vintage, 1993).

238 "the most brilliant young physicist": quoted in Gribbin and Gribbin, *Richard Feynman*, p. 101.

238 "I think Dick": David Goodstein, "Richard P. Feynman, Teacher," in *"Most of the Good Stuff": Memories of Richard Feynman*, ed. Laurie M. Brown and John S. Rigden (New York: Springer Verlag, 1993), pp. 115–24, quotes on pp. 115, 118.

238 "He urged each of us": Laurie M. Brown, "To Have Been a Student of Richard Feynman," ibid., pp. 53–58, quote on p. 54.

238 "I never heard him": quoted in Gribbin and Gribbin, *Richard Feynman*, p. xiv.

238 "this excitement in the house": quoted ibid., p. 9.

239 "discussions turned into laughter": John Wheeler, "The Young Feynman," in Brown and Rigden, *"Most of the Good Stuff,"* p. 19.

239 "When people asked him": Joan Feynman, "The Beginnings of a Teacher," in Brown and Rigden, *"Most of the Good Stuff,"* p. 169.

239 "As I watched": John S. Rigden, "Feynman at La Cañada High School," in Brown and Rigden, *"Most of the Good Stuff,"* p. 157.

239 "done for the excitement": Feynman, *Meaning of It All*, p. 9.

239 "understand and appreciate the great adventure": ibid.

239 "It is a great adventure": ibid., p. 39.

240 "The world is so wonderful": Richard P. Feynman, *The Pleasure of Finding Things Out*, ed. Jeffrey Robbins (Cambridge, Mass.: Perseus, 1999), p. 101.

240 "Where did the stuff of life": Feynman, *Meaning of It All*, p. 12.

241 "I think we should teach them wonders": Feynman, *Pleasure of Finding Things Out*, p. 102.

241 "The most beautiful and deepest experience": Albert Einstein, "My Credo," speech given to the German League of Human Rights, Berlin, 1932.

241 "to penetrate deeper still": Feynman, *Pleasure of Finding Things Out*, p. 144.

241 "makes a straight-cut ditch": Henry David Thoreau, journal entry, September 19, 1850, in *The Journal of Henry D. Thoreau*, ed. B. Tarrey and F. H. Allen (Boston: Houghton Mifflin, 1906), p. 22.

241 "Not explaining science": Carl Sagan, "Describing the World as It Is, Not as It Would Be," in *The Writing Life*, ed. Marie Arana (New York: Public Affairs, 2003), p. 306.

241 "we're moved": ibid., p. 308.

242 "I'm delighted with the width of the world!": Feynman, *Pleasure of Finding Things Out*, p. 203.

242 "I love to teach": ibid.

242 "concurred with the pre-Copernican Theory": Pat Conroy, *The Water Is Wide* (1972; New York: Bantam, 1987), p. 33. For a recent account of the exuberant teaching of disadvantaged children, see Rafe Esquith's *There Are No Shortcuts* (New York: Pantheon, 2003).

242 "What could I teach them": Conroy, *Water Is Wide*, p. 81.

243 "All right, young cats": ibid., pp. 51–52.

243 "life was good": ibid., p. 319.

243 "going to have more": ibid., p. 120.

243 "The chief wonder of education": Adams, *Education*, p. 55.

243 "embrace life openly": Conroy, *Water Is Wide*, p. 251.

244 "Lord, I am a teacher": Pat Conroy, *The Prince of Tides* (New York: Bantam, 1987), p. 662.

Chapter 9: "We Should Grow Too Fond of It"

245 "grows hot and lives within us": William James, *The Varieties of Religious Experience* (1902; New York: Penguin, 1982), p. 197.

246 "monarchical" in their power: "The passions," writes Philip Fisher in his excellent book *The Vehement Passions*, "are best described as thorough. They do not make up one part of a state of mind or a situation. Impassioned states seem to drive out every other form of attention or state of being. . . . [They] are what we could call monarchical states of being." Philip Fisher, *The Vehement Passions* (Princeton, N.J.: Princeton University Press, 2002), p. 43.

246 "We cannot write well": Henry David Thoreau, journal entry, September 2, 1851, in Henry D. Thoreau, *Journal*, vol. 4: 1851–1852, gen. ed. Robert Sattelmeyer, ed. Leonard N. Neufeldt and Nancy Craig Simmons (Princeton, N.J.: Princeton University Press, 1992), pp. 27–28.

246 intoxicates more quickly: Andy Coghlan, "Fizz, Bang, Wallop," *New Scientist*, December 22, 2001, p. 7. The scientists at the University of Surrey who conducted the study speculate that the carbon dioxide in the bubbles speeds up the flow of alcohol into the intestines. Champagne is dangerous in other ways, reports the *Wall Street Journal* (June Fletcher, "Champagne Openers: Popping in Safety," December 28, 2001). Flying corks result in hundreds of visits to the emergency room every New Year's Day. "Corks are very small and they travel at tremendous velocity," noted one ophthalmologist interviewed; they can cause black eyes and detached retinas, among other injuries.

246 the capacity for excess: It is difficult to imagine greater excess in the pursuit of joy than that reflected in a deadly statistic used in advertising for Joy, the Jean Patou perfume. One ounce of Joy, boast its manufacturers, contains the "essence" from more than ten thousand jasmine flowers and twenty-eight dozen Bulgarian roses.

246 "hath such a charm": William Shakespeare, *Measure for Measure*, Act IV, scene 1, lines 14–15.

246 "Surely the fever process": James, *Varieties of Religious Experience*, p. 15.

247 "A Heart—how shall I say?": Robert Browning, "My Last Duchess," in *Browning: A Selection*, ed. W. E. Williams (New York: Penguin, 1954), p. 67.

248 "there was this restless creativity": Mark Leithauser, quoted in *The Georgetowner* (Washington, D.C.), June 27, 2002.

248 "something buoyant": Paul Richard, "The Touch of Class," *Washington Post*, June 19, 2002.

248 "lope into others' pastures": J. Carter Brown's comments about exuberance are drawn from his conversations with the author and correspondence in response to specific questions, March 15, 2000.

248 "His text was beauty": The Reverend Peter J. Gomes, memorial service for J. Carter Brown, Washington National Cathedral, July 17, 2002.

249 "I've put a lot of energy into my students": Richard Feynman, quoted in John Gribbin and Mary Gribbin, *Richard Feynman: A Life in Science* (London: Penguin, 1997), p. 212.

249 Feynman was a magician: Marc Kac, quoted in *No Ordinary Genius,* ed. Christopher Sykes (New York: Norton, 1994), p. 19.

249 "who was always in front": A. A. Milne, *The House at Pooh Corner* (1928; New York: Puffin, 1992), p. 75.

250 "Americans really believe": Leon Wieseltier, *Kaddish* (New York: Knopf, 1998), p. 576.

250 "Rainbows flowered for my father": Wallace Stegner, *Where the Bluebird Sings to the Lemonade Springs* (New York: Random House, 1992), p. xx.

250 "Led by pillars of fire and cloud": ibid., p. 3.

250 "Complete independence": ibid., p. 200.

250 "probably time we settled down": ibid., p. 206.

251 "[t]he principle of Barnum's museum": Christopher Irmscher, *The Poetics of Natural History* (New Brunswick, N.J.: Rutgers University Press, 1999), p. 115.

252 "Ya got trouble": Lyrics are from Meredith Willson's *The Music Man,* first performed in 1957; book, music, and lyrics by Meredith Willson, Frank Music Corporation.

253 "Not only fools": John Kenneth Galbraith, *A Short History of Financial Euphoria* (New York: Penguin, 1993), p. viii.

253 "Some artifact or some development": ibid., pp. 2–4.

253 "the mass escape from reality": ibid., p. 12.

253 "wishful thinking on the part of investors": Robert J. Shiller, *Irrational Exuberance* (Princeton, N.J.: Princeton University Press, 2000), p. xii.

254 "equivalent to no less than three hundred times": Mike Dash, *Tulipomania* (New York: Crown, 1999), p. 220.

254 "In reading the history of nations": Charles MacKay, *Extraordinary Popular Delusions and the Madness of Crowds* (New York: Three Rivers Press, 1980), p. xvii; first published in 1841 as *Memoirs of Extraordinary Popular Delusions.*

254 "Men, it has been well said": ibid., p. xviii.

254 "the world can't show a dye": ibid., p. 93.

255 "Many persons grow insensibly attached": ibid., p. 94.

255 The fact that the flower: Anna Pavord, *The Tulip* (London: Bloomsbury, 2000).

255 a single bulb was exchanged: MacKay, *Extraordinary Popular Delusions,* pp. 94–95.

255 A lone bulb of "Semper Augustus": Dash, *Tulipomania*, pp. 108–9.

256 "I must say I'm not very fond": Sinclair Lewis, *Arrowsmith* (1925; New York: Signet, 1998), p. 241.

256 "being galvanized": ibid., p. 226.

256 "I learned during the war": quoted in obituary for Sir Mark Oliphant, *The Economist*, July 22, 2000, p. 85.

257 "The only reaction that I remember": Richard Feynman, *The Pleasure of Finding Things Out*, ed. Jeffrey Robbins (Cambridge, Mass.: Perseus, 1999), p. 10.

257 "All invasive moral states": James, *Varieties of Religious Experience*, p. 90.

257 "always affords some of the enjoyments": Mme the Baroness de Staël-Holstein, *Germany*, 2 vols., trans. O. W. Wight (New York: Derby & Jackson, 1861), vol. 2, p. 361. She went on to say, "We ought to choose our object by enthusiasm, but to approach it by character; thought is nothing without character; enthusiasm is every thing for literary nations, character is every thing to those which are active; free nations stand in need of both" (p. 362).

258 "It is well that war": Robert E. Lee to James Longstreet after the battle of Fredericksburg, December 13, 1862.

258 "a transformation of the personality": Paul Fussell, *The Great War and Modern Memory* (London: Oxford University Press, 1975), pp. 114–54.

258 It is for some a romantic quest: Eric Auerbach's ideas are discussed further by Fussell, ibid., p. 135.

258 "a blood-red blossom": Alfred Lord Tennyson, *Maud: A Definitive Edition*, ed. Susan Shatto (London: Athlone Press, 1986), p. 156 (Part III, verse 4, line 53).

258 "All men who feel": Theodore Roosevelt, quoted in Edmund Morris, *The Rise of Theodore Roosevelt* (New York: Ballantine, 1970), p. 654.

258 "we were all in the spirit": quoted in Nathan Miller, *Theodore Roosevelt: A Life* (New York: William Morrow, 1992), p. 303.

258 "was just revelling in victory": quoted in Morris, *Rise of Theodore Roosevelt*, p. 656.

258 "Power when wielded": Henry Adams, *The Education of Henry Adams* (1907; Boston: Houghton Mifflin, 2000), p. 417. Mark Twain commented that Roosevelt was "clearly insane . . . and insanest upon war and its supreme glories" (quoted in Morris, *Rise of Theodore Roosevelt*, p. 12).

258 "restless agitation": Adams, *Education*, p. 418.

259 "when in the hottest of a battle": letter from Captain David Embree to his sister, September 1863, in *War Letters: Extraordinary Correspondence from American Wars*, ed. Andrew Carroll (New York: Scribners, 2001), pp. 92–93.

259 "the memories of which will remain": quoted in Niall Ferguson, *The Pity of War* (London: Penguin, 1998), p. 361.

259 "Once you have lain in her arms": quoted ibid.

259 "There is a part of me": Chris Hedges, *War Is a Force That Gives Us Meaning* (New York: Public Affairs, 2002), p. 5.

260 "I've been having a blast": Laura Palmer, "Mystery Is the Precinct Where I Found Peace," in Tad Bartimus, Denby Fawcett, Jurate Kazickas, Edith Lederer, Ann Bryan Mariano, Anne Morrissy Merick, Laura Palmer, Kate Webb, and Tracy Wood, *War Torn: Stories of War from the Women Reporters Who Covered Vietnam* (New York: Random House, 2002), p. 257.

260 "First, when they start shooting at you": ibid.

260 "I never knew": ibid. Psychiatric difficulties of Vietnam veterans who engaged in atrocities are discussed by the social worker Sarah A. Haley in her article "When the Patient Reports Atrocities: Specific Treatment Considerations of the Vietnam Veteran," *Archives of General Psychiatry*, 30: 191–96 (1974).

260 "in strange and troubling ways": William Broyles, Jr., "Why Men Love War," *Esquire*, November 1984, p. 56.

260 "beatific contentment": ibid., p. 59.

261 "joy of slaughter": Joanna Bourke, *An Intimate History of Killing: Face-to-Face Killing in Twentieth-Century Warfare* (London: Granta Books, 1999), p. 31.

261 "I had thought myself more or less immune": quoted ibid.

261 "The everlasting battle": T. E. Lawrence, *Seven Pillars of Wisdom: A Triumph* (1926; New York: Doubleday, 1935), pp. 29–30.

Winston Churchill wrote that "Lawrence was one of those beings whose pace of life was faster and more intense than the ordinary. Just as an aeroplane only flies by its speed and pressure against the air, so he flew best and easiest in the hurricane. He was not in complete harmony with the normal." Winston S. Churchill, "Lawrence of Arabia," in *Great Contemporaries* (1937; Safety Harbor, Fla.: Simon, 2001), p. 123.

261 "I liked the things underneath me": Lawrence, *Seven Pillars*, p. 564.

262 "It began as a border patrol": Richard Wrangham and Dale Peterson, *Demonic Males: Apes and the Origins of Human Violence* (Boston: Mariner, 1996), p. 17.

263 "That day [during World War II]": Chuck Yeager and Leo Janos, *Yeager: An Autobiography* (Toronto: Ballantine, 1985), pp. 84–85.

264 "a subtle change happened": Lawrence, *Seven Pillars*, p. 511.

264 "To rouse the excitement of war": ibid.

265 "God but I wish": quoted in Stanley P. Hirshon, *General Patton: A Soldier's Life* (New York: HarperCollins, 2002), p. 50.

265 "When the cave man": quoted ibid., p. 84.

266 "You have seen what the enthusiasm": quoted in Carlo D'Este, *Patton: A Genius for War* (New York: HarperCollins, 1995), p. 93.

266 "You play games to *win*": quoted ibid., p. 312.

266 "all-pervading, visible personality": Hirshon, *General Patton*, p. 200.

266 "one of those men born": quoted in D'Este, *Patton*, p. 818.
266 "This is a damn fine war": quoted in Hirshon, *General Patton*, p. 411.
266 "Where are the damned Germans": quoted ibid., p. 318.
266 "We're going to go right in": quoted ibid., p. 270.
267 "We'll rape their women": quoted ibid., p. 271.
267 "dashing, courageous, wild": quoted ibid., p. 297.
267 "an extreme case": quoted in D'Este, *Patton*, p. 812.
267 "I think he was about half mad": quoted ibid., p. 815.
267 "I am convinced": quoted ibid.
267 "He will be ranked": obituary in the *New York Times*, December 22, 1945.
268 Mitchell, the son of a U.S. Senator: In addition to Mitchell's own writings, I relied upon biographical information in Emile Gauvreau and Lester Cohen, *Billy Mitchell: Founder of Our Air Force and Prophet Without Honor* (New York: Dutton, 1942); Ruth Mitchell, *My Brother Bill: The Life of General "Billy" Mitchell* (New York: Harcourt Brace, 1953); Isaac Levine, *Mitchell: Pioneer of Air Power* (New York: Duell, Sloan & Pearce, 1958); Alfred F. Hurley, *Billy Mitchell: Crusader for Air Power* (New York: Franklin Watts, 1964); Burke Davis, *The Billy Mitchell Affair* (New York: Random House, 1967); James J. Cooke, *Billy Mitchell* (Boulder, Colo.: Lynne Rienner, 2002). Lieutenant Colonel Johnny R. Jones has compiled a helpful anthology of Mitchell's published and unpublished writings: *William "Billy" Mitchell's Air Power* (Maxwell Air Force Base, Ala.: Airpower Research Institute, 1997).
268 "Those of us in the air": William Mitchell, *Winged Defense: The Development and Possibilities of Modern Air Power—Economic and Military* (1925; New York: Dover, 1988), p. 71.
268 "Napoleon studied the campaigns": "Aeronautical Era," *Saturday Evening Post*, December 20, 1924, p. 103.
269 "The competition will be": Mitchell, *Winged Defense*, p. 3.
269 "Bold spirits that before wanted": ibid., p. 8.
269 "The old discipline": William Mitchell, *Skyways: A Book on Modern Aeronautics* (Philadelphia: Lippincott, 1930), p. 65.
269 "[t]he [Army's] General Staff": William Mitchell, *Memoirs of World War I* (New York: Random House, 1928), p. 195.
270 "I have been asked": Colonel Mitchell's Statements on Government Aviation, *Aviation*, 19 (September 14, 1925), p. 318. The editor of *Aviation* noted that Mitchell's remarks "represent the most daring indictment of the War and Navy departments ever made by an officer."
271 "He erred in believing": Hurley, *Billy Mitchell*, p. 139.
271 "Americans might well regard Mitchell": ibid., p. 140.

271 "began to feel the joy": Ovid, *Metamorphoses*, VIII, trans. Horace Gregory (New York: Viking, 1958), p. 212. Daedalus had warned Icarus of the temptation:

> *Remember*
> *To fly midway, for if you dip too low*
> *The waves will weight your wings with thick saltwater,*
> *And if you fly too high the flames of heaven*
> *Will burn them from your sides. Then take your flight*
> *Between the two.* (pp. 211–12)

272 "Who cares that he fell": Anne Sexton, "To a Friend Whose Work Has Come to Triumph," in *The Complete Poems* (Boston: Houghton Mifflin, 1981), p. 53. Likewise, John Burnside writes, "The things that fall / are what we treasure most." From "Of Gravity and Light: (Icarus)," in *The Light Trap* (London: Jonathan Cape, 2002), p. 35.

272 " 'Gemini Four' ": Alan Shepard and Deke Slayton, *Moon Shot: The Inside Story of America's Race to the Moon* (Atlanta: Turner, 1994), p. 182.

272 "One loses all misgivings": Trevor Norton, *Stars Beneath the Sea: The Extraordinary Lives of the Pioneers of Diving* (London: Century, 1999), p. 213.

273 "no easy matter": David Weeks and Jamie James, *Eccentrics* (London: Weidenfeld and Nicolson, 1995), p. 95.

273 "[p]retty soon I decided": Eric Hansen, *Orchid Fever: A Horticultural Tale of Love, Lust, and Lunacy* (New York: Random House, 2000), p. 36.

273 "One morning she sat him down": ibid., p. 37.

274 "he or she has little choice": ibid., pp. 60–61.

274 "In a person of irritable temperament": June Z. Fullmer, *Young Humphry Davy: The Making of an Experimental Chemist* (Philadelphia: American Philosophical Society, 2000), p. 67.

274 "It seemed as though the refreshing breath": Clifford Beers, *A Mind That Found Itself* (New York: Longmans, Green, 1908), p. 73.

275 "Survival might often depend": James Watson, interview with the author at Cold Spring Harbor Laboratory, December 8, 1993.

275 more complexly express it in the arts and sciences: The complicated relationship between depressed and exalted states, and its bearing on the artistic imagination, is more extensively discussed in my *Touched with Fire: Manic-Depressive Illness and the Artistic Temperament* (New York: Free Press, 1993).

275 "can never permanently get the upper hand": Lucretius, *On the Nature of the Universe*, trans. R. E. Latham (London: Penguin, 1951), pp. 76–77.

276 "have a soft temperament": Ernst Kretschmer, translated into English in J. D.

Campbell, *Manic-Depressive Disease: Clinical and Psychiatric Significance* (Philadelphia: J. B. Lippincott, 1953), pp. 26–27.

276 "the drummer takes the muffling handkerchief": Ben Ratliff, "In the Sorrow, the Seeds of Joy," *New York Times,* September 13, 2001.

277 "two or three within": George Gordon, Lord Byron, *Don Juan,* Canto XVII, stanza 11, in *Lord Byron: The Complete Poetical Works,* ed. Jerome J. McGann (Oxford, U.K.: Clarendon Press, 1986), vol. 5, p. 660.

277 "listening to [Woolf]": Christopher Isherwood, in *Recollections of Virginia Woolf by Her Contemporaries,* ed. Joan Russell Noble (Athens: Ohio University Press, 1972), p. 178.

277 "One would hand her": Nigel Nicolson, ibid., p. 128.

277 "I was aware": Elizabeth Bowen, ibid., p. 49. To tell Virginia Woolf anything, added her friend Rose Macaulay, "was like launching a ship on the shifting waters of a river, which flashed back a hundred reflections, enlarging, beautifying, animating, rippling about the keel, filling the sails, bobbing the craft up and down on dancing waves, enlarging the small trip into some fantastic Odyssean voyage among islands of exotic flowers and amusing beasts and men" (ibid., p. 166).

278 "No pen could convey": Cynthia Asquith, *Portrait of Barrie* (New York: Dutton, 1954), p. 22.

278 "A childlike mirth": Edmund Gosse, "Robert Louis Stevenson: Personal Memories," in *The Robert Louis Stevenson Companion,* ed. Jenni Calder (Edinburgh: Paul Harris, 1980), p. 45.

279 "most robust and ordinary men": Sidney Colvin, "Robert Louis Stevenson," in *The Robert Louis Stevenson Companion,* p. 19.

279 "brought into our lives": Isobel Field (Belle Osbourne), "The Best of All Things at Grez," in R. C. Terry, *Robert Louis Stevenson: Interviews and Recollections* (London: Macmillan, 1996), p. 69.

279 "so gay and buoyant": Lloyd Osbourne, "The Scotch Literary Mediocrity," in Terry, *Robert Louis Stevenson,* p. 81.

279 "excited a passionate admiration": Andrew Lang, "Sealed of the Tribe of Louis," in Terry, *Robert Louis Stevenson,* p. 60.

279 "life carries swiftly before it": Henry James, "Robert Louis Stevenson," in *The Robert Louis Stevenson Companion,* pp. 82–83.

279 "He [had] lighted up": Henry James, letter to Fanny Stevenson, in *Henry James Letters,* ed. Leon Edel, vol. III: 1883–1895 (Cambridge, Mass.: Belknap Press, 1980), p. 498.

279 "When I came to London": quoted in *Letters of J. M. Barrie,* ed. Viola Meynell (New York: Scribners, 1947), p. 250.

279 "It was not in Louis": quoted in *The Robert Louis Stevenson Companion,* p. 47.

279 "he was as restless": quoted ibid., p. 43.

280 "a sort of uncommon celerity": quoted in Terry, *Robert Louis Stevenson,* p. 58.

280 "seems never to rest": Henry Adams, "Queer Birds—Mighty Queer Ones Too," in Terry, *Robert Louis Stevenson,* p. 158.

280 "an insane stork": quoted ibid., p. 160.

280 "He was as active": S. J. Whitnee, quoted in Frank McLynn, *Robert Louis Stevenson: A Biography* (London: Hutchinson, 1993), p. 477.

280 "mutable as the sea": W. E. Henley, early draft of "Apparition," quoted ibid., p. 91.

280 "there were two Stevensons": H. J. Moors, *With Stevenson in Samoa* (London: Small, Maynard and Co., 1910), p. 29.

280 "It is in vain": Robert Louis Stevenson, "Crabbed Age and Youth," in *The Lantern-Bearers and Other Essays* (New York: Farrar, Straus and Giroux, 1988), p. 63; essay first published in 1877.

281 "My childhood": letter from Robert Louis Stevenson to William Archer, 1885, quoted in McLynn, *Robert Louis Stevenson,* p. 17.

281 "The family evil": Robert Louis Stevenson, "A Christmas Sermon" (New York: Scribners, 1900), p. 15.

282 "a profound essential melancholy": Robert Louis Stevenson, "Thomas Stevenson, Civil Engineer," in *The Lantern-Bearers,* pp. 213–15; essay first published in 1887.

282 "At one moment": McLynn, *Robert Louis Stevenson,* p. 83.

282 "ah! what bonds we have": Robert Louis Stevenson, letter to W. Craibe Angus, April 1891, in *The Letters of Robert Louis Stevenson,* ed. Bradford A. Booth and Ernest Mehew, vol. 7: September 1890–December 1892 (New Haven, Conn.: Yale University Press, 1995), p. 110.

282 "I'm getting tired": letter from Robert Louis Stevenson to Charles Baxter, January 16, 1873, in *Letters,* vol. 1: 1854–April 1874, pp. 271–72. In the same letter Stevenson describes his brain as "just like a wet sponge: soft, pulpy, and lying spread out, flat and flacid, over my eyes" (p. 271).

282 "You will understand": letter from Robert Louis Stevenson to Frances Sitwell, September 12, 1873, ibid., vol. 1, p. 298.

283 "If you knew how old I felt": letter from Robert Louis Stevenson to Frances Sitwell, November 21, 1873, ibid., vol. 1, p. 374.

283 "The world is disenchanted": Robert Louis Stevenson, "Ordered South," in *Essays and Poems* (London: J. M. Dent, 1992), p. 5; essay first published in 1874.

283 "Black care was sitting": Robert Louis Stevenson, *The Cevennes Journal: Notes on a Journey Through the French Highlands,* ed. Gordon Golding (Edinburgh: Mainstream, 1978), p. 107.

283 "Insomnia is the opposite pole": letter from Robert Louis Stevenson to Edmund Gosse, July 1881, in *Letters,* vol. 3, p. 202.

284 "devilish little left to live for": letter from Robert Louis Stevenson to Charles Baxter, May 21, 1888, in *Letters,* vol. 6: August 1887–September 1890, p. 191.

284 "Half of the ills": Robert Louis Stevenson, quoted in McLynn, *Robert Louis Stevenson,* p. 387.

284 "Health I enjoy": letter from Robert Louis Stevenson to Henry James, August 19, 1890, in *Letters,* vol. 6, pp. 402–3.

284 "Drinks plenty": Letter from Robert Louis Stevenson to J. M. Barrie, April 2 or 3, 1893, in *Letters,* vol. 8, p. 44.

284 "every guarantee": Robert Louis Stevenson, *The Strange Case of Dr. Jekyll and Mr. Hyde,* Norton Critical Edition (1886; New York: Norton, 2003), p. 47.

284 "certain impatient gaiety": ibid., pp. 47–48.

284 "I thus drew steadily nearer": ibid., pp. 48–49.

285 "younger, lighter, happier": ibid., p. 50.

285 "like a mill race": ibid.

285 "solution of the bonds of obligation": ibid.

285 "tenfold more wicked": ibid.

285 "screwed to the topmost peg": ibid., p. 56.

285 "contempt of danger": ibid., p. 58.

285 "spring headlong": ibid., p. 52.

286 "In these flashing revelations": Herman Melville, *Pierre; or, The Ambiguities* (1852; New York: Signet, 1964), p. 114.

286 "Write with fury": Henry David Thoreau, quoted in Robert D. Richardson, Jr., *Henry Thoreau: A Life of the Mind* (Berkeley: University of California Press, 1986), p. 380.

286 "Keep your early enthusiasm": Louis Pasteur, speech given at the opening of the Pasteur Institute in Paris, November 14, 1888, in René Valery-Radot's *The Life of Pasteur*, vol. 2, trans. R. L. Devonshire (London: Constable, 1911), pp. 221–22.

Chapter 10: "It Is Not Down in Any Map"

287 "It Is Not Down in Any Map": Herman Melville, *Moby-Dick* (1851; Berkeley: University of California, 1979), p. 57.

288 "giddy humor": William Bradford, *Of Plymouth Plantation: 1620–1647,* ed. Samuel Eliot Morison (New York: Knopf, 1952), p. 23.

288 "vast and unpeopled countries": ibid., p. 25.

288 "casualties of the sea": ibid., p. 26.

288 "sore sicknesses and grievous diseases": ibid.

288 "The very hearing of these things": ibid.

288 "All great and honourable actions": ibid., p. 27.

288 "It was granted the dangers were great": ibid.

288 "good and honourable": ibid.

289 "yet might they have comfort": ibid.

289 "This land was an enigma": Willa Cather, *O Pioneers!* (1913; New York: Penguin, 1989), p. 15.

289 "wanted to be let alone": ibid., p. 10.

290 "A pioneer should have imagination": ibid., p. 33.

290 "In a pack on his back": Vachel Lindsay, "In Praise of Johnny Appleseed," in *Johnny Appleseed and Other Poems* (Cutchogue, N.Y.: Buccaneer Books, 1976), pp. 84–85.

290 "Love's orchards climbed": ibid., p. 90.

291 "He saw the fruits unfold": ibid., p. 91. At the beginning of Douglas Dunn's poem "The Apple Tree," he quotes Martin Luther—"And if the world should end tomorrow, I still would plant my apple tree"—and then goes on to speak beautifully of perseverance, of holding on to the past and reaching out to the future:

Tonight I saw the stars trapped underneath the water.
I signed the simple covenant we keep with love.
One hand held out an apple while the other held
Earth from a kirkyard where the dead remember me.

Douglas Dunn, "The Apple Tree," in *New Selected Poems, 1964–2000* (London: Faber and Faber, 2003), p. 73.

291 "at the hither edge": Frederick Jackson Turner, *The Frontier in American History* (1920; New York: Dover, 1996), p. 3.

291 "As they wrested their clearing": ibid., p. 345.

291 "coarseness and strength": ibid., p. 37.

291 "restless, nervous energy": Jackson links restlessness to exuberance, as did Henry T. Buckle in his *History of Civilization* (London: John W. Parker & Son, 1869): "An exuberant and therefore a restless nation" (vol. III, pt. i., p. 9).

291 "flush with enthusiasm": quoted in Turner, *Frontier,* p. 319.

292 "From this hour I ordain myself": Walt Whitman, "Song of the Open Road," in *Leaves of Grass* (1855; New York: Norton, 1965), p. 151. Whitman, says Robert Louis Stevenson, "sees that, if the poet is to be of any help, he must testify to the livableness of life. His poems, he tells us, are to be 'hymns of the praise of things.' They are to make for a certain high joy in living." Robert Louis Stevenson, "Walt Whitman," in *The Lantern-Bearers and Other Essays,* ed. J. Treglown (New York: Farrar, Straus and Giroux, 1988), p. 75; essay first published in 1878.

292 *"Here something was about to go wrong"*: O. E. Rölvaag, *Giants in the Earth: A Saga of the Prairie* (New York: Harper Perennial Classics, 1999), p. 33; first published in the United States in 1927.

293 "This vast stretch": ibid., p. 41.
293 "This formless prairie": ibid., p. 43.
293 "Where Per Hansa was": ibid., p. 257.
293 "Now it had taken possession": ibid., pp. 48, 52–53.
294 "such a zest for everything": ibid., p. 56.
294 "plow[s] and harrow[s]": ibid.
294 "He was never at rest": ibid., pp. 125, 127, 128.
294 "unchangeable—it was useless": ibid., p. 149.
294 "called forth all that was evil": ibid., p. 174.
294 "heeded not the light of the day": ibid., p. 255.
294 "seemed to reflect": ibid., p. 179.
295 "even louder in his optimism": ibid., pp. 148–49.
295 "they would all become wild beasts": ibid., p. 215.
295 "the power to create a new life": ibid., p. 337.
295 "As the mild spring weather set in": ibid., p. 338.
295 "He walked so lightly": ibid., pp. 337–38.
295 "It was as if nothing affected people": ibid., pp. 485–86.
296 Americans see enthusiasm: S. Sommers, "Adults Evaluating Their Emotions: A Cross-Cultural Perspective," in *Emotion in Adult Development*, ed. C. Z. Malatesta and C. E. Izard (Beverly Hills, Calif.: Sage, 1984), pp. 319–38.
296 Optimism is a related and defining American trait: John Leland, "Why America Sees the Silver Lining," *New York Times*, June 13, 2004.
296 high rates of manic-depressive illness: B. H. Roberts and J. K. Myers, "Religion, National Origin, Immigration, and Mental Illness," *American Journal of Psychiatry*, 153: 418 (1988); B. Malzberg, "Mental Disease Among Native and Foreign-Born Whites in New York State, 1949–1951," *Mental Hygiene*, 48: 478–99 (1964); L. Rowitz and L. Levy, "Ecological Analysis of Treated Mental Disorders in Chicago," *Archives of General Psychiatry*, 19: 571–79 (1968).
297 "My ties and ballasts": Walt Whitman, "Song of Myself," in *Leaves of Grass*, p. 61. Whitman's expansiveness found a match in that of his contemporary Melville: "Give me a condor's quill!" he wrote. "Give me Vesuvius' crater for an inkstand!" (*Moby-Dick*, p. 465).
297 "I began to feel that I lived": Charles A. Lindbergh, *The Spirit of St. Louis* (1953; New York: Scribners, 2003), pp. 261–62.
297 he made an ironic exception: "I once thought that there were no second acts in American lives, but there was certainly to be a second act to New York's boom days." F. Scott Fitzgerald, "My Lost City," in *The Crack-up with Other Pieces and Stories* (London: Penguin, 1965), p. 29; essay first published in 1945.
297 "New York had all the iridescence": ibid., p. 22.
297 "The buildings were higher": ibid., p. 28.

297 "bloated, gutted, stupid with cake and circuses": ibid., p. 29.

297 "a widespread neurosis": F. Scott Fitzgerald, "Echoes of the Jazz Age," in *The Crack-up,* p. 16; essay first published in 1931.

298 "Something bright and alien": ibid., p. 17.

298 "We were somewhere in North Africa": Fitzgerald, "My Lost City," p. 29.

298 "had begun to disappear": Fitzgerald, "Echoes of the Jazz Age," p. 16.

298 "I began to realize": F. Scott Fitzgerald, "The Crack-up," in *The Crack-up,* p. 42; essay first published in 1936.

298 "an over-extension of the flank": ibid., p. 48.

299 "often approached such an ecstasy": ibid., pp. 55–56.

299 "It's not much fun": letter from Robert Lowell to Peter Taylor, March 15, 1958, quoted in Ian Hamilton, *Robert Lowell: A Biography* (New York: Random House, 1982), p. 253.

299 "It takes just a moment": Robert Lowell, "Balloon," in *Robert Lowell: Collected Poems* (New York: Farrar, Straus and Giroux, 2003), p. 987.

300 "In all the woods": Anonymous, "The Flora of the Somme Battlefield," *Nature,* 100: 475 (1918). An irrepressible scientific curiosity came through as well: "The innumerable shell-hole ponds present many interesting features to the biologist. In July they were half-full of water, and abounded in water beetles and other familiar pond creatures, with dragonflies flitting around. . . . [O]ften growing out of the water were stout plants . . . and water grasses."

301 "It was the invisible and intangible spirit": Peter Ackroyd, *London: The Biography* (London: Chatto & Windus, 2000), p. 745.

301 "London itself would rise again": ibid., pp. 745–46.

302 "I could no longer take any pleasure": Joyce Poole, *Coming of Age with Elephants: A Memoir* (New York: Hyperion, 1996), pp. 232–33.

302 "There were no roads here": ibid., pp. 265–66.

303 "I admit that the slave *does* sometimes sing": Frederick Douglass, "Lecture on Slavery, No. 1," delivered in Corinthian Hall, Rochester, New York, December 1, 1850, in Frederick Douglass, *Selected Speeches and Writings,* ed. P. S. Foner, abridged and adapted by Y. Taylor (Chicago: Lawrence Hill Books, 1999), pp. 164–70.

303 "I sit down on the wire": Philippe Petit, *To Reach the Clouds: My High-Wire Walk Between the Twin Towers* (New York: North Point Press, 2002), p. 194.

304 "gods of the billion constellations": ibid., pp. 194–95.

304 Fight terror: Two weeks after the September 11 attacks on the World Trade Center in New York and the Pentagon in Washington, *The Economist* harked back to the pioneering origins of America: "America's response to the first sustained terrorist attack on its soil is balanced, level-headed and sensible. . . . The can-do pioneers who tamed a wild continent and then helped to win three world

confrontations have not disappeared after all." Lexington, "America the Sensible," *The Economist,* October 27, 2001, p. 34.

304 "Let us print": Petit, *To Reach the Clouds,* p. 241.

305 "our nation's strength": Michael Collins, *Carrying the Fire: An Astronaut's Journeys* (1974; New York: Bantam, 1989), p. 476.

306 "Man has always gone": ibid., p. 487.

306 "inexorable motion of human beings": Alan Bean, talk given at the National Museum of Naval Aviation, Pensacola, Fla., June 15, 2003. Bean's comments are borne out by the level of public interest in space. A poll published by Britain's ITN archive (the world's largest collection of television news) revealed that Neil Armstrong's 1969 moonwalk continued to be the most requested piece of film footage ever shown on television ("Moonwalk Footage Is Brits' Favorite Clip," *Washington Post,* September 3, 2003). In the first six weeks of 2004, NASA's Mars websites registered more than 4.6 billion hits ("Space Sites Garner High Hits," *USA Today,* February 17, 2004).

306 "I am alone now": Collins, *Carrying the Fire,* pp. 408–9.

306 "As I scurry about": ibid., p. 417.

307 a piece of lunar rock: "We have completed the rock section for the Washington Cathedral. This section is from Apollo 11 rock number 10057 and is specific piece number 230. Using the average radius, thickness, and the average specific gravity of lunar rocks, we have calculated the weight of this section to be 7.18 grams. The rock, collected by Astronauts Neil Armstrong and Edwin Aldrin, is a basalt, probably from a lava flow. From isotopic analysis, we believe this rock to have crystallized approximately 3.6 billion years ago. The mineral pyroxferroite, unknown from Earth, was discovered in this rock." Memo from TL/Acting Lunar Sample Curator to NASA Headquarters, June 10, 1974 (from the archives of Washington National Cathedral).

307 "Is not God": "Is not God in the height of heaven? and behold the height of the stars, how high they are!" (Job 22:12).

307 "My life flows on": Robert Lowry, "How Can I Keep from Singing," in *Bright Jewels for the Sunday School* (New York: Bigelow & Main, 1869), p. 16.

Robert Louis Stevenson, in his prayer "For the Renewal of Joy"(from Vailima Prayers, reprinted in Fanny and Robert Louis Stevenson, *Our Samoan Adventure,* ed. Charles Neider [New York: Harper & Brothers, 1955], p. 247), wrote:

> Look down, call upon the dry bones, quicken, enliven; re-create in us the soul of service, the Spirit of peace; renew in us the sense of joy.

Acknowledgments

I am indebted to the following individuals for generously agreeing to answer my questions about the role of exuberance in their lives: Dr. Samuel Barondes, the late J. Carter Brown, Dr. Andrew Cheng, Dr. Robert Farquhar, Dr. Carleton Gajdusek, Dr. Robert Gallo, Senator George McGovern, Katy Payne, Dr. Joyce Poole, Hope Ryden, Jean Schulz, Judy Sladky, Dr. James Watson, the late Senator Paul Wellstone, and Dr. Ellen Winner. Senator Wellstone died in an airplane accident in late 2002, before I was able to complete my follow-up interview with him. With regret, I decided that it would be best not to include his earlier, incomplete remarks in this book. Paul Wellstone was a magnificently exuberant man, and his contagious enthusiasm for life, ideas, and politics was of importance not only to those he represented in Minnesota but to the entire nation. He is greatly missed by those of us who were fortunate to work with him on mental health advocacy causes.

J. Carter Brown, for many years the director of the National Gallery of Art in Washington, D.C., also died during the course of my writing this book. Because he had written out responses to my questions and we had, on many occasions and at great length, discussed the essential role of exuberance in his life, I felt comfortable including his observations here.

Several people were helpful to me while I was doing research for this book: Duncan Blanchard, atmospheric scientist and biographer of Wilson "Snowflake" Bentley; David Dugan, who provided me with useful information about Michael Faraday's lectures at the Royal Institution; Dr. Ellen Gerrity, Senator Wellstone's legislative assistant for mental health and addiction issues; Randi and Hart Johnson and Keith Charles, who were gracious when I visited St. Paul, Minnesota, for the Peanuts on Parade Festival; the Reverend Stuart Kenworthy of Christ Church, Georgetown; Marla Krauss, Special Collections Librarian for Vertebrate Zoology, American Museum of Natural History in New York; Alain Moreau, for his initial and elegant art design for the chapter illustrations; Jinny Nathans, Archivist, American Meteorological Society in Boston, for information about Wilson "Snowflake" Bentley; Diane Ney, Records Manager, Washington National Cathedral, for providing me access to the correspondence between Richard Feller, Clerk of the Works, and Rodney Winfield, the artist who designed the Cathedral's stained-glass Space Window; and Marian O'Keefe and the staff of the Barnum Museum in Bridgeport, Connecticut. Financial support from the John D. and Catherine T. MacArthur Foundation has been generous and greatly appreciated.

I am indebted to those who read early versions of my manuscript and made helpful suggestions: Pat Conroy, Dr. Carleton Gajdusek, Dr. Robert Gallo, Donald Graham, William Graham, Dr. Jerome Kagan, Matt Ridley, Dr. Jeremy Waletzky, and Dr. James Watson. William Collins has, as always, typed my manuscripts with unbelievable accuracy, celerity, and grace under endless time pressures. Ioline Henter has been extraordinarily helpful in locating references, tracking down quirky topics, and ferreting out information of all kinds. Carol Janeway, my editor, has been her usual remarkable self. I am deeply indebted to her, as well as to Stephanie Koven Katz and Ellen Feldman at Knopf. Christopher Mead and Bradley Clements have also been enormously helpful. I am fortunate in my colleagues in the Department of Psychiatry at Johns Hopkins, as well as in the School of English at the University of St. Andrews in Scotland, especially Douglas Dunn, Robert Crawford, and Phillip Mallett. More than anyone else, however, I owe a profound debt to Silas Jones for his help, caring, and friendship.

I have been blessed with kind and generous friends who have seen me through some very difficult as well as wonderful times: Ray and Joanne De Paulo, Bob and Kay Faguet, Bob and Mary Jane Gallo, Chuck and Gwenda Hyman, Carol Janeway, Joanne Leslie, Alain Moreau, Bob Packwood, Norm Rosenthal, Jeff and Kathleen Schlom, Richard Sideman, and Jim and Liz Watson. Jeremy Waletzky has been a friend beyond imagining. My family, as always, has been a tremendous source of love and support: my mother, Dell Jamison; my father, Marshall Jamison; Julian and Sabrina, Eliot, and Leslie Jamison; Danica and Kelda Jamison; Kin Bing Wu; my cousin James Campen; and my brother, Dean Jamison.

My husband, Richard Wyatt, died while I was writing this book. He was delighted by the idea of my writing about exuberance, and he encouraged me in every conceivable way. He supported my ideas with enthusiasm, made many imaginative suggestions, and never let a day go by without expressing his love and encouragement. I admired him enormously: he was an excellent scientist and physician, as well as a gentle, immensely curious, and quietly exuberant man. I miss him more than I can say.

Illustrations

Permissions Acknowledgments

Grateful acknowledgment is made to the following for permission to reprint previously published material:

Alfred A. Knopf: Excerpt from "Dream Variations" from *The Collected Poems of Langston Hughes* by Langston Hughes. Copyright © 1994 by the Estate of Langston Hughes. Reprinted by permission of Alfred A. Knopf, a division of Random House Inc.

The American Association for the Advancement of Science: Excerpt from "The Untold Story of HUT78" from *Science,* 248: 1499–1507 (1990). Copyright © 1990 by AAAS. Reprinted by permission of the American Association for the Advancement of Science.

Blackwell Publishing Ltd.: Excerpts from "Individual Distinctiveness of Brown Bears" by R. Fagen and J. M. Fagen from *Ethology,* 102: 212–26 (1996). Reprinted by permission of Blackwell Publishing Ltd.

Cambridge University Press: Excerpts from *Cecilia Payne-Gaposchkin: An Autobiography and Other Recollections,* 2d ed., by Cecilia Payne-Gaposchkin, edited by Katherine Haramundanis. Reprinted by permission of Cambridge University Press.

The Central African Journal of Medicine: Excerpt from "Provine" by A. M. Rankin and P. J. Philip from *The Central African Journal of Medicine,* 9: 167–70 (1963). Reprinted by permission of *The Central African Journal of Medicine.*

Crumb Elbow Publishers: Excerpt from "Dedicatory Oda" from *Hillaine Belloc: A Collection of Poems* edited by Edward Thompson. Reprinted by permission of Crumb Elbow Publishers.

David Higham Associates Limited: Excerpts from "Canticle of the Sun: Dancing on Easter Morning" from *Collected Poems: 1943–1987* by John Heath-Stubbs (Carcanet Press). Reprinted by permission of David Higham Associates Limited.

Dutton Children's Books and Egmont Books Limited: Excerpts from *The House at Pooh Corner* by A. A. Milne, illustrations by E. H. Shepard. Copyright © 1928 by E. P. Dutton, renewed © 1956 by A. A. Milne. Copyright under the Berne Convention. All rights reserved. Reprinted by permission of Dutton Children's Books, A Division of Penguin Young Readers Group, A Member of Penguin Group (USA) Inc., 345 Hudson Street, New York, NY 10014 and Egmont Books Limited, London.

Farrar, Straus and Giroux, LLC: Excerpts from *Carrying the Fire: An Astronaut's Journey* by Michael Collins. Copyright © 1974 by Michael Collins. Excerpt from "Balloon" from *Collected Poems* by Robert Lowell. Copyright © 2003 by Harriet Lowell and Sheridan Lowell. Excerpts from *To Reach the Clouds* by Philippe Petit. Copyright © 2002 by Philippe Petit. Reprinted by permission of Farrar, Straus and Giroux, LLC.

Hal Leonard Corporation: Excerpt from the song lyric "Ya Got Trouble" from Meredith Willson's *The Music Man* by Meredith Willson. Copyright © 1957, 1966 (Renewed) by Frank Music Corp. and Meredith Willson Music. Excerpt from the song lyric "Rock Island" from Meredith Willson's *The Music Man* by Meredith Willson. Copyright © 1957, 1958 (Renewed) by Frank Music Corp. and Meredith Willson Music. All rights reserved. Reprinted by permission of Hal Leonard Corporation.

HarperCollins Publishers Inc.: Excerpts from *General Patton: A Soldier's Life* by Stanley P. Hirshson. Copyright © 2002 by Stanley P. Hirshson. Excerpts from *Giants in the Earth* by O. E. Rölvaag. Copyright 1927 by Harper & Row, Publishers, Inc. Renewed 1955 by Jennie Marie Berdahl Rölvaag. Reprinted by permission of HarperCollins Publishers Inc.

Houghton Mifflin Company: Excerpt from "To a Friend Whose Work Has Come to Triumph" from *All My Pretty Ones* by Anne Sexton. Copyright © 1962 by Anne Sexton, © renewed 1990 by Linda G. Sexton. Excerpts from *The Life and Letters of John Muir* edited by W. F. Bade. Copyright © 1923, 1924 by Houghton Mifflin Company, renewed 1951, 1952 by John Muir Hanna. All rights reserved. Reprinted by permission of Houghton Mifflin Company.

International Creative Management, Inc.: Excerpts from "Why Men Love War" by William Broyles from *Esquire* (November 1984). Copyright © 1984 by William Broyles. Reprinted by permission of International Creative Management, Inc.

Lawrence Erlbaum Associates, Inc., Publishers: Excerpts from "Early Inspiration" from *Creativity Research Journal*, 7: 341–49 (1994). Reprinted by permission of Lawrence Erlbaum Associates, Inc., Publishers.

New Directions Publishing Corp. and David Higham Associates Limited: Excerpts from *The Crack-up* by F. Scott Fitzgerald. Copyright © 1945 by New Directions Publishing Corp. Excerpt from "Fern Hill" from *The Poems of Dylan Thomas* by Dylan Thomas. Copyright © 1945 by The Trustees for the Copyrights of Dylan Thomas. Reprinted by permission of New Directions Publishing Corp. and David Higham Associates Limited.

New Directions Publishing Corp. and Pollinger Limited: Excerpt from "I Am Cherry Alive" from *Selected Poems: Summer Knowledge* by Delmore Schwartz. Copyright © 1959 by Delmore Schwartz. Reprinted by permission of New Directions Publishing Corp. and Pollinger Limited and the proprietor.

Index

Page numbers 309 and higher refer to notes.

ALSO BY KAY REDFIELD JAMISON

NIGHT FALLS FAST
Understanding Suicide

Suicide has become one of the most common killers of Americans between the ages of fifteen and forty-five. *Night Falls Fast* is critical reading for parents, educators, and anyone wanting to understand the epidemic. An internationally acknowledged authority on depressive illnesses, Dr. Jamison has also struggled with manic-depression, she tried at age twenty-eight to kill herself. Weaving together a historical and scientific exploration of the subject with personal essays on individual suicides, she brings not only her remarkable compassion and literary skill but also all of her knowledge and research to bear on this devastating problem. This is a book that helps us to understand the suicidal mind, to recognize and come to the aid of those at risk, and to comprehend the profound effects on those left behind.

Psychology/0-375-70147-8

AN UNQUIET MIND
A Memoir of Moods and Madness

One of the foremost authorities on manic-depressive illness, Dr. Kay Redfield Jamison has also experienced it firsthand. While pursuing her career in academic medicine, Jamison found herself succumbing to the exhilarating highs and catastrophic depressions that afflicted many of her patients, as her disorder launched her into ruinous spending sprees, episodes of violence, and an attempted suicide. In *An Unquiet Mind*, she examines manic-depression from the dual perspectives of the healer and the healed, revealing both its terrors and the cruel allure that at times prevented her from taking her medication. She has emerged with a memoir of enormous candor, vividness, and wisdom, a rare book that may have the power to transform lives—and even save them.

Psychology/Memoir/0-679-76330-9

VINTAGE BOOKS

Available at your local bookstore, or call toll-free to order:
1-800-793-2665 (credit cards only)